# PCR 2

# The Practical Approach Series

SERIES EDITORS

**D. RICKWOOD**
*Department of Biology, University of Essex*
*Wivenhoe Park, Colchester, Essex CO4 3SQ, UK*

**B. D. HAMES**
*Department of Biochemistry and Molecular Biology*
*University of Leeds, Leeds LS2 9JT, UK*

Affinity Chromatography
Anaerobic Microbiology
Animal Cell Culture (2nd edition)
Animal Virus Pathogenesis
Antibodies I and II
Basic Cell Culture
Behavioural Neuroscience
Biochemical Toxicology
Bioenergetics
Biological Data Analysis
Biological Membranes
Biomechanics—Materials
Biomechanics—Structures and Systems
Biosensors
Carbohydrate Analysis (2nd edition)
Cell–Cell Interactions
The Cell Cycle
Cell Growth and Division
Cellular Calcium
Cellular Interactions in Development
Cellular Neurobiology
Centrifugation (2nd edition)
Clinical Immunology
Computers in Microbiology
Crystallization of Nucleic Acids and Proteins
Cytokines
The Cytoskeleton
Diagnostic Molecular Pathology I and II
Directed Mutagenesis
DNA Cloning: Core Techniques
DNA Cloning: Expression Systems
Drosophila
Electron Microscopy in Biology
Electron Microscopy in Molecular Biology
Electrophysiology
Enzyme Assays
Essential Developmental Biology
Essential Molecular Biology I and II

Experimental Neuroanatomy
Extracellular Matrix
Fermentation
Flow Cytometry (2nd edition)
Gas Chromatography
Gel Electrophoresis of Nucleic Acids (2nd edition)
Gel Electrophoresis of Proteins (2nd edition)
Gene Targeting
Gene Transcription
Genome Analysis
Glycobiology
Growth Factors
Haemopoiesis
Histocompatibility Testing
HPLC of Macromolecules
HPLC of Small Molecules
Human Cytogenetics I and II (2nd edition)
Human Genetic Disease Analysis
Immobilised Cells and Enzymes
Immunocytochemistry
*In Situ* Hybridization
Iodinated Density Gradient Media
Light Microscopy in Biology
Lipid Analysis
Lipid Modification of Proteins
Lipoprotein Analysis
Liposomes
Lymphocytes
Mammalian Cell Biotechnology
Mammalian Development
Medical Bacteriology
Medical Mycology
Medical Parasitology
Medical Virology
Microcomputers in Biochemistry
Microcomputers in Biology
Microcomputers in Physiology
Mitochondria
Molecular Genetic Analysis of Populations
Molecular Genetics of Yeast
Molecular Imaging in Neuroscience
Molecular Neurobiology
Molecular Plant Pathology I and II
Molecular Virology
Monitoring Neuronal Activity
Mutagenicity Testing
Neural Transplantation
Neurochemistry
Neuronal Cell Lines
NMR of Biological Macromolecules
Non-isotopic Methods in Molecular Biology
Nucleic Acid Hybridization
Nucleic Acid and Protein Sequence Analysis
Nucleic Acids Sequencing
Oligonucleotides and Analogues
Oligonucleotide Synthesis
PCR 1
PCR 2
Peptide Antigens
Peptide Hormone Action
Peptide Hormone Secretion

Photosynthesis: Energy Transduction
Plant Cell Biology
Plant Cell Culture
Plant Molecular Biology
Plasmids (2nd edition)
Pollination Ecology
Postimplantation Mammalian Embryos
Preparative Centrifugation
Prostaglandins and Related Substances
Protein Architecture
Protein Blotting
Protein Engineering
Protein Function
Protein Phosphorylation
Protein Purification Applications
Protein Purification Methods
Protein Sequencing
Protein Structure
Protein Targeting
Proteolytic Enzymes
Radioisotopes in Biology
Receptor Biochemistry
Receptor–Effector Coupling
Receptor–Ligand Interactions
Ribosomes and Protein Synthesis
RNA Processing I and II
Signal Transduction
Solid Phase Peptide Synthesis
Spectrophotometry and Spectrofluorimetry
Steroid Hormones
Teratocarcinomas and Embryonic Stem Cells
Transcription Factors
Transcription and Translation
Tumour Immunobiology
Virology
Yeast

# PCR 2

## A Practical Approach

Edited by

M. J. McPHERSON

*Department of Biochemistry and Molecular Biology, University of Leeds*

B. D. HAMES

*Department of Biochemistry and Molecular Biology, University of Leeds*

and

G. R. TAYLOR

*Regional DNA Laboratory, Leeds*

at
OXFORD UNIVERSITY PRESS
Oxford New York Tokyo

*Oxford University Press, Walton Street, Oxford OX2 6DP*
*Oxford New York*
*Athens Auckland Bangkok Bombay*
*Calcutta Cape Town Dar es Salaam Delhi*
*Florence Hong Kong Istanbul Karachi*
*Kuala Lumpur Madras Madrid Melbourne*
*Mexico City Nairobi Paris Singapore*
*Taipei Tokyo Toronto*
*and associated companies in*
*Berlin Ibadan*

*Oxford is a trade mark of Oxford University Press*

*Published in the United States*
*by Oxford University Press Inc., New York*



*A catalogue record for this book is available from the British Library*

*Library of Congress Cataloging in Publication Data*
*PCR 2, a practical approach.*
*(The Practical approach series)*
*Includes bibliographical references and index.*
*1. Polymerase chain reaction.*
*I. McPherson, M. J. II. Hames, B. D.*
*III. Taylor, G. R. IV. Series.*
*[DNLM: 1. DNA Replication. 2. Gene*
*Amplification. QH 450 P7833]*
*QP606.D46P66 1995 574.87'3282 90–14354*

*ISBN 0 19 963425 4 (Hbk)*
*ISBN 0 19 963424 6 (Pbk)*

*Typeset by Footnote Graphics, Warminster, Wilts*
*Printed in Great Britain by Information Press Ltd, Eynsham, Oxon.*

# Preface

PCR is one of the most widely used basic molecular biology techniques due to its remarkable speed, specificity, flexibility, and resilience. As a testament to these characteristics PCR has found a firm place in undergraduate practicals, which are notorious for requiring simple, resilient, preferably quick, and nearly failure-proof techniques! In research, PCR can now be regarded as an indispensable tool, and one which has led to an unparalleled increase in research productivity in all areas of molecular biology. Now, the cloning of a gene or other sequence does not generally represent the major time-consuming component of a research project, therefore more effort can be directed towards the analysis of the gene and the biological significance of its expression patterns and products. PCR has undoubtedly been one of the major factors influencing the rapid advances now being witnessed in genome mapping and sequencing projects.

Since many applications of PCR have commercial potential, the PCR process and the polymerases that drive PCR are the subjects of patents now held by Hoffman-LaRoche. Fortunately for the academic researcher, the patent situation seems to be relatively straightforward; if you purchase reagents from one of the companies licensed by Hoffman-LaRoche to sell *Taq* polymerase then you are essentially free to use PCR for whatever non-profit-making research you wish. The situation is different when there is some commercial application, for example in the field of diagnostic testing.

The basic PCR technique provides the *in vitro* DNA amplification cornerstone upon which further features can be built to expand the range of PCR-based molecular biology applications. Since the publication of *PCR 1: A Practical Approach*, the number of applications of PCR has, not surprisingly, continued to increase along with the number of acronyms of the general format ???-PCR, where ? can be any letter. The purpose of this book, *PCR 2: A Practical Approach*, is to address some of these exciting new topics. *PCR 2* is **not** a revised version of *PCR 1*, but sets out to address different applications of PCR and so is **complementary** to the original volume.

This volume starts with an essential guide, Optimizing PCR, which not only describes the important features of the process, as they are currently understood, but also provides rational guidance for optimizing reaction conditions to improve your PCRs; it should be of benefit to everyone whatever their PCR applications. Chapters 2 and 3 describe the use of non-standard phosphoramidites for primer synthesis (Chapter 2) and the 5′ non-isotopic labelling of primers (Chapter 3); such modifications can expand the range of subsequent manipulations of PCR products. Chapters 4 and 5 describe solid phase approaches to PCR and product manipulation. Chapter 4 outlines the use of solid phase supports for RNA purification, cDNA synthesis and cDNA handling during PCR while Chapter 5 discusses solid phase sequencing of PCR products. Chapter 6 deals with the cloning of full-length cDNA molecules and multigene families. The next two Chapters, 7 and 8, describe alternative approaches to mRNA quantitation which offer different advantages for measuring transcript levels, an area of increasing importance for the detailed analysis of gene expression. Chapter 9 describes procedures for the direct expression of proteins from PCR products, thus allowing *in vitro* synthesis of not only the initial DNA template, but also of the encoded product. The remainder of the volume is largely concerned with genomic DNA analysis. Genome mapping is comprehensively covered in Chapter 10 which illustrates a variety of complementary approaches. The use of arbitrary priming as a tool for genomic fingerprinting is discussed in Chapter 11 which also covers the increasingly important area of analysis of differential RNA expression patterns by arbitrarily primed PCR. Chapters 12 and 13 describe approaches for the identification of known and unknown mutations, respectively, in genomic DNA. Linear amplification as a tool for genomic footprinting of ligand/DNA interaction is described in Chapter 14. Finally Chapter 15 describes ligation-mediated PCR, a generally applicable tool for the PCR amplification from any template where the sequence of only one end of the target region is known; specifically this chapter describes the use of this approach to genomic DNA sequencing and footprinting.

We thank all the authors for their contributions and for their care in compiling such useful practical accounts of their specialist areas of PCR technology. We hope that the reader will find that *PCR 2: A Practical Approach*, represents an informative and useful bench manual to complement the earlier volume in the series.

*Leeds* M. J. M.
October 1994 B. D. H.
G. R. T.

# Contents

# Contributors

ALEX ANDRUS
Applied Biosystems, Division of Perkin Elmer Co., 850 Lincoln Center Drive, Foster City, CA 94404, USA.

JÜRGEN BROSIUS
Institut für Experimentelle Pathologie, Zentrum für Molekularbiologie der Entzündung (ZMBE), Westfälische Wilhelms-Universität, Von-Esmarch-Str. 56, 48149 Münster, Germany.

JEAN BAPTISTE DUMAS MILNE EDWARDS
GENSET, 11, passage Delaunay, 73011 Paris, France.

PAUL A. GARRITY
Howard Hughes Medical Institute/UCLA, 5-748 MacDonald Building, 10833 LeConte Ave, Los Angeles, CA 90024-1662, USA.

KARSTEN HENCO
EVOTEC Biosystems GmbH, Grandweg 64, 22529 Hamburg.

THOMAS HULTMAN
Department of Biochemistry and Biotechnology, Royal Institute of Technology, Teknikringen 34, S100 44 Stockholm, Sweden.

CHRISTINE ICARD-LIEPKALNS
Laboratoire de Génétique Moléculaire de la Neurotransmission et des Processus Neurodégénératifs, CNRS, 91198 Gif sur Yvette Cedex, France.

ANDREAS IMMELMANN
Qiagen, Institut für Molekularbiologische Diagnostik GmbH, Max-Volmer-Str. 4, 40724 Hilden, Germany.

JEAN-PIERRE JOST
FMI, PO Box 2534, CH-4002 Basel, Switzerland.

JIE KANG
Qiagen, Institut für Molekularbiologische Diagnostik GmbH, Max-Volmer-Str. 4, 40724 Hilden, Germany.

DAVID E. KELLOGG
Clontech, 4030 Fabian Way, Palo Alto, CA 94303, USA.

K. K. KIDD
Department of Genetics, Yale University School of Medicine, 333 Cedar Street, New Haven, CT 06510, USA.

JOCHEN E. KÜHN
Institut für Virologie, Universität Köln, Fürst Pückler Str. 56, 50935 Köln, Germany.

M. R. A. LALLOZ
Department of Haematological Medicine, King's College Hospital, Denmark Hill, London SE5 9RS, UK.

SCOTT A. LESLEY
Promega Corporation, 2800 Woods Hollow Road, Madison, WI 53711-5399, USA.

JACQUES MALLET
Laboratoire de Génétique Moléculaire de la Neurotransmission et des Processus Neurodégénératifs, CNRS, 91198 Gif sur Yvette Cedex, France.

A. F. MARKHAM
Department of Molecular Medicine, CSB, St James's University Hospital, Beckett Street, Leeds LS9 7TF, UK.

MICHAEL McCLELLAND
California Institute of Biological Research, 11099 N. Torrey Pines Road, La Jolla, CA 92037, USA.

MICHAEL J. McLEAN
Zeneca Ltd, Cambridge Research Biochemicals, Gadbrook Park, Rudheath, Northwich, Cheshire CW9 7RA, UK.

K. MICHAELIDES
Haemostasis Research Group, Clinical Sciences Centre, Royal Postgraduate Medical School, Hammersmith Hospital, Du Cane Road, London W12 0NN, UK.

J. F. J. MORRISON
Department of Molecular Medicine, CSB, St James's University Hospital, Beckett Street, Leeds LS9 7TF, UK.

PAUL R. MUELLER
Division of Biology, 156-29, California Institute of Technology, Pasadena, California 91125, USA.

CLIVE R. NEWTON
Zeneca Pharmaceuticals, Mereside, Alderley Park, Macclesfield, Cheshire SK10 4TG, UK.

MIGUEL PEINADO
California Institute of Biological Research, 11099 N. Torrey Pines Road, La Jolla, CA 92037, USA.

MANUEL PERUCHO
California Institute of Biological Research, 11099 N. Torrey Pines Road, La Jolla, CA 92037, USA.

DAVID RALPH
California Institute of Biological Research, 11099 N. Torrey Pines Road, La Jolla, CA 92037, USA.

PHILIPPE RAVASSARD
Laboratoire de Génétique Moléculaire de la Neurotransmission et des Processus Neurodégénératifs, CNRS, 91198 Gif sur Yvette Cedex, France.

GUALBERTO RUANO
Bios Laboratories, Five Science Park, New Haven, Connecticut 06511, USA.

HANS PETER SALUZ
HKI, Beutenbergstrasse 11, 07749 Jena, Germany.

PETER SCHÄFER
Universitätskrankenhaus Eppendorf, Institut für Medizinische Mikrobiologie, Martinstr. 52, 20251 Hamburg, Germany.

W. SCHMIDT
Institut für Experimentelle Hämatologie und Transfusion Medizin der Universität Bonn, Sigmund Freud Str. 25, 53105 Bonn 1, Germany.

R. SCHWAAB
Institut für Experimentelle Hämatologie und Transfusion Medizin der Universität Bonn, Sigmund Freud Str. 25, 53105 Bonn 1, Germany.

PAUL D. SIEBERT
Clontech, 4030 Fabian Way, Palo Alto, CA 94303, USA.

STEFAN STAMM
Max-Planck-Institut für Psychiatrie, Neurobiochemie, Am Klopferspitz 18A, 81252 Planegg-Martinsried, Germany.

E. G. D. TUDDENHAM
Haemostasis Research Group, Clinical Sciences Centre, Royal Postgraduate Medical School, Hammersmith Hospital, Du Cane Road, W12 0NN, UK.

MATHIAS UHLÉN
Department of Biochemistry and Biotechnology, Royal Institute of Technology, Teknikringen 34, S100 44 Stockholm, Sweden.

JOHAN WAHLBERG
Department of Biochemistry and Biotechnology, Royal Institute of Technology, Teknikringen 34, S100 44 Stockholm, Sweden.

JOHN WELSH
California Institute of Biological Research, 11099 N. Torrey Pines Road, La Jolla, CA 92037, USA.

BARBARA WOLD
Division of Biology, 156-29, California Institute of Technology, Pasadena, California 91125, USA.

# Abbreviations

| | |
|---|---|
| A | adenine/adenosine/adenylate |
| *A* | absorbance |
| AAT | $\alpha_1$-antitrypsin |
| AP | alkaline phosphatase |
| ARMS | amplification refractory mutation system |
| ASA | allele-specific amplification |
| ASO | allele-specific oligonucleotide |
| ASP | allele-specific PCR |
| ATP | adenosine 5′-triphosphate |
| bp | base pair(s) |
| C | cytosine/cytidine/cytidylate |
| CE | capillary electrophoresis |
| CF | cystic fibrosis |
| CFTR | cystic fibrosis transmembrane regulator |
| cDNA | complementary DNA |
| cDNA-PCR | complementary DNA-polymerase chain reaction |
| COP | competitive oligonucleotide priming |
| c.p.m. | counts per minute |
| CTAB | *N*-cetyl-*N,N,N*-trimethylammonium bromide |
| dATP | deoxyadenosine 5′-triphosphate |
| dCTP | deoxycytidine 5′-triphosphate |
| ddNTP | 2′,3′-dideoxyribonucleotide 5′-triphosphate |
| DGGE | denaturing gradient gel electrophoresis |
| DMS | dimethyl sulphate |
| DMSO | dimethyl sulphoxide |
| DMT | dimethoxytrityl |
| dsDNA | double-stranded DNA |
| ssDNA | single-stranded DNA |
| mtDNA | mitochondrial DNA |
| DNase | deoxyribonuclease |
| dNTP | deoxyribonucleoside 5′-triphosphate |
| DTT | dithiothreitol |
| EDTA | ehylenediamine-tetraacetic acid |
| EtBr | ethidium bromide |
| FISH | fluorescent *in situ* hybridization |
| G | guanine/guanosine/guanylate |
| *g* | gravity |
| G6PD | glucose 6-phosphate dehydrogenase |
| HPLC | high-performance liquid chromatography |

| | |
|---|---|
| HRP | horseradish peroxidase |
| IL-2 | interleukin-2 |
| kb | kilobase |
| LMP | low melting point |
| LMPCR | ligation-mediated PCR |
| MMLV | Moloney murine leukaemia virus |
| MOPS | 3-(*N*-morpholino)-propane sulphonic acid |
| nt | nucleotide |
| NTP | nucleoside 5′-triphosphate |
| OPC | oligonucleotide purification cartridge |
| PAGE | polyacrylamide gel electrophoresis |
| PBS | phosphate-buffered saline |
| PCR | polymerase chain reaction |
| PEST | primer extension sequence test |
| PFGE | pulse-field gel electrophoresis |
| Poly(A)RNA | polyadenylated RNA |
| RACE | rapid amplification of cDNA ends |
| RAP-PCR | RNA arbitrarily-primed PCR |
| RAPD | random amplified polymorphic DNA |
| RNase | ribonuclease |
| hnRNA | heterogenous nuclear RNA |
| mRNA | messenger RNA |
| rRNA | ribosomal RNA |
| RP-HPLC | reverse-phase HPLC |
| RT | reverse transcriptase |
| RT-PCR | reverse transcription-polymerase chain reaction |
| SDS | sodium dodecyl sulphate |
| SDS-PAGE | SDS-polyacrylamide gel electrophoresis |
| SLIC | single strand ligation of cDNA |
| SNuPE | single nucleotide primer extension |
| SSCP | single-stranded conformation polymorphism |
| SSPE | standard saline–phosphate–EDTA |
| STS | sequence-tagged sites |
| *Taq* | *Thermus aquaticus* |
| TAE | Tris–acetate–EDTA buffer |
| TBE | Tris–borate–EDTA buffer |
| TE | Tris–EDTA buffer |
| TEAA | triethylammonium acetate |
| TEMED | *N,N,N′,N′*-tetramethylenediamine |
| TGGE | temperature gradient gel electrophoresis |
| $T_m$ | melting temperature |
| TFA | trifluoroacetic acid |
| tRNA | transfer RNA |
| UV | ultraviolet |

| | |
|---|---|
| WBB | washing and binding buffer |
| X-gal | 5-bromo-4-chloro-3-indolyl-β-D-galactoside |
| YAC | yeast artificial chromosome |
| YLS | yeast lysis buffer |
| YRB | yeast resuspension buffer |

# 1

# Optimizing PCR

K. K. KIDD and G. RUANO

## 1. Perspectives

Although this entire volume is intended to provide explicit, practical guidelines and protocols for PCR in its myriad manifestations, we wish to start with a series of perspectives on various aspects of the several processes that are collectively called the polymerase chain reaction. No individual protocol will be optimal for all PCR reactions, nor will any single, simple set of variables to be optimized necessarily produce a functioning protocol for a specific case. Therefore, in order to optimize PCR, the researcher must have a mental picture of the underlying processes and be able to grasp how the various aspects of a protocol affect the chemical reactions. The relevance of most of the underlying physico-chemical principles can be understood mechanistically to allow a researcher to optimize a PCR reaction for a specific application. In this chapter, we present some of these principles and our perspective on the process without resorting to precise physico-chemical characterizations and quantification. If such a detailed treatment is required, it can be found in Ruano *et al.* (1).

PCR of genomic DNA is a complex series of chemical reactions whose relative contributions to the overall process vary between early, middle, and late cycles, as shown in *Figure 1*. The crucial chemical variable is the net synthesis of product during thermal cycling; because of this synthesis, the molecular balance between product, template, thermostable DNA polymerase, primers, and deoxynucleotides changes with each cycle. Similarly, the annealing interactions between complementary DNA molecules, whether primers with genomic template, primers with product, or product with product, are under continuous flux during PCR cycling.

PCR is a process with two primary phases:

(a) a screening phase during the first few cycles when the desired DNA fragment is selected by specific primer binding,
(b) an amplification phase during the subsequent cycles when the copy number of the desired DNA fragment increases exponentially.

Similar principles apply to simpler templates, such as plasmids and cosmids,

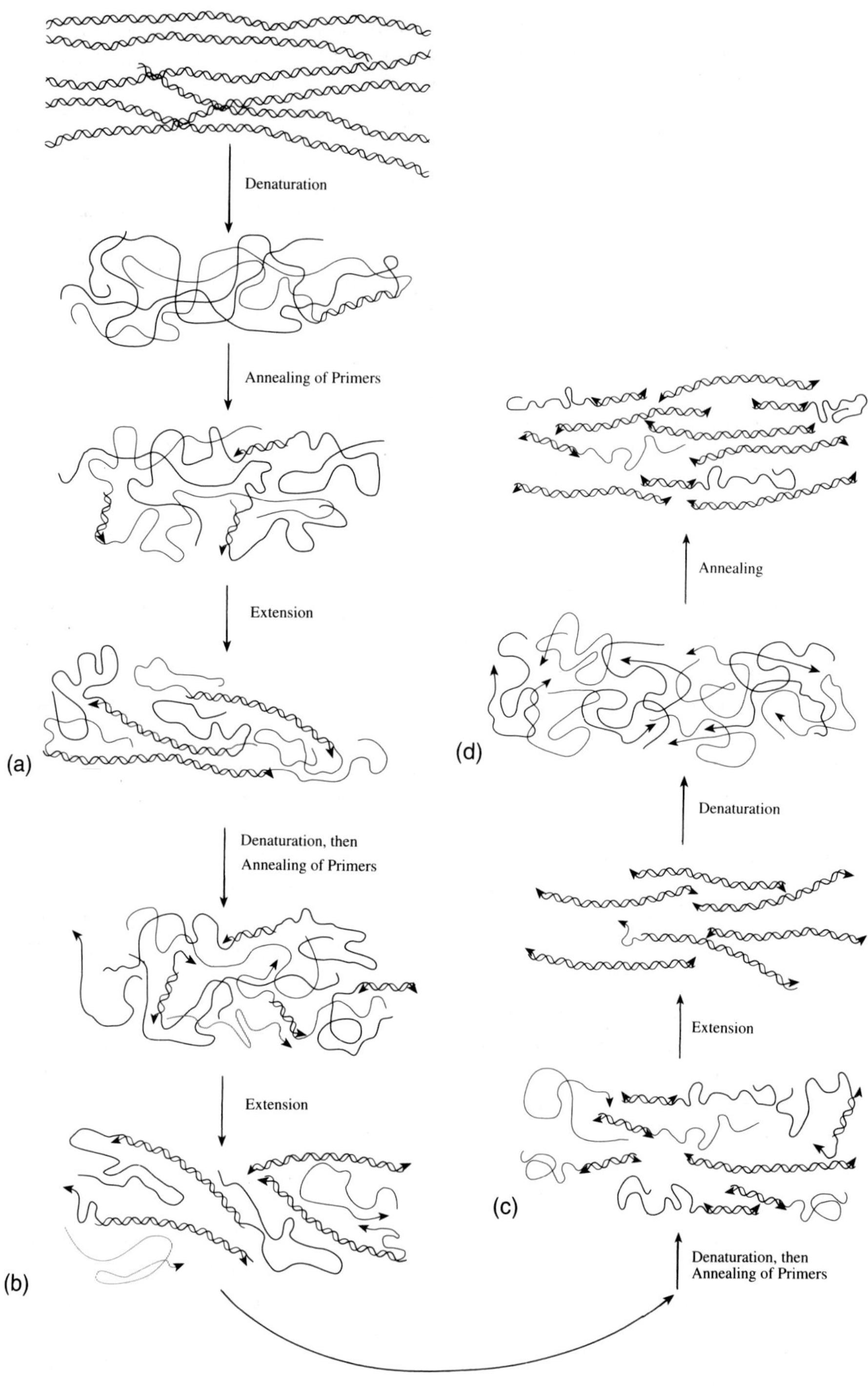
Denaturation
Annealing of Primers
Extension
(a)
Denaturation, then
Annealing of Primers
Extension
(b)
Denaturation, then
Annealing of Primers
(c)
Extension
Denaturation
(d)
Annealing

**Figure 1.** (a) The first cycle of PCR. The double-stranded DNA is denatured. When the temperature is lowered, primers anneal to their homologous sequences in the middle of long stretches of single-stranded DNA. If the temperature is too low, some primers will anneal to regions that do not have perfect homology. (Primers and newly synthesized fragments are shown with arrow-heads at the 3′-end.) A large number of primer extensions occurs producing a heterogeneous mix of short to moderately long stretches of double-stranded DNA. Some newly synthesized fragments correspond to the target (solid lines) but others represent non-target sequences (broken lines). (b) The early cycles. There continue to be primer extensions (from the template DNA) of both target sequences and other sequences. The target products from previous rounds can also be replicated by priming of the complementary oligonucleotide, starting the exponential phase of the amplification. Non-target sequences accumulate linearly with each cycle. (c) The middle cycles. The reaction mixture consists primarily of fragments with perfect complementarity, mostly at the ends, to the primers. This is the exponential phase of the reaction. (d) The late cycles. During the late cycles, product strands are in such high concentration that the complementary strands reanneal, removing many of the potential primer binding sites and restricting the amount of further synthesis. At this point product from spurious primer annealing could begin to be synthesized at proportionately higher levels than the target sequence.

---

or previously amplified products. The screening stage is vastly simplified for these simple templates which are thus likely to amplify with a broader set of conditions than genomic DNA.

All cycles of PCR begin by denaturing both the template and any previously synthesized product. As the temperature is lowered, primers anneal to the template. The annealing step of the early cycles requires primers to scan the genomic template for their correct target. Following primer annealing, the DNA polymerase anchors itself to the primer–template complex, abstracts free dNTPs from the medium, and extends along the template strand. During the middle cycles, previously synthesized product is the preferred template. Finally, in the late cycles, amplified products in high concentration self-hybridize, thereby blocking the primers from their complementary sites (*Figure 1d*).

The thermodynamic driving force for PCR is the molar excess of reagents with respect to template. The molar ratios of PCR reagents with respect to template are highest in the first few cycles but decline as PCR product accumulates. This ratio is reduced by the creation of amplified target molecules but, surprisingly, not by consumption of reagents.

## 1.1 The first few cycles

The very first cycle in PCR consists of primer extension reactions in which a long single strand of DNA is synthesized from every location on the pool of template molecules where the 3′-end of a primer binds sufficiently long for the extension reaction to be initiated (*Figure 1a*). Having an ultimately successful outcome requires, firstly, that the primers being used bind very well in the desired locations but poorly, if at all, in any other location of the template DNA. Secondly, among the many different extension products that are likely to result, only the desired product should produce a template to which one of the primers can bind to initiate an extension reaction synthesizing the complementary strand in the subsequent cycles. The efficient binding of the primer to the desired sequence is a function of primer design (sequence and length) and reaction conditions (especially magnesium ion concentration and temperature). The specificity and purity of the final product depend upon the relative frequency with which primer extension reactions occur from the desired versus undesired positions in the template DNA.

Essentially any interaction of primer with genomic template will lead to an extension product whose specificity may be difficult to control during the first few cycles, given the vast molar excess of possible annealing sites in the template with respect to the desired site. What renders PCR of a defined target feasible is, as observed by Mullis and Faloona (2), the dependence of the reaction on the specificity of two primers which must bind to sites on either complementary strand of DNA. Further constraints are that these primers stably hybridize to these sites at a given stringency (determined by the annealing temperature and $Mg^{2+}$ concentration) and that these sites are not more than, say, 10 kb apart.

Starting with the second cycle of PCR (*Figure 1b*), a primer can anneal to a site on one of the newly synthesized strands and use it to synthesize the fully complementary strand. From this point there exist, in the reaction mixture, strands at which both 3′-ends are precisely complementary to the primers in the reaction mix whether or not the original template DNA had precisely complementary sequences. Thus, initiation of a primer extension reaction in a relatively non-specific way can nonetheless result in a target molecule that has precise complementarity at both ends to the primers. Depending upon the intended use of the product of a PCR reaction, such additional molecules at an early stage of the process will result in a very heterogeneous product. In a 'worst case' scenario, the desired target in the template DNA is at a lower copy number than an undesired sequence which allows priming to occur. Under these conditions, even though the initial priming may be less efficient, more new molecules are produced from the 'false' template than from the 'true' template during the first few cycles and the ultimate product consists almost exclusively of the 'false' template. Priming during PCR of repetitive sequences (3) and priming in mitochondrial sequences (4) are examples.

A variety of techniques has been developed to shift the dynamic balance towards the desired product:

- *Booster PCR* (5) uses low primer concentrations during the first few cycles to minimize priming from the wrong sequence, followed by a primer boost to allow exponential amplification of the already enriched correct template.

- *Hot-start PCR* (6–8) involves completion of the full reaction mixture at a high temperature so that priming from the wrong sequences will not occur under the low stringency conditions of the first transition from low ('room') temperature to the first denaturation step.
- *Touchdown PCR* (9) is a technique in which the annealing temperature is gradually decreased, especially during the first few cycles, so that the first priming will occur only at sequences with perfect homology to the primers.

Booster PCR and hot-start PCR are discussed in more detail in Section 4.8. Touchdown PCR is discussed in more detail in Section 4.11.2.

## 1.2 The exponential amplification phase

During most cycles of the amplification stage of PCR, the templates are the perfectly demarcated segments amplified in previous cycles (*Figure 1c*). The number of those segments is determined by the stringency of the screening in the first few cycles. The task of scanning for homologous primer-annealing sites is simplified drastically during amplification as the fraction of the total available template contributed by genomic DNA becomes negligible. The complexity of screening is effectively reduced by the great excess of the newly amplified material bearing perfectly complementary sites to both primers.

Once template molecules exist with precise complementarity at their ends to primers in the reaction mixture, most standard reaction mixtures have sufficient primers and sufficient enzyme molecules so that all template molecules will be replicated at each cycle. Thus, whatever molecules are present will all be doubled in number each cycle and their ratio will remain unaltered for many cycles while all increase in number. The only requirements for this phase to proceed are:

- sufficient primer and enzyme molecules
- conditions that allow full primer annealing, full extension (e.g. adequate concentrations of free nucleotides), and full denaturation of the product molecules before initiating the next cycle.

## 1.3 The plateau phase

As the reaction products accumulate, all the enzyme present becomes totally occupied and the primer to template ratio decreases, promoting self-annealing of the strands (10) (*Figure 1d*). Also, since the full products are far longer than the primers, such reannealing can start at much higher temperatures than would allow primer annealing and so will tend to occur unavoidably as reaction mixtures cool from the denaturing step. When self-annealing becomes significant or the amount of enzyme is limiting, the reaction begins to saturate and ceases to be exponential.

The amplification phase is self-limiting for any target. Thermal cycling beyond this stage diverts amplification into spurious targets which were not preferred sites during the first few cycles. Because the spurious products are in lower concentration, a much smaller percentage of them will reanneal to themselves than will the products in higher concentration. Thus, there will be differential dampening of the exponential increase such that the most prevalent molecules will increase at a slower rate and the less common molecules will continue their exponential increase. Usually, the most prevalent molecule will be the desired product and the others will be 'contaminants' from undesired 'mispriming' during the first few cycles.

In some cases of extreme product excess, concatamerization of product is possible. This problem is most often seen in re-amplification of aliquots of PCR products. In these cases, so much product may be used as template that dimerization and higher order polymerization of PCR products occur. The results are bands of much higher molecular weight than the desired product. This artefact can be prevented by reducing the number of cycles (Section 4.11.1) and by careful calculation of the primer to template ratio (Section 4.9).

## 1.4 The strategy for successful PCR

Although high reagent to template ratios could be maintained by increasing the initial amounts of thermostable DNA polymerase and primers, these modifications most often result in loss of specificity because primers are extended promiscuously. Extra bands and smears appear. Adjusting reaction and cycling conditions specifically for screening in one stage and for amplification in another would make it possible to:

- maintain constant primer to template and enzyme to template ratios compatible with high specificity during the screening phase;
- boost these reagents later, once amplification has rendered the simple amplified piece the preferred template over complex genomic DNA;
- devise 'touchdown' thermal cycling profiles with higher annealing temperatures and shorter extension times for the first few cycles than for the rest of PCR in order to maximize the stringency of screening (5).

Based on the above considerations, the following steps should be pursued:

(a) Primer sequences need to be those that are exact or very nearly exact matches for the desired target to be amplified and should have virtually no homology to any other sequences in the template mixtures.

(b) The reaction conditions during the first few cycles should be those that minimize the possibility of primers initiating an extension reaction from any part of the template mixture except the desired sequences.

(c) The reaction conditions after the first few cycles should be those that allow all newly synthesized molecules to be perfectly replicated at high efficiency so that they double in number at each cycle.

(d) The reaction should be stopped before the major product becomes saturated and can no longer increase exponentially while the contaminant molecules continue to increase.

### 1.5 Contamination

Contamination is a serious consideration when carrying out PCR reactions. To prevent contamination of samples by previously amplified products, considerable thought and care should be devoted to the actual logistics of the experiments. A set of micropipettors should be devoted exclusively to preparation of the reaction mixtures at a designated PCR workstation set apart from the rest of the laboratory. Products from completed amplifications should be handled with a separate set of micropipettors at a different location in the laboratory. The key concern is 'carry-over' of amplified target to unamplified reaction mixtures (11). Negative controls (i.e. complete reaction mixtures lacking template DNA) should be included in each experiment as an assay for overt contamination.

## 2. Example of a standard protocol

Although the focus of this section pre-supposes that no existing protocol will be suitable for a particular application, it is useful to start with a specific example protocol and then discuss how a protocol can be optimized, based on the results obtained in an initial reaction. The protocol given (*Protocol 1*) is one that has been used successfully for amplification from total genomic DNA of a 1100 bp region near the human HOX2B gene.

**Protocol 1.** A sample protocol for PCR

*Equipment and reagents*

- Thermal cycler (e.g. PE 9600 from Perkin-Elmer)
- Genomic DNA (100 ng/μl)
- dNTP stock solutions (5 mM stock solutions of each dNTP)
- *Taq* polymerase (5 U/μl from Perkin-Elmer)
- 10 × PCR buffer stock (Perkin-Elmer buffer to a final concentration of 15 mM $MgCl_2$, 500 mM KCl, 100 mM Tris–HCl, pH 8.4, 0.1% gelatin, 1% Triton X-100 (Perkin-Elmer))
- Primers CMC6FIN3 and BS2UP2 (see below for sequences) at 10 μM

*Method*

The reaction is best described as follows:

Site: Human *Sac*I polymorphic site, detected by clone BS3, about 4 kb upstream of HOX2B locus on 17q (12).

Primers: CMC6FIN3: 5′ . . . CCTAAGTTAATTTGCTCACTG
BS2UP2: 5′ . . . CCCATACCAATACAACGCAAG

Product: 1100 bp fragment (Genbank accession number U15407) cut into 600 bp and 500 bp by *Sac*I when the *Sac*I site is present.

**Protocol 1.** *Continued*

1. Prepare a 25 μl reaction mixture containing
   - Genomic DNA 1 μl
   - 10 × reaction buffer 2.5 μl
   - Each primer 1 μl
   - dNTP stock solution 1 μl
   - *Taq* polymerase 0.1 μl
2. Cycle in PE 9600 thermal cycler as follows:
   - 94°C for 5 min, initially, followed by 30 cycles of
   - 94°C for 30 sec
   - 60°C for 30 sec
   - 72°C for 1.0 min
3. Hold at 4°C.

## 3. Primer design

Several pointers to good primer design are as follows:

- to prevent self-annealing, primers should not be complementary. This precaution is critical at the extreme 3′-ends where any complementarity may lead to considerable primer-dimer formation.
- None of the primers should have a 3′ T; as T is the least discriminating nucleotide at the 3′-position, primers with a 3′ T have greater tolerance of a mismatch (13).
- It is advisable for each primer to have at least one A or T within the 3′ most triplet to prevent mismatch tolerance of primers with consecutive Gs or Cs.
- The dissociation temperature of primers in a PCR pair should be roughly the same.

Primer design is most elegant from individual thought, but software is also available from many commercial and academic sources to assist in the process (14). Most software packages for DNA sequence analysis now include menus for PCR primer design.

## 4. Optimization of reaction conditions

### 4.1 Primer sequence

Primers that are 18 or more nucleotides long should, on a random expectation, be unique in a complex eukaryote genome. Primers as short as 10 nucleotides are likely to be unique in any purified cloned DNA up to (and even above) cosmid size. However, DNA is not a single random sequence of nucleotides; gene families, repetitive elements, simple duplications, and complex repeat sequences (e.g. alpha satellite sequences in human DNA) all present problems. Depending on the organism and what is already known about the template sequences, primer sequences should be checked against databases to eliminate obvious errors such as unintentionally designing a human primer that is part of an *Alu* element or homologous to mitochondrial DNA (4).

In the example protocol above (*Protocol 1*), the two primers are not perfectly matched for annealing temperature, but are close enough. In this case, each was designed separately for a different use, but later tried together because they were already available.

## 4.2 Annealing temperature

$T_d$, the temperature at which one-half of the primers are annealed to their target sequence, is roughly calculated for oligonucleotides 20 bases long or less by following the equation (15):

$$T_d = 4(G + C) + 2(A + T),$$

where $A$, $T$, $G$, and $C$ are the numbers of those bases in the oligonucleotide. More elaborate calculations can be performed (16), but this simple rule suffices for most applications.

Sometimes it is necessary to try several annealing temperatures over a range, if accurate data or dissociation temperatures are not available for the primers. In addition, if too many spurious products are amplified as a result of primer(s) annealing to the wrong template, increasing the annealing temperature may improve the result (see also Section 4.11.2).

## 4.3 Reaction buffer

One of the key variables in PCR is the $Mg^{2+}$ concentration; it is relevant to both the specificity and yield (17). Other aspects of the ionic environment provided by the buffer are also critical. The $Mg^{2+}$ concentration affects the reaction differently at high and low concentrations. Higher concentrations of $Mg^{2+}$ stabilize double-stranded DNA and prevent complete denaturation of the product at each cycle, reducing yield. Excess $Mg^{2+}$ could also stabilize spurious annealing of primer to incorrect template sites, resulting in larger amounts of undesired products and lower specificity (17). On the other hand, very low $Mg^{2+}$ concentrations (less than 0.5 μM) impair the extension reaction as $Mg^{2+}$ is a required co-factor for enzymatic activity of most DNA polymerases—certainly of all of the common thermostable polymerases. Some $Mg^{2+}$ ions will also be chelated by the dNTPs in the reaction mixture. Thus, there will be a $Mg^{2+}$ concentration for which yield and specificity will be optimum, all other conditions being constant. Indeed, for some recalcitrant regions (such as regions with a very high GC content) only a very narrow range of $Mg^{2+}$ concentration allows acceptable yield of sufficiently specific product.

Quite apart from the $Mg^{2+}$ concentration, the ionic environment provided by the buffer is also critical. KCl buffers have been the most widely used and probably are adequate for the majority of PCR assays. However, some loci may be difficult to amplify in this milieu, regardless of the $Mg^{2+}$ concentration selected. PCR of GC-rich loci shows increased specificity when amplified in NaCl-based buffers, probably because of more complete denaturation (18). Buffers containing ammonium sulphate were used in early PCR research (19) and are now often prescribed for newly isolated thermostable DNA polymerases such as *Pfu* and Vent. We have found that ammonium sulphate buffers reduce the amount of incompletely extended products generated during PCR with *Taq* DNA polymerase. These artefacts are detected as low molecular weight bands on high resolution denaturing acrylamide gels of radiolabelled PCR products.

Another issue concerns the pH of the reaction solution. Most reaction solutions are buffered with 10–50 mM Tris–HCl at pH 8.3, measured at 20°C. However, the pH will decrease with increasing temperature. In the range of temperatures most frequently used for PCR, the pH will cycle between pH 7.8 and 6.8 (20). It is possible that, for amplification of some loci, this pH range may be suboptimal and therefore empirical testing of different starting pH levels will be necessary.

## 4.4 Deoxynucleotide concentration

We recommend that 200 μM of each deoxynucleotide be used. The initial excess of deoxynucleotides with respect to 100 ng genomic template ($10^{10}$ to $10^{11}$) is such that their concentration is essentially constant. Given an initial concentration of 200 μM, the final dNTP concentration will constitute a $10^4$ excess even with 50% degradation during thermal cycling (1). It is very important that the concentrations of all deoxynucleotides are equal to prevent misincorporation errors (20).

## 4.5 Thermostable DNA polymerase

DNA polymerase in excess may synthesize DNA from spurious primer–template interactions (21). The suggested amount of *Taq* DNA polymerase in most protocols (0.5 U per 25 μl reaction) provides the enzyme as the lowest molar concentration of all reaction components, 1 nM. After a $10^6$-fold amplification of target starting at 1 fM, there are roughly the same number of target molecules as enzyme molecules. This limiting concentration is required to control the specificity of the amplification reaction. Concentrations of more than 2.5 nM of enzyme (1.25 U per 25 μl reaction) are to be avoided.

For some of the more recently available thermostable DNA polymerases, the optimal concentration will differ from that recommended above for *Taq* DNA polymerase. The user is advised to follow manufacturers' guidelines at first. If those are inappropriate for a specific reaction, the amount of enzyme may have to be adjusted empirically by performing reactions with different amounts of enzyme and checking the results obtained. Comparison of different thermostable DNA polymerases is simplified by using buffers known to be compatible with most commercial preparations of enzyme.

### 4.6 Enzyme stabilizers

Some manufacturers include gelatin or Triton X-100 in the enzyme storage buffer. Others provide both or, in some cases, neither. It is useful to prepare buffers which always provide these components. In order to stabilize the enzyme during thermal cycling, each reaction buffer should contain gelatin at a final concentration of 0.01% (w/v) and Triton X-100 at a final concentration of 0.1% (v/v).

### 4.7 Concentration of primers

The initial excess of primers with respect to template ($10^7$) is such that their concentration is essentially constant; 95% of the initial primer molecules remain unconsumed after 30 cycles. For most PCR applications, the two primers should have the same concentration. The final concentration of each primer should be 0.1 μM, which is equivalent to 2.5 pmol (16.25 ng of a 20-mer) in a 25 μl reaction mixture. This concentration may have to be adjusted up to 0.5 μM, depending on the locus being amplified. Previous protocols called for as much as 3 μM primer (2), but such a high concentration is not only unnecessary but actually disadvantageous as primer excess can lead to formation of primer dimers and to excessive mispriming. As discussed in more detail in Section 4.9, the relative concentration of primers is as relevant as absolute concentration (22). Thus, for very dilute templates, even lower primer concentrations may give the best results.

### 4.8 DNA template

Whole genomic human DNA at 125 ng per 25 μl reaction is used for the titrations discussed below. This corresponds to 0.063 attomoles of DNA ($\sim 10^{-19}$ mol) and a concentration of 2.5 fM. When amplifying targets from simpler DNA templates, the amount of template should be calculated on a molar basis. This calculation will allow determination of the amount of template DNA required to give the same number or concentration of targets in the template and/or of the number of cycles required to synthesize a given amount of product.

The fidelity of amplification depends on the error rate of the DNA polymerase, but it also depends on the number of initial copies of template DNA.

Random errors introduced in the first few cycles will be propagated in later cycles. The final fraction of product molecules with any specific error will be inversely proportional to the number of initial template molecules. This situation corresponds to 'jackpot' mutations seen in *E. coli*. If the number of original starting molecules is less than 100 and if polymerization fidelity is essential, it may be advisable to do the first few cycles of PCR manually with T7 or T4 polymerase. Although these non-thermostable enzymes will have to be replenished after every denaturing step, they are known to have error rates which are one to two orders of magnitude lower than that of *Taq* polymerase (23). After the first few cycles, regular automated PCR may be started with *Taq* polymerase.

Amplifications from very dilute DNA samples, i.e. 100 copies of template or less (such as single cells or paraffin-embedded tissues), have highlighted the dichotomy between screening and amplification. The screening phase features a higher excess of reagents with respect to the few molecules of available template. Dilute template makes the screening phase of PCR more difficult as the collision frequency of primer and template is markedly reduced to the extent that primer-to-primer contacts are much more common. This situation leads to the formation of primer-dimers and other artefacts. Booster PCR (5, 24) was a first attempt at tackling this problem. As described in Section 1.1, in this procedure the first cycles of PCR were performed with diluted amounts of both primer and template, to achieve a constant $10^7$-fold molar excess of primer to template. For greater amplification, primers were boosted during later cycles. This procedure was devised to amplify the desired target several-fold during the screening phase; those initial product molecules would then become the principal template for subsequent amplification during later cycles. Booster PCR was successful in amplifying single molecules of DNA at higher yield and with fewer primer-dimers than standard PCR (5, 25, 24). It has also been employed to amplify selectively GC-rich exons in the p53 and ras oncogenes (26).

Hot-start PCR is a related method, which has recently become popular in diagnostics and forensics for amplification of dilute DNA samples (6–8). All variants of this technique share deliberate exclusion of at least one essential reagent (dNTPs, $Mg^{2+}$ ions, *Taq* polymerase or primers) from the reaction mixture. The missing component is added only after the mixture has reached a temperature of 70 °C or more. The salient effect of hot-start PCR may be to render DNA template amenable to selective amplification during the first cycle of PCR. Hot-start precludes any synthesis of extraneous products arising from non-selective primer/template hybridization at temperatures between 30 °C and 60 °C, a range in which *Taq* polymerase retains partial activity. Such spurious products would be synthesized during the first cycle of PCR if the entire reaction mixture is initially prepared at room temperature and then ramped to 94 °C. Beginning with the second cycle of PCR, the spurious products could be amplified at the expense of the intended target.

## 4.9 Primer to template ratio

One of the most important concepts in PCR is that of the optimal primer to template ratio. If the ratio is too high, primer-dimers are formed, as occurs in conditions of very dilute template or excess primer (5). If the ratio is too low, product will not accumulate exponentially, since newly synthesized strands will renature after denaturation.

Primers are in a $10^7$ molar excess with respect to template, using the amounts delineated for the titration in Section 5.1.2. For most applications, regardless of template concentration, the primer concentration cannot be raised much higher than 0.5 μM (12.5 pmol per 25 μl reaction) because of primer-dimer formation. What determines the primer to template ratio is, therefore, the amount and complexity of the template provided to the reaction.

## 4.10 Number of target loci in a given template

In most of this chapter, we consider that the existence of multiple annealing sites in the template is undesirable and minimized. However, there are times when more than a single target is to be amplified from the template. In some PCR applications, multiple targets are amplified simultaneously with multiple primer pairs (multiplex PCR) or with a single primer having multiple annealing sites in the template (e.g. *Alu* PCR, RAPD-PCR (random amplified polymorphic DNA, also known as arbitrarily primed PCR)). In these applications it may be necessary to double the concentrations of every reagent in the cocktail. Without adjustment for the multiplicity of sites, an effective decrease occurs in the primer to template ratio and so the yield of the resulting products will be poor.

## 4.11 Cycle profile for amplification

### 4.11.1 Number of cycles

We recommend 30 cycles of amplification for titrations, but additional cycles may be required in some cases if the yield is very low (see Section 4.7). The number of cycles will depend on the molar primer to template ratio (see Section 5.5). For a ratio of $10^6$–$10^7$, 30 cycles should prove adequate to generate sufficient material for visualization on ethidium bromide stained gels, and 25 cycles for visualization of radiolabelled PCR product by autoradiography. For ratios two orders of magnitude above or below this value, respectively, increase or decrease the number of cycles by 10. To minimize the decay of reagents and thus allow for the possibility of further amplification cycles, samples should be refrigerated at 4 °C after each run.

### 4.11.2 Annealing temperature

To enhance the specificity of amplification, the annealing temperature should be at the lowest dissociation temperature ($T_d$) of either primer. The suggested annealing time is 1 min but this may be reduced to 30 sec for small volume reactions (10–25 μl) that equilibrate quickly to the low temperature, especially when PCR is performed in ultra-thin tubes.

The strategy of Touchdown PCR involves starting the cycling with a very high annealing temperature and then lowering the annealing temperature with successive cycles. In the specific example given in the initial paper (9) the annealing temperature was lowered 1 °C every second cycle, going from 65 °C to 55 °C over 20 cycles, and then held at 55 °C for another 10 cycles. Whenever there is still more annealing of the primers to the correct template than to any incorrect template, the correct product is preferentially amplified. Since this preference occurs in the earliest cycles that allow any annealing, the ratio of correct to incorrect product will be maintained when more permissive temperatures are reached in later cycles. Obviously many variants of the Touchdown protocol are possible, limited largely by the program flexibility of the specific thermal cycler used.

It is possible to use the same thermal profile to amplify several loci, even though their calculated annealing temperatures may be different. Templates whose calculated annealing temperature is above that for the standardized profile may require the lowest $Mg^{2+}$ concentration. Alternatively, those with annealing temperatures below the specified one may require extra $Mg^{2+}$ to compensate for overtly stringent temperatures.

### 4.11.3 Polymerization

*Taq* DNA polymerase is highly processive and extends at 2–4 kb per min (35–70 bases per sec) (20). Hence, for targets 1 kb or less, 1 min at 72 °C is sufficient for amplification. Indeed, the ramp time from the annealing temperature to the denaturing temperature is usually sufficiently long to allow full synthesis of targets less than 500 bp. For fragments longer than 1 kb, a conservative estimate would be to add roughly 1 min at the extension temperature per additional kb. Once the amplification is working well, it should be possible to use a shorter extension time determined by trying a few alternatives.

Note that for 20-mer primers with a GC content of 60% or more, the calculated $T_d$ is in the range used for polymerization. This is another situation in which the annealing and polymerization steps can be combined into one step, at $T_d$, for 1–2 min, depending on the thermal cycler, reaction volume, and length of the segment being amplified.

### 4.11.4 Denaturation

For targets 1 kb or less, denature for 1 min at 94 °C. If using ultra-thin tubes, denaturation time may be reduced to 30 sec. For longer fragments, add roughly 1 min denaturation time (or 30 sec if using ultra-thin tubes) per kb. A long denaturation period may be especially important at the very beginning to make sure the template DNA is fully denatured; indeed, for GC-rich regions of genomic DNA templates, this may be essential (8).

### 4.12 Addition of denaturation reagents to the reaction mixture

The effective stringency of the annealing step can be increased by adding denaturants to the PCR cocktail. Possibilities include formamide (27), DMSO (28), tetramethylammonium chloride (29), and detergents (30). Single-stranded DNA binding proteins from *E. coli* (31) and from T4 phage gene 32 (32) have also been shown to improve both the specificity and yield of PCR. These proteins appear to improve the specific annealing of primer to template and so reduce mispriming.

## 5. Detailed optimization protocol

Buffers used for optimization of the PCR protocol should be prepared in stock solutions concentrated 10-fold with respect to the desired final concentration in the reaction mixture. A suitable selection of buffers for optimization of $Mg^{2+}$ concentration and ionic composition is given in *Table 1*. These buffers are also commercially available as a kit from BIOS Laboratories. A buffer kit for optimization of pH is available from Invitrogen. The detailed optimization procedure is described in *Protocol 2*.

**Table 1.** Buffers for optimization of $Mg^{2+}$ concentrations and ionic composition

| Series | Common components[a] | $MgCl_2$ concentration (mM)[a] |
|---|---|---|
| I | 50 mM KCl | A = 0.75 |
| | 10 mM Tris–HCl pH 8.4 | B = 1.50 |
| | 0.01 gelatin | C = 2.25 |
| | 0.1% Triton X-100 | D = 3.00 |
| | | E = 3.75 |
| II | 40 mM NaCl | F = 0.75 |
| | 10 mM Tris–HCl pH 8.4 | G = 1.50 |
| | 0.01 gelatin | H = 2.25 |
| | 0.1% Triton X-100 | I = 3.00 |
| | | J = 3.75 |
| III | 10 mM $(NH_4)_2SO_4$ | K = 0.75 |
| | 10 mM KCl | L = 1.50 |
| | 10 mM Tris–HCl pH 8.4 | M = 2.25 |
| | 0.01 gelatin | N = 3.00 |
| | 0.1% Triton X-100 | O = 3.75 |

[a] The concentrations given are those in the PCR reaction mixture *after* diluting the 10 × stock. Thus the buffer stock solutions are 10-fold concentrated compared to the concentrations listed.

## **Protocol 2.** Optimization protocol

### *Equipment and reagents*

- 10 × buffer stocks for optimization (see *Table 1*)
- dNTP mix (5 mM each of dGTP, dATP, dTTP and dCTP in water)
- Thermal cycler: e.g. PE 9600 or any reliable thermal cycling instrument
- Primer solutions 10 μM (i.e. 10 picomoles/μl). For example, to prepare a 10 μM solution for a 20-mer oligonucleotide, dissolve 6.5 μg in 100 μl distilled water.[a]
- DNA template (500 ng/μl). Prepare the template DNA by careful extraction to avoid polymerase inhibitors co-extracting with DNA. (Residual phenol and chloroform, as well as haem degradation products in the case of blood, interfere with PCR.) Quantify the template DNA to allow the appropriate number of cycles to be calculated. The key consideration is the number of template copies, not mass, and the number of loci to be amplified per template. The formula for the number of molecules in picomole equivalents given weight $m$ (ng) of template molecules with length $l$ (bp) and total number of loci per template $t$ is 1.54 $mt/l$.
- Standard white mineral oil (Mallinckrodt). This can be light or heavy oil
- *Taq* DNA polymerase (5 U/μl)
- 10 × loading dye (20% (w/v) Ficoll 400, 10 × TAE, or 10 × TBE, 5 mg/ml Bromphenol Blue, 5 mg/ml xylene cyanol)
- Agarose gel. *Either* TAE *or* TBE (1 ×) can be used as gel and tank buffers. Prepare the gel using the following guidelines[b]; 3% agarose gel for target DNA less than 500 bp in size; 2% gel for target DNA between 500 bp and 1000 bp; 1.5% gel for DNA target between 1–2 kb.
- 50 × TAE stock solution pH 8.5.[c] This contains, per litre, 242 g Tris base, 57.1 ml glacial acetic acid, 37.2 g $Na_2EDTA{\cdot}2H_2O$.
- 10 × TBE stock solution pH 8.0.[c] This contains, per litre, 108 g Tris base, 55 g boric acid, 40 ml 0.5 M EDTA.
- 0.5 μg/ml ethidium bromide

### A. *Aliquoting of buffers*

1. Carry out the optimization in 15 reaction mixtures of 25 μl each. Start by labelling 15 microcentrifuge tubes A–O.
2. Gently stir each tube of buffer A–O (*Table 1*) to resuspend any precipitated components. Do so gently to prevent bubble formation. Pipette 2.5 μl of each buffer into the respective microcentrifuge tube.
3. For each tube, layer 25 μl mineral oil over the buffer to prevent condensation during thermal cycling.[d] Wax sealer pellets can be added to each tube containing buffer instead of mineral oil. Melt the pellets by heating the tubes to 60°C. After cooling down the tubes to room temperature, the buffer solution should be sealed by wax.

### B. *Preparation of the master mixture*

4. Add the following reagents to a microcentrifuge tube labelled 'master mixture'.
    - Primer 1 — 4 μl
    - Primer 2 — 4 μl
    - dNTPs stock solution — 16 μl
    - DNA template — 4 μl
    - Distilled $H_2O$ — 327 μl
    - *Taq* DNA polymerase (5 U/μl) — 1.6 μl[e]

Stir the mixture to ensure that all of the components are in homogeneous solution.

C. *Assembly of reaction mixtures*

5. Add 22.5 μl of the master mixture to each of the 15 buffer tubes A–O.[f] Centrifuge the tubes briefly in a microcentrifuge to bring the contents to the tube bottom.

D. *Thermal cycling*

6. Thermally cycle the samples using the following profile:
   - 1 min denaturation at 94°C per kb length
   - 1 min annealing at $T_d$
   - 1 min polymerization at 72°C per kb length

   These time recommendations may be reduced to 30 sec each when using ultra-thin tubes.
7. Repeat the cycling profile for a total number of cycles dependent on the primer to template ratio[g] as follows:

| *No. of cycles* | *Primer to template ratio* |
|---|---|
| 20 | $10^5$ |
| 30 | $10^7$ |
| 40 | $10^9$ |
| 50 | $10^{11}$ |

E. *Sample preparation for electrophoresis*

8. Remove a 12.5 μl aliquot from each amplification vial by sampling the reaction mixture directly through the mineral oil or wax layer. Wipe off the mineral oil or residual wax from the pipette tip with a clean tissue.
9. Add the 12.5 μl aliquots to 5 μl of 5 × loading dye in tubes prelabelled A–O and mix well. Refrigerate the vials containing the remaining reaction mixture (step 8) at 4°C, pending further gels if necessary.

F. *Agarose gel electrophoresis*

10. Load all of the samples on to the gel and electrophorese at 75 V for 4 h.
11. After electrophoresis, stain the gel with ethidium bromide (0.5 μg/ml) in gel tank buffer. Note that the tracking dyes (Bromophenol Blue and xylene cyanol) must never be allowed to run off the gel. The full extent of products and artefacts should be visualized, including primer-dimers which migrate slightly below the Bromophenol Blue front.
12. Evaluate the yield and the specificity of the 15 reactions. Use this to determine the optimal buffer, then prepare this buffer in bulk.

**Protocol 2.** *Continued*

G. *PCR of DNA templates under optimal conditions*

**13.** Prepare the master mixture (see step B4) using the optimal buffer determined in the above titration experiment. Prepare all reactions from a common stock to ensure reproducibility.

**14.** Mix together the following reagents per 100 μl of total reaction mixture:

| | |
|---|---|
| • 10 × optimal buffer | 10 μl |
| • Primer 1 | 1 μl |
| • Primer 2 | 1 μl |
| • dNTPs stock solution | 4 μl |
| • *Taq* DNA polymerase | 2 U |
| • Distilled $H_2O$ | bring to 100 μl |

Stir the mixture to ensure that all of the components are in homogeneous solution.

**15.** Add the master mixture to individual DNA templates (for genomic DNA, the suggested DNA concentration in the final reaction is 5 ng/μl). Always include a control reaction containing a DNA known to amplify.

**16.** Amplify and analyse by electrophoresis as in steps 6–11.

[a] The formula for the amount (μg) of oligonucleotide *n* bases long to be dissolved in 100 μl distilled water to yield a 10 μM solution is 0.325*n*.

[b] If the gel of a product less than 1 kb is to be blotted, we recommend using equal proportions of regular and highly sieving agaroses (e.g. NuSieve, FMC Co.) for preparing the agarose gel used for electrophoresis. The softer consistency of highly sieving agaroses allows for faster and more complete DNA transfer than afforded by regular agarose gels, particularly if the agarose concentration exceeds 1.5%.

[c] *Either* TBE *or* TAE buffer will be required depending on the agarose gel electrophoresis conditions chosen.

[d] Some instruments have features such as overlying heat blocks which preclude condensation automatically and require no oil in the sample. Consult the manufacturer's instructions.

[e] Recommended enzyme amounts are for *Taq* DNA polymerase. Consult manufacturer's recommendations for other enzymes.

[f] Wax-sealed tubes could have the reaction mixture overlaid on the wax.

[g] The primer to template ratio can be calculated as follows. For a mass *g* (ng) of an oligonucleotide *n* bases long, a mass *m* (ng) of a double-stranded template *l* base pairs long, and a total number *t* of loci per template (*t* is 1 for single copy loci but higher for multiple-copy targets), then the molar ratio of primer to template is *2gl/nmt*.

## 6. Troubleshooting

The following analysis is based on the premise that one is attempting to obtain PCR product with a pair of newly designed primers for which one has no prior experience. Obviously, different scenarios present different problems but the points made below will be applicable to many of those alternative scenarios.

## 6.1 No PCR product; an empty lane

### 6.1.1 Was the reaction set up properly?

Before testing each individual component, verify that all components were pipetted into the reaction in the correct amounts by repeating the experiment at least once. Check the micropipettor for accuracy before repeating the experiment. If no product is again the result, then consider the components in order:

(a) *Polymerase*. Was there a positive control of the thermostable polymerase? If not, perform a control amplification with a combination of primers, template, buffer, and cycle profile that is known to work.

(b) *Nucleotides*. Was the nucleotide concentration used adequate for the reaction? The positive control for enzyme function (above) will also control for nucleotides.

(c) *Buffer*. Was a reasonable buffer used? This may be the most critical variable that is unique to this particular template and pair of primers. See Section 5 for optimizing buffer composition.

(d) *DNA template*. Was there a control proving that other primers will amplify from this template? If genomic DNA is involved, re-precipitate and wash the DNA, then retest it. Alternatively, test several independent extractions of the genomic DNA as very small amounts of phenol and/or chloroform left in the DNA solution will inhibit a PCR reaction. Excessive template can also inhibit the reaction by binding all of the primers.

(e) *Primer pair*. Assuming that the primers were not confused so that inappropriate primers were used, the complete absence of any product is probably not the fault of the primers, but if nothing else seems to be wrong, it could be 'faulty' primers. The synthesis of one of the primers could have been poor so that little of the product is of full length primer. Check the absorbance of the primers and/or run them on a sequencing gel to check yield and purity. Alternatively, there may have been a sequence error in ordering a primer or entering the sequence into the synthesizer. Check the records. Finally, the sequence used to design one of the primers may have been erroneous and the primer has reduced or no homology to the template. Selecting other sequence for the primers is the only recourse in this case (see Section 3).

(f) *Water*. Was enough added to properly dilute the buffer?

(g) *Cycling profile*. Did the thermal cycling machine really cycle or did it stay in the hold position overnight? Check this; operators *have* been known to fail to push the start button! More likely possibilities are that the annealing temperature was too stringent (too high) or that the denaturing temperature was not high enough. On thermal cyclers with a slow response of the heat block, wells near the periphery may be slow to heat up. If the reaction volume is large (100 μl) and the amplified sequence high in GC pairs, the reaction may not reach a high enough temperature to denature the template.

### 6.1.2 Is the detection method sensitive enough?

It is possible that 'large' amounts of specific product are present, but are below the level of easy detection using ethidium bromide staining (approximately 10 ng DNA). Try destaining the gel and/or using a longer exposure in order to photograph it. Alternatively, use of a fluorescent-labelled primer will enhance detection of the product. If a suitable radioactive probe is available, another procedure is to blot the gel and then detect the PCR product by hybridization. Such a probe is easily prepared by 5′-labelling of one of the oligonucleotide primers. The author's experience in screening a YAC library by PCR showed that blotting and probing the PCR product from the many pools was extremely sensitive at identifying the proper pools and sub-pools containing the YAC being sought.

## 6.2 Too many bands

(a) *Primers not sufficiently specific.* Although a last resort when troubleshooting, it is always possible the primers are good matches to several sites, not just the desired target (see Section 3).

(b) *Annealing temperature too low, causing mispriming.* As a first step, because it is simple to check, try using much higher annealing temperatures, especially in the first few cycles (see Sections 4.2 and 4.11.2).

(c) *Too many cycles.* Compare the number of bands and their relative amounts after fewer cycles. If there is proportionately more of the desired product at the earlier cycles, there may be much more template to start with than expected.

(d) *Excessive* $Mg^{2+}$*, enzyme, primer, or dNTPs.* If altering the cycling temperatures and profile do not eliminate the extra bands, review the calculations to see if reaction concentrations are in the ranges that generally work. If they are, then try the optimization protocol (see Section 5).

## 6.3 Lots of primer-dimer

(a) Were the 3′-ends of the two primers complementary?

(b) Was the annealing temperature low enough to allow annealing to the template?

(c) Was the primer concentration excessive?

## 6.4 Product of the wrong size

(a) What was the basis for the expected size? Some published sizes for expected products are only very rough estimates, so check the basis for size estimation.

(b) Could the primers have homology with repetitive DNA sequence in the template? (3).

(c) Could the primers have homology with mtDNA? (4).

## 6.5 PCR product with one template but not another

(a) Cloned DNA is a much less complex template than genomic DNA.

(b) The primers may be annealing to polymorphic sequences which render amplification allele specific.

(c) The DNA preparation used for the template that failed to work may be faulty. Confirm that there is an adequate amount of the correct DNA template present and that it is clean and free of *Taq* DNA polymerase inhibitor. One possible problem is that DNA samples reconstituted in TE buffer may have the $Mg^{2+}$ concentration effectively lowered by EDTA chelation, which will inhibit PCR. Similarly, DNA extracted from blood may retain haem degradation products that interfere with PCR (8).

# References

1. Ruano, G., Brash, D. E., and Kidd, K. K. (1991). *Amplifications*, **7**, 1.
2. Mullis, K. B. and Faloona, F. A. (1987). In *Methods in enzymology*, Vol. 155 (ed. R. Wu), pp. 335. Academic Press, London
3. Batzer, M. A., Gudi, V. A., Mena, J. C., Foltz, D. W., Herrera, R. J., and Deinginger, P. L. (1991). *Nucleic Acids Res.*, **19**, 3619.
4. Zullo, S., Kennedy, J. L., Gelernter, J., Polymeropoulos, M. H., Tallini, G., Pakstis, A. J., Shapiro, M. B., Merrill, C. R., and Kidd, K. K. (1993). *PCR Methods and Applications*, **3**, 39.
5. Ruano, G., Fenton, W., and Kidd, K. K. (1989). *Nucleic Acids Res.*, **17**, 5407.
6. D'Aquila, R. T., Gechtel, L. J., Videler, J. A., Eron, J. J., Gorczyoa, P., and Kaplan, J. C. (1991). *Nucleic Acids Res.*, **19**, 3749.
7. Erlich, H. A., Gelfand, D., and Sninsky, J. J. (1991). *Science*, **252**, 1643.
8. Ruano, G., Pagliaro, E. M., Schwartz, T. R., Lamy, K., Messina, D., Gaensslen, R. E., and Lee, H. C. (1992). *Biotechniques*, **13**, 266.
9. Don, R. H., Cox, P. T., Wainwright, B. J., Baker, K., and Mattick, J. S. (1991). *Nucleic Acids Res.*, **19**, 4008.
10. Sardelli, A. D. (1993). *Amplifications*, **9**, 1.
11. Kwok, S. and Higuchi, R. (1987). *Nature*, **339**, 237.
12. Murphy, P. D., Ferguson-Smith, A. C., Miki, T., Feinberg, A. A., Ruddle, F. H., and Kidd, K. K. (1987). *Nucleic Acids Res.*, **15**, 6311.
13. Gelfand, D. H. and White, T. J. (1990). In: *PCR protocols: a guide to methods*

and applications (ed. M. A. Innis, D. H. Gelfand, J. J. Sninsky, and T. J. White), pp. 129. Academic Press, San Diego.

14. Rychlik, W. and Rhoads, R. E. (1989). *Nucleic Acids Res.*, **17**, 8543.
15. Wallace, R. B. and Miyada, C. G. (1987). In *Methods in enzymology* (ed. S. L. Berger and A. R. Kimmel), Vol. 152, pp. 432. Academic Press, London.
16. Rychlik, W., Spencer, W. J., and Rhoads, R. E. (1990). *Nucleic Acids Res.*, **18**, 6409.
17. Oste, C. (1989). *Amplifications*, **1**: 10.
18. Holm, T., Terry, C., and Georges, M. (1991). *Crime Lab Dig.*, **18**, 187.
19. Kogan, S. C., Doherty, M., and Gitschier, J. (1987). *N. Engl. J. Med.*, **317**, 985.
20. Innis, M. A., Myambo, K. B., Gelfand, D. H., and Brow, M. A. D. (1989). *Proc. Natl Acad. Sci. USA*, **85**, 9436.
21. Saiki, R. K., Gelfand, D. H., Stoffel, S., Scharf, S. J., Higuchi, R., Horn, G. T., Mullis, K. B., and Erlich, H. A. (1988). *Science*, **239**, 487.
22. Ruano G., Kidd, K. K., and Stephens, J. C. (1990). *Proc. Natl Acad. Sci. USA*, **87**, 6296.
23. Keohavong, P. and Thilly, W. G. (1989). *Proc. Natl Acad. Sci. USA*, **86**, 9253.
24. Ruano, G. and Kidd, K. K. (1989). *Amplifications*, **3**, 12.
25. Ruano, G., Brash, D. E., and Kidd, K. K. (1991). *Amplifications*, **7**, 1–4, October 1991.
26. Phillips, D. J., Benson, J. M., Pruckler, J. M., and Hooper, W. C. (1992). *PCR Methods and Applications*, **2**, 45.
27. Sarkar, G., Kapeiner, S., and Sommer, S. S. (1990). *Nucleic Acids Res.*, **18**, 7465.
28. Seto, D. (1990). *Nucleic Acids Res.* **18**, 5905.
29. Hung, T., Mak, K., and Fong, K. (1990). *Nucleic Acids Res.*, **18**, 4953.
30. Bachmann, B., Luke, W., and Hunsmann, G. (1990). *Nucleic Acids Res.*, **18**, 1309.
31. Wanner, R., Tilmans, I., and Mischke, D. (1992). *PCR Methods and Applications*, **1**, 193.
32. Schwarz, K., Hansen-Hagge, T., and Bartram, C. (1990). *Nucleic Acids Res.*, **18**, 1079.

# 2

# Use of speciality phosphoramidites in PCR

MICHAEL J. McLEAN

## 1. Introduction

In the decade or so since the introduction of automated oligonucleotide synthesizers, the ability to use these machines for the incorporation of unnatural residues into nucleic acids has led to the chemical synthesis of a wide variety of phosphoramidite derivatives. When included in an oligomer synthesis, these entities provide a source of site-specific, stable modifications of the backbone, thus enabling a rigorous determination of their biochemical effects.

This chapter will focus on the applicability of several such reagents for use in PCR. In particular, reagents will be discussed which permit:

- an increase in the efficiency of primer synthesis
- the generation of double-stranded PCR products with single-stranded tails
- selective strand separation of PCR products
- post-PCR covalent cross-linking of double-stranded DNA

Of the bewildering array of phosphoramidites published in recent years, only a relatively small number are commercially available (although this number is increasing rapidly) and thus readily accessible to those molecular biologists without the help of a synthetic chemist. Of these, perhaps the most widely used has been the reagent which introduces an amino group at the 5′-end of an oligomer (1), thus providing a reactive centre for the attachment of fluorophores, enzymes, haptens and other groups to the DNA (see Chapter 3 for details of methods of non-radioactive labelling of oligonucleotides). Other, simple, modifications that are well known include the phosphorylation of the 5′-terminus for post-PCR ligation to other duplexes or for selective exonuclease digestion of one strand (2, 3), and thiolation of the 5′-end as an alternative reactive group for non-isotopic tagging.

One group of (necessarily) more complicated molecules which, at the time of writing, are not available commercially are the ‘Universal bases’ reported by Brown *et al.* (4, 5). These compounds include a purine analogue which binds with comparable affinity to both thymine and cytosine, and a pyrimidine analogue which similarly recognizes both adenine and guanine. Because of the degeneracy of the genetic code, DNA sequences deduced from amino acid sequences cannot be uniquely defined and this complicates the design of primers for their amplification. It is proposed, however, that the use of these analogues in primer synthesis should avoid or reduce the need to make multiple primers (4, 5).

Other reagents which at first glance may appear to have no immediate application in PCR include the *N*-aminoethyl adenine analogue reported by Cosstick (6), which promotes covalent cross-linking, and a group of molecules which deserve special mention, the 'versatile' phosphoramidites of Xu *et al.* (7, 8). Using these compounds, a masked nucleoside derivative is introduced into the oligomer which can be selectively modified during the deprotection procedure to generate species such as 4-methoxythymidine, 6-thiodeoxyguanosine, etc. Such modifications may have applications in PCR (for example to reduce or augment the strength and/or specificity of hybridization) which have yet to be realized and attempted.

The aim of this chapter is to highlight the use of speciality phosphoramidites which do have a clear applicability to PCR.

## 2. Reagents for the preparation of two oligomers per synthesis (TOPS™)

Because of the short cycle times employed by DNA synthesizers, it is feasible to prepare four 30-mers within the working day and to start the synthesis of another four oligomers which will be completed late in the evening. This means that the synthesizer is lying idle for approximately 12 hours each day, until the operator sets up further syntheses the next morning. Many core facilities often have a severe backlog of synthesis requests and so the synthesizer is not being used to its full capacity. This is particularly so for the growing number of people who have one-column machines.

The majority of oligonucleotides are used in pairs, principally as primers for PCR, and yet are synthesized separately and subsequently combined. Making pairs of oligomers co-temporaneously should generate considerable savings in time and labour.

### 2.1 The TOPS concept

The concept behind the TOPS reagents is shown schematically in *Figure 1*. Instead of synthesizing two oligomers separately, they are entered into the synthesizer memory as one long sequence punctuated by the TOPS compound, the reservoir of which is attached to a spare port on the instrument. When synthesis is complete, the oligomer is cleaved from the support by

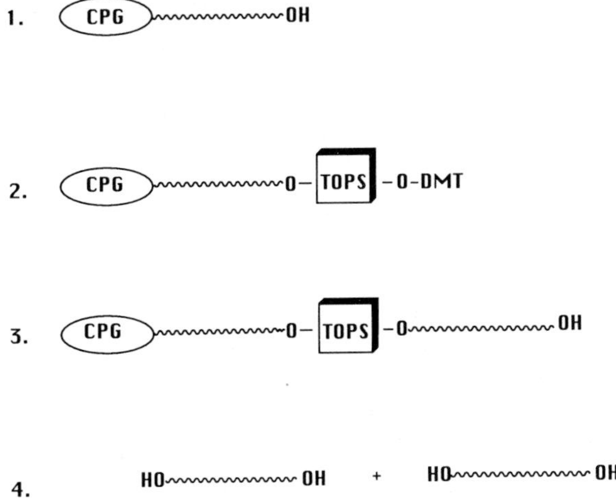


**Figure 1.** Schematic representation of the TOPS concept. (1) The first primer is synthesized. (2) The TOPS reagent is added to the 5′-end of the first primer. (3) The second primer is synthesized on the end of the first. (4) Base deprotection also leads to cleavage of the link between the two oligomers, generating two primers free in solution.

treatment with concentrated aqueous ammonia and the protecting groups on the heterocyclic bases are removed by prolonged treatment of the ammoniacal solution at elevated temperature. During this process, the link between the two oligomers and the TOPS compound will also be cleaved, generating the two primers free in solution.

To achieve the aim of making several oligomers on one support and in a single user-defined synthesis, a reagent has to be designed that meets several important criteria:

(a) The reagent should be compatible with standard synthesis protocols as used on automated instruments and, in particular, the coupling of it to the 5′-hydroxyl of the first oligomer must be as efficient as that achieved by nucleoside phosphoramidites. Even though phosphoramidite chemistry provides extremely high coupling yields, there is nevertheless a fall-off in yield of final product which is proportional to the length of the oligomer. This means that there will always be slightly less of the second oligomer than the first and, if the coupling of the TOPS reagent is inefficient, this difference could become significant.

(b) Whatever method is chosen to effect chain scission to provide the separate oligomers, this process must go to completion before the mixture can be used for the desired purpose. This is particularly pertinent when the mixture is to be used as a source of primers for PCR, as any uncleaved oligomer will also prime synthesis from one end of the target DNA, leading to amplified fragments that are longer than expected. Another consequence may be a reduction in hybridization stringency, giving rise to spurious PCR products.

(c) It is essential that oligomers produced by this process bear free hydroxyls at their 3′-termini, as the overwhelming majority of applications for synthetic oligonucleotides require that this position is available for further reaction.

(d) For some applications, a useful modification of the 5′-position of the first oligomer is its phosphorylation. For PCR primers, modification of the 5′-position rarely influences their properties, but other applications require that enzymatic activity can proceed unhindered at this terminus.

(e) Since the whole purpose of the TOPS approach is to save time and effort, the method used should involve the operator in as little post-synthesis chemistry as possible. Ideally, the cleavage of the link should be performed at the same time as the obligatory deprotection of the nucleoside bases.

The compounds we prepared in order to overcome these problems are shown in *Figure 2*. In the synthesis of oligonucleotides using these compounds, the phosphoramidite reacts with the 5′-hydroxyl of an oligomer tethered to a support, and the succinate esters are stable to the conditions

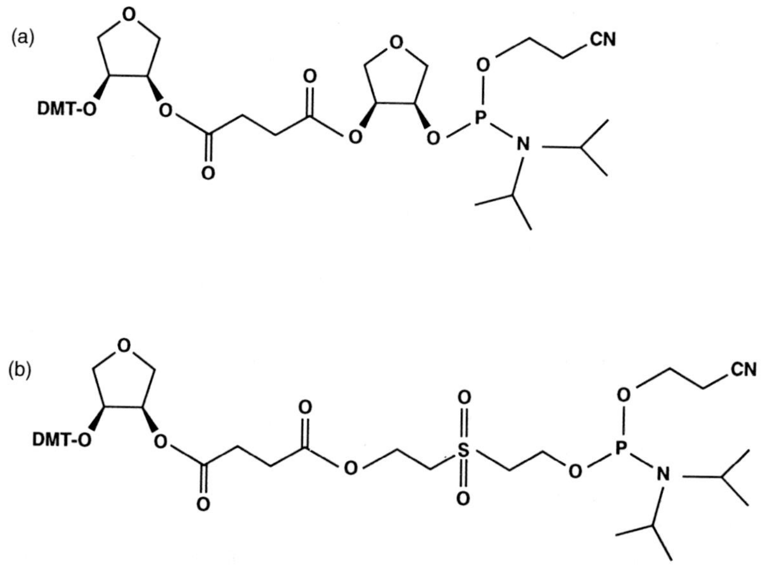

(a) DMT-O O O O O O O P O CN N

(b) DMT-O O O O O O S O O O P O CN N

**Figure 2.** Chemical structures of the TOPS phosphoramidites. (a) Universal TOPS phosphoramidite; (b) Universal TOPS Phosphate ON phosphoramidite.

employed in subsequent, repetitive steps in the synthesis cycle. After oxidation of the phosphite intermediate, deprotection of the secondary hydroxyl enables reaction with an incoming phosphoramidite, i.e. a second oligomer sequence is constructed on the first.

At the end of the synthesis, the conditions used for base deprotection also cause hydrolysis of the succinate esters, leaving two oligomers with terminal phosphodiester moieties. Under the prevailing basic conditions, a further reaction takes place which involves an attack at the phosphate by the vicinal *cis*-hydroxyl, ejecting DNA as a leaving group resulting in oligonucleotides with terminal hydroxyls. In the case of the Universal TOPS Phosphate ON reagent (see *Figure 2*) this reaction occurs only at the 3′-end of the second oligomer. Assisted by the electron-withdrawing sulphone, β-elimination from the sulphonyldiethanol results in the release of the first oligomer with a phosphate on the 5′-hydroxyl.

The procedures for synthesizing a pair of oligonucleotides using the Universal TOPS phosphoramidite are described in *Protocol 1*.

**Protocol 1.** Making a pair of oligomers using the TOPS compound

*Reagents*

- Universal TOPS phosphoramidite (Cambridge Research Biochemicals)
- Nucleoside phosphoramidites and ancillary reagents (e.g. from ABI, Cruachem, or Millipore)
- 40% methylamine (Aldrich)

*Method*

1. Dissolve the TOPS phosphoramidite in 1 ml anhydrous acetonitrile, using the procedures normally followed in oligonucleotide synthesis.
2. Swirl the contents of the vial to ensure complete dissolution (because the reagent is a glass it may take several minutes to dissolve).
3. Install the vial at one of the spare amidite ports on the synthesizer.
4. Type in the sequences of the desired oligomers as one long sequence punctuated by the TOPS reagent. For example, the oligomers 5′-AGCT-3′ and 5′-TAGTTC-3′ would be entered as:

   5′-AGCT 5 TAGTTC-3′

   with the TOPS amidite in bottle position 5.
5. For optimum efficiency of synthesis, extend the wait time during coupling of the TOPS compound from the usual 30 sec to 5 min. The best way to do this is to modify an existing cycle which can then be recalled from the memory each time the TOPS reagents are used. An example of the modification is shown below.

**Protocol 1.** *Continued*

| *Step No.* | *Function No.* | *Name* | *Step time* | *Step active for bases A* | *G* | *C* | *T* | *5* | *6* | *7* | *Safe step* |
|---|---|---|---|---|---|---|---|---|---|---|---|
| – | – | – | – | – | – | – | – | – | – | – | – |
| – | – | – | – | – | – | – | – | – | – | – | – |
| – | – | – | – | – | – | – | – | – | – | – | – |
| 30 | −50 | Group 3 off | 1 | Yes | Yes | Yes | Yes | Yes | Yes | Yes | Yes |
| 31 | 4 | Wait | 30 | Yes | Yes | Yes | Yes | Yes | Yes | Yes | Yes |
| 32 | 4 | Wait | 270 | No | No | No | No | Yes | Yes | Yes | Yes |
| 33 | 16 | Cap prep | 10 | Yes | Yes | Yes | Yes | Yes | Yes | Yes | Yes |

In this cycle, the new step is number 32.

6. Perform oligonucleotide synthesis.
7. For machines with the capacity for automated cleavage of the oligomer from the support, add 1 ml 40% methylamine to the vial containing the oligomer solution. Cap the vial tightly and incubate at 60°C for 8–16 h.
8. For manual cleavage of the oligomer from the support, replace the concentrated ammonia solution usually used with 2 ml 40% methylamine. Elute the oligomer into a glass screw cap vial and incubate the tightly capped vial at 60°C for 8 h.
9. Lyophilize the solution and dissolve the residue in 1 ml of sterile distilled water. The oligomers are now ready for use in PCR.

The following experiment was performed to test the efficiency in PCR of primers prepared by the TOPS approach. A 103-residue oligonucleotide was synthesized using the Universal TOPS reagent. Cleavage of this oligomer was designed to give a pair of PCR primers, 48 nt and 54 nt in length, which were then used to amplify a portion of the pBR322-encoded β-lactamase gene (9). Each primer incorporated a restriction enzyme recognition site, *Asp*718 (a *Kpn*I isoschizomer) in the 48-mer and *Nar*I in the 54-mer, for cloning the DNA encoding the mature protein. PCR using this primer pair in three 100 μl reactions with pBR322 template DNA gave the expected 848 bp product (see *Figure 3*). The integrity of the ends of this product was demonstrated by digestion using *Asp*718 and *Nar*I, followed by gel purification and ligation to concatamers as shown in *Figure 3*. Redigestion with the same enzymes reduced the concatamers to the original monomeric product. An aliquot was removed after PCR, after the subsequent restriction enzyme digestions, and after ligation. The aliquots were analysed by agarose gel electrophoresis. The final product was successfully cloned into an M13mp8 derivative using a synthetic *Nar*I/*Eag*I linker duplex during the construction of a phage–enzyme (10) and conferred ampicillin resistance to transfected *E. coli* TG1 cells. Expression and secretion of the *bla*/M13 gene 3 fusion protein was driven by the M13 gene 3 promoter and secretory leader sequence (11), respectively.

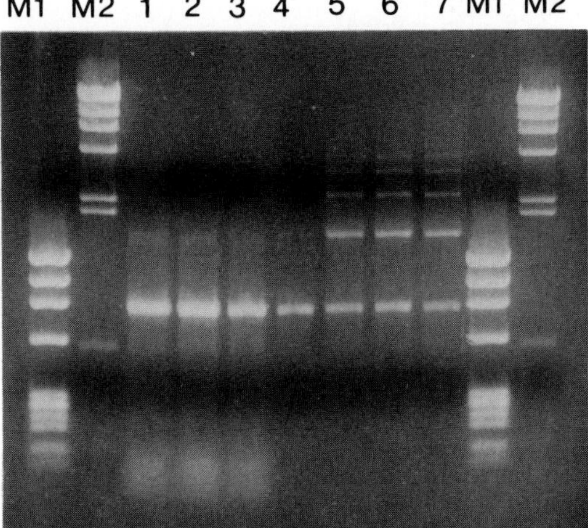


**Figure 3.** Testing the efficiency of PCR using primers prepared by the TOPS approach. The figure shows electrophoretic analysis of the products using a 1.2% ethidium bromide stained agarose gel. Lane 1, 10% aliquot of pBR322 encoded *bla* gene PCR product; lane 2, 10% aliquot of *NarI* digest of PCR product; lane 3, 10% aliquot of *NarI* plus *Asp*718 digested PCR product; lane 4, 10% aliquot of gel-purified *NarI* plus *Asp*718 digested PCR product; lane 5, 10% aliquot of self-ligation of gel-purified *NarI* plus *Asp*718 digested PCR product after 1 h; lane 6, as lane 5 after ligation for 4 h; lane 7, as lanes 5 and 6 after ligation for 16 h. M1 and M2 are size markers, *Hae*III digested ϕX174 DNA and *Hin*dIII digested lambda DNA, respectively.

## 2.2 The introduction of redundancies or modified residues at the 3′-end of oligomers

For some applications (such as ARMS, see Chapter 12), it may be desirable to use a mixture of primers which differ from each other only in the identity of the base at the 3′-end. Clearly, these could be prepared by repeated synthesis of the necessary oligomers with subsequent combination or by manually mixing the appropriate base-derivatized solid supports and refilling an empty synthesis column. It is sometimes also desirable to construct an oligomer with a modified base (for example, inosine) at the 3′-terminus for which there is no appropriately derivatized support.

A simple extension of the TOPS concept answers both these problems; the method is shown schematically in *Figure 4*. Any available derivatized support is installed on the synthesizer and the TOPS reagent is coupled to it, thus providing a 'universal' 3′-end. Next, the modified residue, or mixture of nucleotides (using the ability of most synthesizers to add any combination of the four bases at any position during synthesis) is coupled to the TOPS reagent and synthesis progresses as normal. At the end of the synthesis,

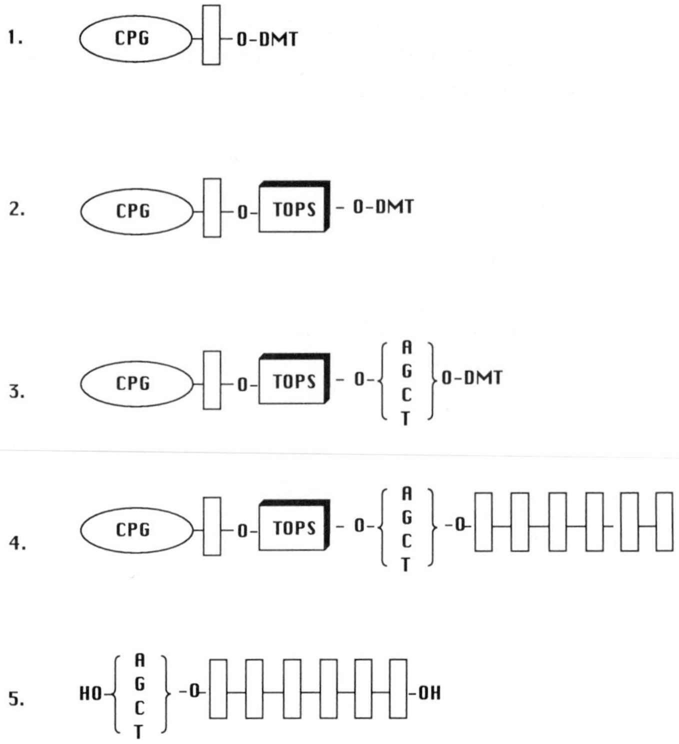

**Figure 4.** Schematic representation of the use of TOPS reagents to introduce modified residues at the 3′-end of an oligomer. (1) Synthesis is initiated with any standard column. (2) Addition of the TOPS reagent. (3) The mixture of bases (or modified residue) is coupled to the TOPS reagent. (4) Oligomer synthesis continues. (5) Alkali-cleavage of the link gives the modified oligomer in solution.

cleavage of the link generates the desired 3′-modification together with the single nucleoside which was used to initiate synthesis. This approach is described in practical detail in *Protocol 2*.

## Protocol 2. Introduction of redundancies at the 3′-end

### *Reagents*

- Universal TOPS phosphoramidite (Cambridge Research Biochemicals)
- Nucleoside phosphoramidites and ancillary reagents (e.g. from ABI, Cruachem, or Millipore)
- 40% methylamine (Aldrich)

### *Method*

1. Perform steps 1–3 as in *Protocol 1*.

2. Type in the sequence of the desired oligomer, extending the sequence at the end to include the TOPS reagent and the column to be used. For example, to make the oligomers:

   5′-CCCCT-3′
   CCCCG
   CCCCA
   CCCCC

   type in the sequence as:

   5′-CCCC(TGCA) 5 T-3′

   with the TOPS reagent in bottle position 5 and (for example) a T-column on the instrument.
3. Perform steps 5–9 as in *Protocol 1*.

## 3. Phosphoramidites for the preparation of primers with non-amplifiable tails

The PCR process can be adapted to introduce additional sequences to the ends of amplified products by means of tailed primers. To achieve this, an oligonucleotide is prepared in which the 3′-portion is complementary to the region of DNA to be amplified, and the 5′-portion contains the additional sequence information to be used in some post-PCR manipulation. This primer is used to initiate DNA synthesis in the usual way, and the new strand so produced becomes the template for the reverse primer in the second (and subsequent) round(s) of PCR. In this way, the single-stranded tail is incorporated into the double-stranded DNA product where it can then be (for example) cut by restriction enzymes for directional cloning. A further variant of this type of approach is found in the Chemical Genetics system, described in Chapter 10.

If the tails are not recognized as a template for DNA synthesis, or if the polymerase is prevented from accessing them as templates, then repeated rounds of PCR will generate double-stranded amplification products with single-stranded tails which can be used as the basis for an assay for the detection of specific PCR products. The single-stranded tail at one end of the DNA segment can be used to bind to a complementary sequence covalently attached to a well of a microtitre plate, and a tail of different sequence at the other end can be used to bind to a complementary oligomer bearing a signal-generating moiety. Detailed protocols for performing these reactions are given in Chapter 12.

The method used for preventing the tails being incorporated into double-stranded product relies on the insertion into the primer of modified phosphoramidites which are not recognized as templates by DNA polymerases. Thus, when the polymerase reaches these positions during DNA synthesis, nucleotide

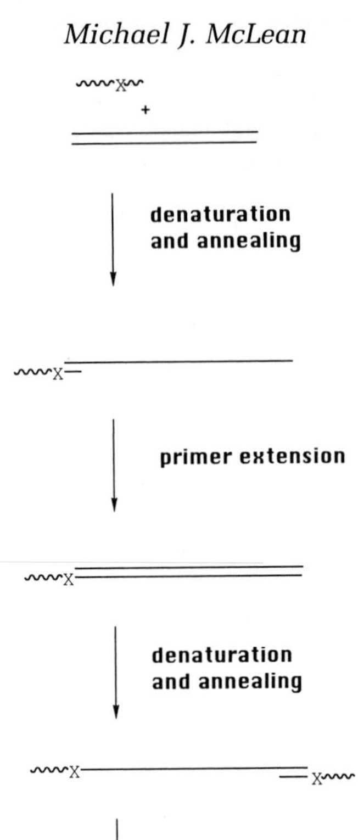


**Figure 5.** Schematic for the generation of amplified products with single-stranded tails. 'X' corresponds to the position of the modified residue which prevents the passage of *Taq* polymerase.

incorporation is prohibited and synthesis stops. This process is shown schematically in *Figure 5*.

Two analogues which are particularly effective at preventing *Taq* polymerase read-through are shown in *Figure 6*. *Protocol 3* decribes the preparation of primers with non-amplifiable tails using one of these: $C_{16}$-Spacer phosphoramidite. The non-base-pairing nucleoside analogue, naphthosine, is slightly 'leaky' in this regard (12) and at least two contiguous residues are needed for complete blockade of polymerase processivity.

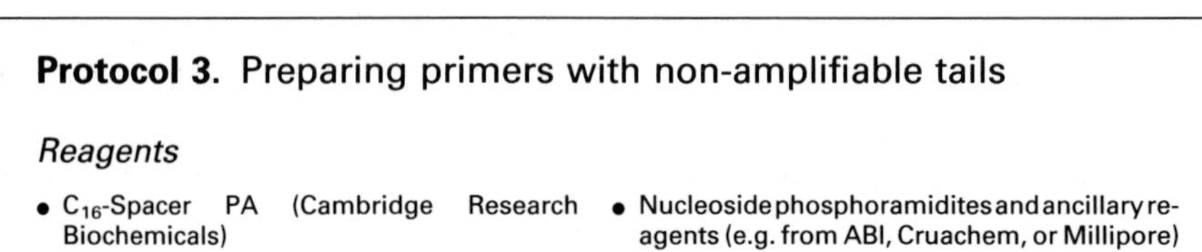

**Protocol 3.** Preparing primers with non-amplifiable tails

*Reagents*

- $C_{16}$-Spacer PA (Cambridge Research Biochemicals)
- Nucleoside phosphoramidites and ancillary reagents (e.g. from ABI, Cruachem, or Millipore)

*Method*

1. Dissolve the $C_{16}$-Spacer PA in 1.3 ml anhydrous acetonitrile using the procedures usually followed in oligonucleotide synthesis.
2. Swirl the contents of the vial to ensure complete dissolution (because the reagent is a gum, it may take several minutes to dissolve).
3. Install the vial at one of the spare amidite ports on the synthesizer.
4. Type in the sequence of the desired oligomer with the $C_{16}$-Spacer at the break point between the sequence-specific primer and the non-amplifiable tail.
5. Perform oligonucleotide synthesis.

(a)

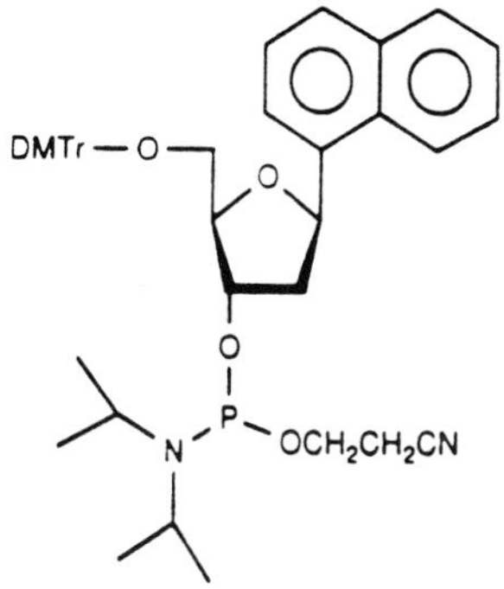

DMTr—O

O

O

P

N

$OCH_2CH_2CN$

(b) DMTr-O O P O CN N

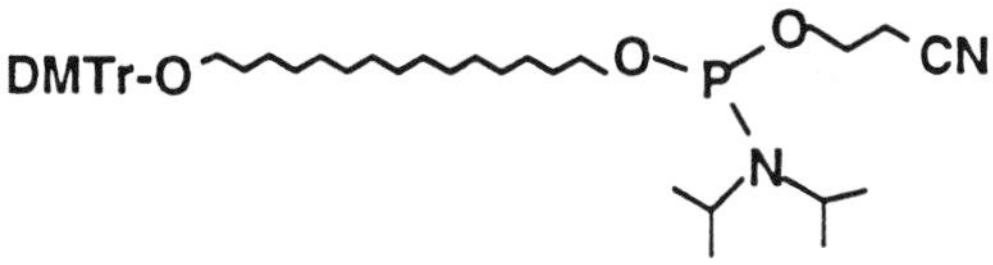

**Figure 6.** Chemical structures of naphthosine phosphoramidite (a) and $C_{16}$-Spacer phosphoramidite (b).

## 4. Reagents for the introduction of biotin

The increased use of biotinylated oligomers for the non-isotopic detection of nucleic acids and, in particular, for sequencing applications (13) has led to the development of a number of biotin phosphoramidite reagents that permit automated on-column synthesis. These give considerable improvements in terms of convenience, time saved, and yield, compared to conventional off-column biotinylations.

The presence of biotin at the 5′-end of one of a pair of PCR primers enables the solid phase capture of the amplified product via the strong interaction of biotin and immobilized avidin. Subsequent treatment of this complex under denaturing conditions (chemical or physical) specifically releases into solution the DNA strand which does not possess the biotin, while retaining the biotinylated strand on the solid phase. After washing out the non-biotinylated strand and the denaturants, the remaining DNA is available in a single-stranded form, ready for sequencing. Detailed protocols for these procedures are presented in Chapters 5 and 13.

*Figure 7* shows the structures of some of the reagents available for auto-

(I)

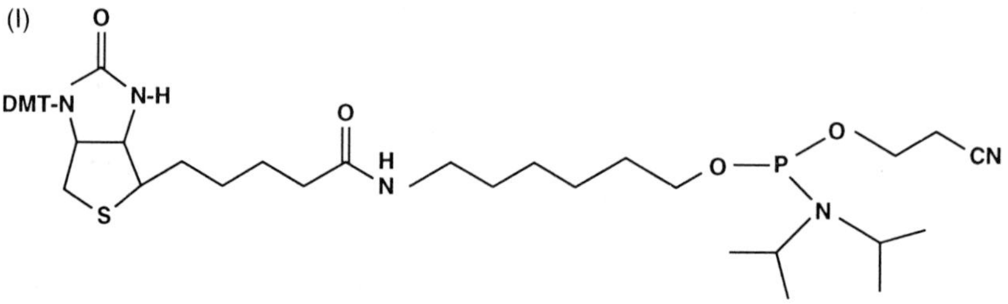

O

DMT-N N-H

O

H N

S

O—P O CN

N

(II)

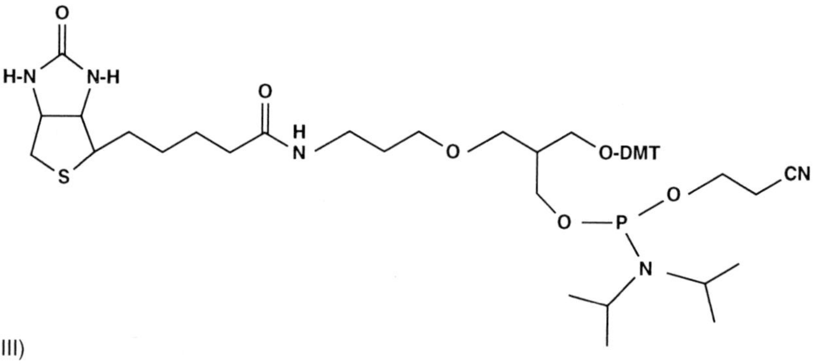

O

H-N N-H

O

H N O O-DMT

S

CN

O

O —P

N

(III)

O

H-N N-H

O

O-DMT

H N

S

O

O—P CN

N

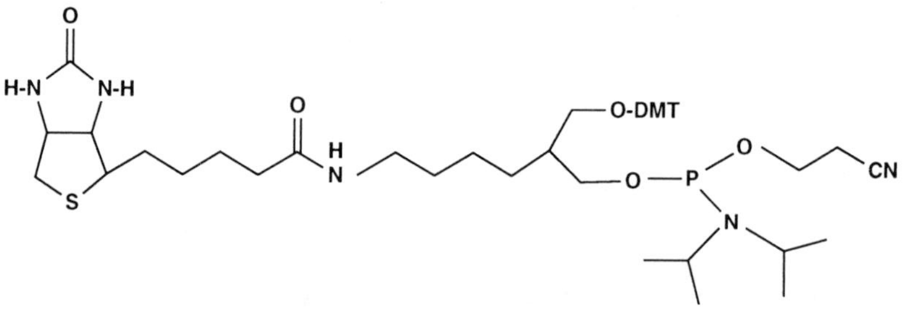

**Figure 7.** Chemical structures of three biotin phosphoramidites. Reagent I, DMT-biotin-$C_6$-PA from CRB; Reagent II, biotin phosphoramidite, from Amsterdam International; Reagent III, biotin phosphoramidite from Glen Research.

mated biotinylation of oligomers. Whereas the use of any of these reagents will give high yields of biotinylated oligomers suitable for use in PCR, there are some important chemical differences between them which should be borne in mind when considering the particular application in which the biotinylated oligomer is to be used.

Reagent I in *Figure 7* (14) introduces a single biotin at the 5′-end of oligomers. It possesses a dimethoxytrityl group (DMT) on N-1 of the ureido ring of biotin. One of the reasons given for this feature (15, 16) is that it prohibits phosphitylation at this position by another biotin phosphoramidite during the pre-activation with tetrazole or during coupling to the oligomer (or both).

Reagents II and III in *Figure 7* (17, 18) each contain a DMT-protected pendant hydroxyl, thus permitting the addition of more than one biotin unit per oligomer. This feature also allows these reagents to be incorporated into the chain of the oligonucleotide, rather than at the 5′-end, if so desired. However, the lack of protection at the ureido ring nitrogen in these molecules can lead to the occurrence of unwanted side reactions leading to chain branching (thus giving a heterogeneous product) and a lowered yield of biotinylated oligomer, necessitating purification of the primer.

This effect is shown quite clearly in *Figure 8*, for which 25-mer oligonucleotides were synthesized with one, three, and six biotins, using reagents II (lanes 1–3) and III (lanes 4–6). The unpurified oligomers were radiolabelled at their ends and analysed by denaturing PAGE. In lane 7 is the same oligomer mono-biotinylated with reagent I, and lane 8 shows an attempt to add more than one biotin to the 5′-end using this reagent. It can be seen that there is a large amount of 25-mer in lanes 1–6, due to failure of these reagents to couple quantitatively; these failure sequences are markedly absent in lanes 7 and 8, where the full length biotinylated product is marked by an arrow.

For some applications, the advantage of being able to introduce more than one biotin will outweigh the disadvantages of the heterogeneity produced, especially since a single HPLC purification will effect a marked improvement in purity. However, for sequencing with streptavidin-coated magnetic beads, sensitivity is not an issue and a single biotin is perfectly adequate.

## 5. Psoralen reagents for use in DGGE and TGGE

The detection of point mutations can be achieved by the use of denaturant gradient gel electrophoresis (DGGE) and temperature gradient gel electrophoresis (TGGE), both of which rely on the reduced electrophoretic mobility of DNA fragments at their melting points. Detailed methods for performing these analyses are given in Chapters 7 and 13. Commonly, one end of the fragment to be analysed is rendered more resistant to denaturation by the incorporation into the amplified product of a long GC-rich tail. The synthesis

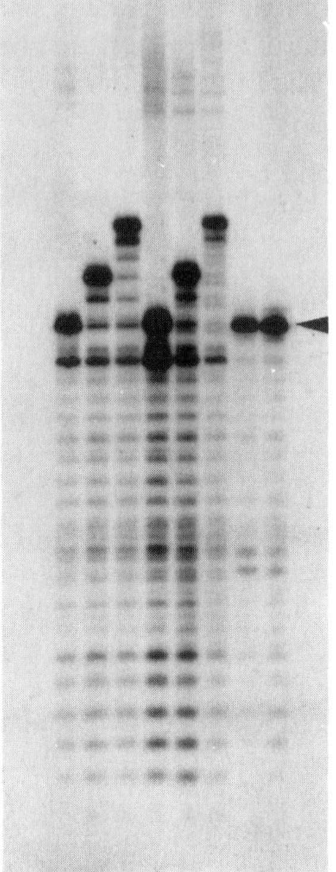


**Figure 8.** Gel electrophoresis of 3′-end-labelled, multiply biotinylated oligomers. Lanes 1–3: oligomers with 1, 3, and 6 biotins, respectively, prepared using reagent II of *Figure 7*. Lanes 4–6: oligomers with 1, 3, and 6 biotins, respectively, prepared using reagent III of *Figure 7*. Lanes 7 and 8: oligomers with 1 and 3 (nominally) biotins, respectively, prepared using reagent I of *Figure 7*. The full length, monobiotinylated product is marked with an arrow.

of primers with such long tails is both expensive and time-consuming, and the overall approach is not universally applicable to all sequences.

An alternative route is provided by the incorporation of a psoralen molecule at the 5′-end of one of the primers (19). After amplification, the psoralen can 'loop back' and intercalate between the last two base pairs of the duplex. If the DNA is now irradiated with long wavelength (365 nm) ultraviolet light, the psoralen photoreacts with the pyrimidine bases to give a covalent adduct (20, 21). This approach is described in *Protocol 4*. The preferred reaction is with thymidine, and so the best intercalation/reaction site is provided by the sequence TpA to give the bis adduct, thus covalently clamping one end of the two strands together. Clearly, one cannot always guarantee the presence of this dinucleotide step at the end of the desired product, and it is advisable to incorporate this step as a 5′-tail in the synthesis of the primer. Appligene SA provide the 5′-modification of oligomers with psoralen as a custom synthesis service.

**Protocol 4.** UV-induced cross-linking of amplified DNA

*Equipment and reagents*

- 5′-psoralen-modified oligomers (Appligene SA)
- UV transilluminator (365 nm, 200 W, e.g. from Flowgen Instruments Ltd)

*Method*

1. Chill the solution of amplified DNA, contained in an open microcentrifuge tube, in an ice bath.
2. Invert the transilluminator and support it on either side so that the lamp is approximately 5 cm from the top of the liquid in the tube.
3. Irradiate at 365 nm for 5 min.
4. Remove the light source and cap the tube. The DNA will be quantitatively cross-linked and is ready for use in DGGE or TGGE (see Chapters 7 and 13.

# References


1. Connolly, B. A. (1987). *Nucleic Acids Res.*, **15**, 3131.
2. Higuchi, R. G. and Ochman, H. (1989). *Nucleic Acids Res.*, **17**, 5865.
3. Copley, C. G. and Boot, C. (1992). *Biotechniques*, **13**, 3.
4. Lin, P. K. T. and Brown, D. M. (1989). *Nucleic Acids Res.*, **17**, 10373.
5. Lin, P. K. T. and Brown, D. M. (1991). *Nucleosides and Nucleotides*, **10**, 675.
6. Cosstick, R. and Douglas, M. E. (1991). *J. Chem. Soc., Perkin Trans. I*, 1035.
7. Xu, Y.-Z., Zheng, Q., and Swann, P. F. (1992). *Tetrahedron*, **48**, 1729.
8. Xu, Y.-Z., Zheng, Q., and Swann, P. F. (1992). *J. Org. Chem.*, **57**, 3839.
9. Sutcliffe, J. G. (1978). *Proc. Natl Acad. Sci. USA*, **75**, 3737.
10. McCafferty, J., Jackson, R. H., and Chiswell, D. J. (1991). *Protein Eng.*, **4**, 955.
11. van Wezenbeek, P. M. G. F., Hulsebos, T. J. M., and Schoenmakers, J. G. G. (1980). *Gene*, **11**, 129.
12. Newton, C. R., Holland, D., Heptinstall, L. E., Hodgson, I. J., Edge, M. D., Markham, A. F., and McLean, M. J. (1993). *Nucleic Acids Res.*, **21**, 1155.
13. Hultman, T., Stahl, S., Hornes, E., and Uhlén, M. (1989). *Nucleic Acids Res.*, **17**, 4937.
14. Pon, R. T. (1991). *Tetrahedron Lett.*, **32**, 1715.
15. Alves, A. M., Holland, D., and Edge, M. D. (1989). *Tetrahedron Lett.*, **30**, 3089.
16. Pieles, U., Sproat, B. S., and Lamm, G. M. (1990). *Nucleic Acids Res.*, **18**, 4355.

17. Misiura, K., Durrant, I., Evans, M. R., and Gait, M. J. (1990). *Nucleic Acids Res.*, **18**, 4345.
18. Catalogue of Glen Research, 1991.
19. Catalogue of Appligene, SA (1992–3), p. 82.
20. Gamper, H. B., Cimino, G. D., and Hearst, J. E. (1987). *J. Mol. Biol.*, **197**, 349.
21. Hearst, J. E. (1988). *Ann. Rev. Biochem.*, **39**, 291.

# 3

# Chemical methods for 5′ non-isotopic labelling of PCR probes and primers

ALEX ANDRUS

## 1. Introduction

Synthetic oligonucleotides are now accessible to most laboratories conducting molecular biology experiments. The decreases in preparation time, costs, and the expertise required have enabled the annual production and use of several million oligonucleotides (1). PCR undoubtedly consumes more of these sequences than any other application. Many PCR experiments require the covalent attachment (labelling, tagging, conjugation) of small molecules to perform ancillary functions, in addition to base-specific annealing to complementary nucleic acid and 3′ chain extension by polymerase. The labels (reporter groups) conduct or facilitate detection, quantification, isolation, sizing, and location of the amplified sequence. Covalent attachment of molecules, such as fluorescent dyes, biotin, and proteins, can be made at virtually any site on the oligonucleotide, including the 5′- and 3′-termini, through the sugars and bases, and by internucleotide phosphate modification (2). The sequence- and base-specific hybridization properties must be preserved, as well as the chain extension capacity for PCR primers. The methods described in this chapter will focus only on examples of specific, chemical 5′-labelling of oligonucleotides, for use as PCR primers and probes of PCR products (3). Post-synthesis protocols such as cleavage/deprotection, analysis, purification, and 5′-labelling operations are described. Enzymatic labelling methods conducted before, during, or after PCR have been reviewed elsewhere (4).

Labelled oligonucleotides can be broadly categorized by function and application (*Table 1*). The label examples discussed in this chapter will be of the detection or capture variety. A distinction can be made between oligonucleotides used as primers for the PCR and those used as probes for annealing to PCR products. Products synthesized from labelled primers, present in excess in the PCR, will bear the same label, such as biotin or fluorescent dye.

**Table 1.** The function and application of labelled oligonucleotides

| Function | Label | Application |
|---|---|---|
| Detection | Fluorescent dyes, enzymes | Laser induced fluorescence, chemiluminescence, bioluminescence, colorimetry, electrochemiluminescence |
| Capture | Biotin, solid support, intercalators, haptens | Immobilization, isolation, antibody binding |
| Solubility change | PEG, cholesteryl, triglycerides, peptides | Hydrophobicity, hydrophilicity |
| Chemical reaction | Psoralen, EDTA, phosphate | Cross-linking, chain-cleavage, ligation |

Section 2 describes the synthesis, purification, and analysis of oligonucleotides incorporating a labelling attachment site at the 5′-end. Such aminolink oligonucleotides then allow the coupling of non-isotopic labels at a subsequent step. The rest of this chapter describes specific methods whereby a variety of labels can be incorporated. In addition, Section 3.2 describes direct 5′ fluorescent dye labelling of oligonucleotides during automated synthesis.

# 2. Synthesis, purification, and analysis of oligonucleotides

## 2.1 Automated oligonucleotide (primer/probe) synthesis

Almost all synthetic oligonucleotides, including PCR primers and probes, are prepared with automated DNA synthesizers (the syntheses described in this chapter were conducted on Applied Biosystems Models 391 PCR Mate, 392 and 394) (5). Synthesis is initiated with the 3′-nucleoside attached to an insoluble support material: controlled-pore glass (CPG) or polystyrene. Nucleoside monomers, with a trivalent phosphoramidite group, are the reactive form of the latent internucleotide phosphodiester. At least four different synthesis scales are commonly conducted. At 20 pmol of primer per experiment, the smallest scale of synthesis (40 nmol) provides enough for about 500–1500 PCR experiments. Yields after synthesis and purification approximate 50% of the theoretical maximum defined by the solid support scale (*Table 2*).

Attachment sites for labels, or the labels themselves in some recent examples, can be incorporated in the oligonucleotide on the automated synthesizer. The reactive functionality for labelling, usually an amine or thiol nucleophile, can be coupled in protected form with special reagents, many of which are commercially available. Nucleophilic functionality for labelling is most conveniently and efficiently introduced at the 5′-end of an oligonucleotide.

**Table 2.** Automated DNA synthesis; approximate crude yield of unlabelled 20-mer

| **Scale of synthesis** | $A_{260}$ [a] | **mg** [b] | **nmol** |
|---|---|---|---|
| 40 nmol | 5–10 | 0.16–0.33 | 20 |
| 0.2 μmol | 20–30 | 0.66–1.0 | 100 |
| 1 μmol | 100 | 3.3 | 500 |
| 10 μmol | 800 | 26 | 4000 |

[a] $A_{260}$ = 1 absorbance unit/ml at 260 nm in a 1 cm path length.
[b] 1 $A_{260}$ = 33 μg single-stranded DNA.

Derivatization of the 5′-terminus also least affects annealing and 3′-extension. Between the reactive amine or thiol and the oligonucleotide is an inert spacer, or linker. The type or length of the linker sometimes influences binding, detection, or solubility. There are several commercially available reagents for adding an amino group, such as AminoLink (5), to the 5′-terminus of an oligonucleotide. The structure of AminoLink and the product, an aminolinked oligonucleotide, is shown in *Figure 1*. The procedures for Aminolink oligonucleotide synthesis on the DNA synthesizer are described in *Protocol 1*. An even more reactive nucleophile, a thiol group, can be appended to the 5′-end through a variety of methods (6, 7). Suitable thiol protecting groups require an extra deprotection step. The 5′ thiol-linked oligonucleotide is prone to sulphur oxidation. These drawbacks may negate the reactivity advantage of thiol, relative to amino functionality, for labelling.

## **Protocol 1.** Aminolink oligonucleotide synthesis

### *Equipment and reagents*

- Automated DNA synthesizer
- Phosphoramidite synthesis reagents
- AminoLink, 250 mg (Applied Biosystems P/N 400808)
- Acetonitrile (Burdick & Jackson, <100 p.p.m water, P/N 015-4)

### *Method*

1. Dilute the AminoLink reagent with 3.3 ml dry acetonitrile and install on the DNA synthesizer.
2. Include the AminoLink bottle position in the oligonucleotide sequence as the 5′-terminus base.
3. Synthesize the aminolink oligonucleotide at a scale from 40 nmol to 10 μmol, according to the manufacturer's instructions.

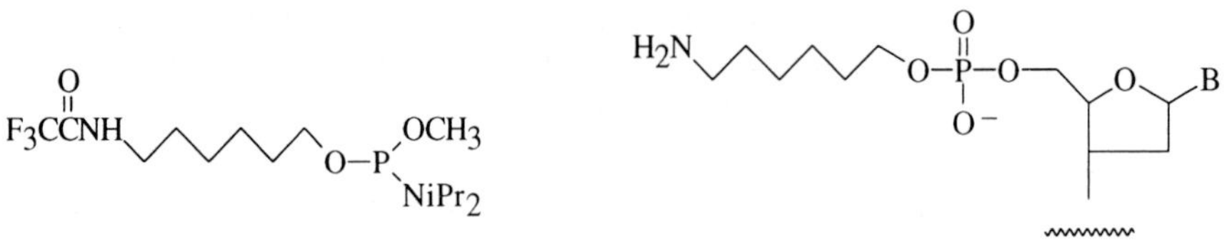

AminoLink

Aminolink Oligonucleotide

**Figure 1.** AminoLink reagent and 5′ aminolink oligonucleotide product.

## 2.2 Cleavage and deprotection of oligonucleotides

Post-synthesis processing, completed before analysis and purification, includes cleavage from the support, removal of protecting groups, desalting, and yield quantification by absorbance. These procedures are described in *Protocol 2*. Cleavage of the 3′-ester linkage to the solid support with concentrated ammonium hydroxide is complete within 1 h at room temperature and is often an automated feature on synthesizers. Deprotection of an oligonucleotide entails removing the protecting groups from the phosphates, the exocyclic amines, and the 5′-hydroxyl. Incomplete deprotection of synthetic oligonucleotides will adversely affect analysis and purification, as well as PCR probe and primer activity.

### **Protocol 2.** Cleavage, deprotection, and quantification of oligonucleotides

*Equipment and reagents*

- Concentrated ammonium hydroxide (J.T. Baker Co., P/N 9072-01). Store at 4°C and discard one month after being opened.
- Screw cap vials with Teflon-lined caps (Wheaton, Inc., P/N 240408, size 13–425)
- Variable temperature heating block or water bath
- UV/VIS spectrophotometer
- 0.1 M TEAA (triethylammonium acetate) buffer, pH 7.

*Method*

1. After synthesis of the oligonucleotide, deliver 1–2 ml concentrated ammonium hydroxide to the column and let stand for 1 h.
2. Transfer the cleaved, partially deprotected solution to a tightly-sealed vial and heat in a hot block or water bath at 55°C for 8 h (standard base-protecting groups; $A^{bz}$, $G^{ibu}$, $C^{bz}$). The support material should not be transferred from the synthesis column to a vial and heated in ammonium hydroxide for a combined cleavage/deprotection operation. Such a procedure will dissolve the controlled-pore glass (CPG) and liberate impurities which remain bound to the support during the 1 h, room temperature cleavage step.

3. Chill the crude, deprotected oligonucleotide solution on ice before opening the vial. Dilute a measured aliquot to 1 ml for quantification by absorbance at 260 nm on a spectrophotometer.[a]
4. Concentrate the oligonucleotide to dryness in a vacuum centrifuge, or under a stream of inert gas, and store as a dry pellet at −20°C. Alternatively, it may be dissolved in neutral, aqueous buffer (e.g. TEAA) and stored at −20°C. Unlabelled oligonucleotides are stable at −20°C in concentrated ammonia for at least several months, but many labels, such as fluorescent dyes, are not. Exposure to strong light, oxidizing agents, and pH extremes should be avoided.

[a] According to Beer's Law, $A = \epsilon cl$ ($A$ = absorbance, $\epsilon$ = molar absorption coefficient in $M^{-1}\ cm^{-1}$, $c$ = concentration in M (mol/litre) and $l$ = path length in cm), so that the mass and concentration of a DNA sample can readily be calculated from its absorbance of light. The average absorbance maximum of the four bases, and most oligonucleotides, is approximately 260 nm with an average molar absorption coefficient near 10 000 $M^{-1}\ cm^{-1}$ per base. One $A_{260}$ unit is the amount of DNA in 1 ml of a solution which has an absorbance of 1 at 260 nm in a 1 cm light path, and represents approximately 33 μg of single-stranded oligonucleotide. For example, 1 mg of an oligonucleotide is about 30 $A_{260}$ units. Conversely, for rapid calculation purposes, 1 μmol of oligonucleotide will absorb, in $A_{260}$ units, 10 times the number of bases. For example, 0.2 μmol of an 18-mer absorbs 36 $A_{260}$ units.

The crude oligonucleotide mixture resulting from *Protocol 2,* dissolved in ammonium hydroxide after deprotection, contains a variety of failure oligonucleotides along with salts, protecting group by-products, and other impurities. Desalting the sample (8) removes some contaminants, gives a more accurate quantitation by UV absorption, and allows exchange of the ammonium counter-ion. Other cations, such as sodium, may be better enzymatic cofactors or confer different solubility effects. Desalting methods include ethanol precipitation, size exclusion/gel filtration media, and the oligonucleotide purification cartridge (OPC, Applied Biosystems). Ethanol precipitation is quick and efficient, and can desalt large quantities of oligonucleotide. The shorter (failure) oligonucleotides remain in the supernatant after ethanol precipitation, so partial size fractionation is achieved. A suitable procedure is described in *Protocol 3.*

**Protocol 3.** Oligonucleotide precipitation

*Equipment*

- Vacuum centrifuge
- Microcentrifuge
- Ethanol
- 3 M sodium acetate

*Method*

1. In a microcentrifuge tube dissolve the oligonucleotide in 30 μl water (20 μl at large scale) and 5 μl of 3 M sodium acetate per $A_{260}$ unit of oligonucleotide.

**Protocol 3.** *Continued*

2. Add 100 μl chilled ethanol (absolute or 95%) per $A_{260}$ unit of oligonucleotide, and mix. Isopropanol may be substituted for ethanol to ensure complete precipitation for very short oligonucleotides (<15-mers).
3. Store at freezer or refrigerator temperatures (−20°C to 4°C) for about 30 min, then centrifuge at high speed (10 300 g) for 5 min.
4. Remove the supernatant with a micropipettor and discard, being careful not to disturb the pellet. Small quantities (less than several $A_{260}$ units or 100 μg) may not be visible.
5. Add another 100 μl chilled ethanol, mix briefly, and centrifuge for 1–5 min.
6. Remove the supernatant again and discard, being careful not to disturb the pellet. Dry the oligonucleotide pellet by vacuum centrifugation.
7. Resuspend the desalted oligonucleotide in aqueous buffer and quantify by absorbance at 260 nm.

## 2.3 Analysis, purification, and quantification

Labelled oligonucleotides may be analysed and purified by the same methods as unlabelled oligonucleotides:

- high performance liquid chromatography (HPLC): purification and analysis
- polyacrylamide gel electrophoresis (PAGE): purification and analysis
- gel capillary electrophoresis: analysis
- oligonucleotide purification cartridge (OPC): purification.

Although unlabelled PCR primers or probes may not require rigorous analysis or purification, their labelled counterparts warrant extra care due to the added complexity and costs. As a general rule, the sequence and base content of an oligonucleotide do not affect analysis, purification, or synthesis (9). The exceptions are primarily under non-denaturing conditions, such as reverse-phase HPLC, when sequences occasionally show anomalous behaviour, probably due to hydrogen-bonding secondary structures. Sequences containing four or more contiguous G bases are most often responsible for anomalous or irreproducible behaviour during analysis or purification. Oligonucleotide length difference is usually the discriminating factor which allows efficient separations.

### 2.3.1 High performance liquid chromatography (HPLC)

HPLC is a high resolution and high sensitivity method for the analysis and purification of labelled oligonucleotides (8, 9). The advantages of HPLC are unattended automation, quantification of the separated products, and easy recovery of purified product. There are many HPLC modes for evaluation and purification of oligonucleotides, but the two most popular are reverse-phase and ion-exchange. Reverse-phase (RP) columns separate by hydrophobicity differences, while ion-exchange columns separate by charge differences. HPLC is useful for analysis and purification of oligonucleotides up to about 50 bases in length. Longer oligonucleotides are best analysed by electrophoretic methods such as polyacrylamide slab gel (PAGE) or gel capillary electrophoresis. Of all the methods, HPLC has the largest purification capacity. Most labels impart added hydrophobicity so that labelled oligonucleotides are easily separated from unlabelled oligonucleotides and impurities by reverse-phase HPLC. For example, biotinylated oligonucleotides elute later than their corresponding unlabelled, 5′-hydroxyl sequences (*Figure 2*). This example shows the typical elution pattern of benzamide, a protecting group by-product, at 4.9 min, followed by failure sequences starting at about 13 min, a small amount of unlabelled 5′-hydroxyl 18-mer at 18.5 min and the 5′-biotin-labelled 18-mer product at 22.0 min. Fluorescent dyes, most commonly attached through the 5′-terminus, also impart additional hydrophobicity to the oligonucleotide and retard elution.

### 2.3.2 Polyacrylamide gel electrophoresis (PAGE)

PAGE is versatile and effective for analysis and purification of labelled oligonucleotides. The most convenient and familiar format for PAGE is the slab gel apparatus using a cross-linked polyacrylamide gel matrix between

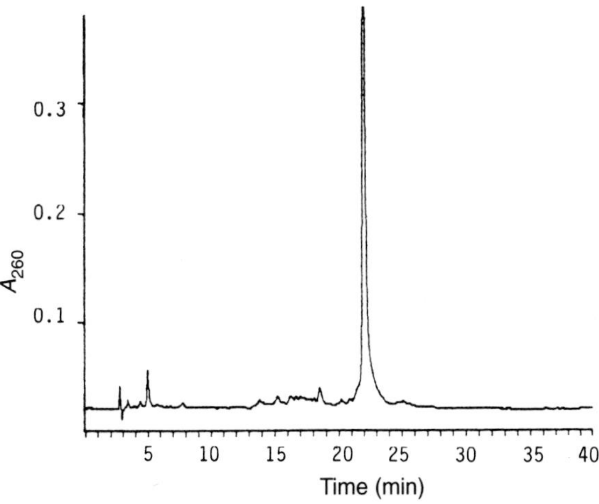


**Figure 2.** Reverse phase HPLC chromatogram of crude 18-mer 5′ biotin–TCA CAG TCT GAT CTC GAT 3′ synthesized at 0.2 μmol scale. HPLC: model 152 (Applied Biosystems); column: Aquapore RP-300, 250 × 4.6 mm; mobile phase: A: 0.1M TEAA; B: $CH_3CN$; gradient: 8–20% B 0–24 min, then 40% B 24–34 min; flow rate 1 ml/min.

two glass plates (8, 10). Polyanionic oligonucleotides migrate through the gel matrix, separating on the basis of size. Electrophoretic mobility varies inversely with length. Denaturing conditions are usually maintained with urea (7 M) to minimize secondary structure formation. Coloured tracking dyes, such as bromophenol blue and xylene cyanol blue, are also loaded on the gel to help determine location of the sample. Oligonucleotides are visualized by UV shadowing and purified by excision of the product band from the gel and elution into an aqueous media. Desalting to remove urea, buffer, and gel debris is necessary.

### 2.3.3 Gel capillary electrophoresis

Gel capillary electrophoresis is a powerful new method for oligonucleotide analysis (11, 12). Capillary electrophoresis (CE), already established as an important analytical tool for other biomolecules such as proteins, peptides, and high molecular weight double-stranded nucleic acids, has been extended to single-stranded oligonucleotides. Capillaries filled with polymer gels give

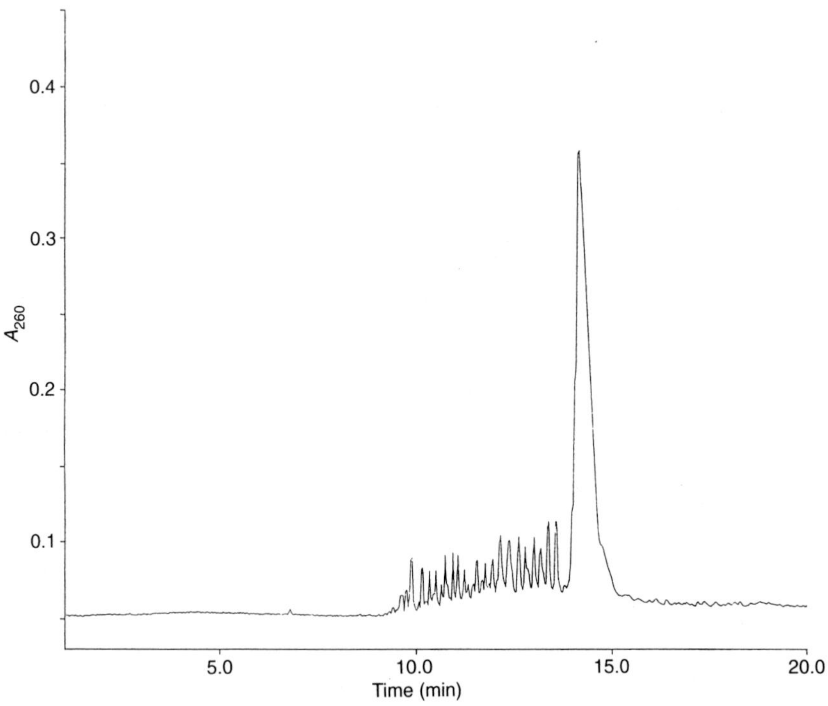


**Figure 3**. Gel capillary electrophoretic separation of oligonucleotides. The sample applied was crude 28-mer 5′ (6-FAM) GAA ACT GGC CTC CAA ACA CTG CCC GCC G 3′ synthesized at 0.2 μmol scale. Electrophoresis conditions: model 270A (Applied Biosystems), 15 kV, 18 μA, 40 °C.

the predictable electrophoretic elution order of small oligonucleotides followed by the largest, usually the product. The advantages of gel CE are short analysis times, excellent resolution, in-capillary detection, small sample requirements, low maintenance, and automation. Single-base resolution can often be attained beyond 100 bases. The analysis, called an electropherogram, is quantitative and can be displayed, stored, integrated, and printed like HPLC chromatograms. The combination of the gel materials and heating of the capillary confers a significant denaturing effect, minimizing secondary structure artefacts. Gel capillaries sustain multiple injections, depending on the gel matrix and handling conditions. Labelled oligonucleotides are excellent substrates for gel CE, resolving well from unlabelled species (*Figure 3*). In this example, a 6-FAM fluorescent dye (see Section 3) labelled oligonucleotide elutes at 14.8 min, after the unlabelled by-products at 10–14 min.

### 2.3.4 Oligonucleotide purification cartridge (OPC)

The OPC was designed specifically for rapid, easy purification of synthetic oligonucleotides (13, 14). It is a syringe- or vacuum manifold-mounted cartridge containing an adsorbent with a specific affinity for oligonucleotides with hydrophobic labels, including 5′-DMT (dimethoxytrityl). Labelled oligonucleotides are efficiently purified and separated from unlabelled oligonucleotides with a simple protocol on OPC. The entire operation requires 15–20 min, and many OPC purifications can be conducted in parallel on a simple vacuum manifold. Examples of OPC purification are given later in *Protocols 5–7*.

# 3. Fluorescent dye labelling

## 3.1 Fluorescent dye NHS coupling

Fluorescent dye labelled oligonucleotides are useful for a variety of PCR experiments (15, 16). For example, 5′ fluorescent dye labelled primers give quantitative PCR product analysis with real-time laser detection during electrophoresis (17). Many different fluorescent dyes are available in active ester form, such as *N*-hydroxysuccinimide (NHS), for coupling to aminolink–oligonucleotides (18). *Protocol 4* describes a representative method for coupling.

**Protocol 4.** Fluorescent dye NHS and aminolink–oligonucleotide coupling reaction

*Equipment and reagents*

- Fluorescent dye NHS ester/DMSO[a]
- 0.25 M sodium carbonate/sodium bicarbonate buffer pH 9.0. Adjust 0.25 M $NaHCO_3$ to pH 9.0 with 0.25 M $Na_2CO_3$
- Purified or desalted aminolink–oligonucleotide (see Sections 1 and 2)
- Size-exclusion column (e.g. PD-10 column, Pharmacia P/N 17-0851-01) or gel filtration matrix (e.g. Sephadex G-25 medium, Pharmacia P/N 17-003-02)
- 0.1 M TEAA (triethylammonium acetate) pH 7.0

**Protocol 4.** *Continued*

*Methods*

1. Dry 5–50 nmol (1–10 $A_{260}$ units, 33–330 μg of a 20-mer) of 5′-aminolink–oligonucleotide in a 0.4 ml or 1.5 ml microcentrifuge tube.
2. Add 40 μl 0.25 M sodium carbonate/sodium bicarbonate buffer pH 9.0 and 4 μl of the fluorescent dye NHS ester/DMSO solution. Mix, protect from light with a foil wrap, and let stand at 37°C for 1 h.
3. Equilibrate the PD-10 column with 12 ml 0.1 M TEAA, then load with the crude dye mixture.
4. Elute the column with another 5 ml 0.1 M TEAA. The colourless void volume is about 2.4 ml and should be discarded. The partially purified dye-labelled oligonucleotide elutes next as a coloured band in about 1 ml 0.1 M TEAA. This fraction will also contain any unreacted amino-link–oligonucleotide. The excess dye impurities will follow.
5. Purify the dye-labelled oligonucleotide further by PAGE or reverse-phase HPLC (see Section 2) to remove unlabelled oligonucleotides.

[a] Four fluorescent dye NHS (5 mg in 100 ml DMSO) reagents are available (Applied Biosystems: FAM, P/N 400985; JOE, P/N 400986; ROX, P/N 400980; TAMRA, P/N 400981).

## 3.2 Fluorescent dye phosphoramidite labelling

Reagents are available for direct 5′ fluorescent dye labelling of oligonucleotides on the DNA synthesizer. Fluorescent dye phosphoramidites are used like nucleoside monomers on DNA synthesizers (19). They obviate intermediate aminolink coupling and streamline post-synthesis purification. Excess dye reagent is automatically removed on the synthesizer, and labelling efficiency is generally higher than solution coupling of the fluorescent dye NHS with aminolink–oligonucleotides. Fluorescent dye phosphoramidite labelling is described in *Protocol 5*.

**Protocol 5.** Fluorescent dye phosphoramidite coupling during synthesis of fluorescent dye-labelled oligonucleotide

*Equipment and reagents*

- Fluorescent dye phosphoramidite (6-FAM, P/N 401527, HEX, P/N 401526, Applied Biosystems)
- Concentrated ammonium hydroxide (see *Protocol 2*)
- 8% acetonitrile in 0.1 M TEAA: 8 ml acetonitrile, 5 ml 2 M triethylammonium acetate (Applied Biosystems P/N 400613), 87 ml water
- Oligonucleotide purification cartridge (OPC, Applied Biosystems P/N 400771)

A. *Synthesis, cleavage, and deprotection*

1. Dissolve the fluorescent dye phosphoramidite in dry acetonitrile (<100 p.p.m. $H_2O$) to make a 0.1 M solution (0.05 M for 40 nmol scale synthesis).

2. Install the diluted reagent on DNA synthesizer at any monomer position.
3. Define the dye-labelled oligonucleotide sequence with the monomer position of the dye phosphoramidite as the 5′-terminus base.
4. Follow the manufacturer's recommended procedures for synthesis, which may include a longer coupling time for the fluorescent dye phosphoramidite, e.g. 120 sec.
5. Cleave and deprotect the dye-labelled oligonucleotide in concentrated ammonium hydroxide under the recommended conditions.

B. *Purification of the dye-labelled oligonucleotides*

1. Pass 5 ml acetonitrile through the OPC to waste, then 5 ml 2 M TEAA.
2. Remove ammonium hydroxide from the crude dye-labelled oligonucleotide by vacuum centrifugation or under a stream of inert gas and dissolve the oligonucleotide in 1 ml 0.1 M TEAA.
3. Pass the solution through the OPC at a rate of about one drop per sec.
4. Collect the eluate from the OPC and pass it through a second time.
5. Pass 5 ml 8% acetonitrile in 0.1 M TEAA through the OPC to waste then 5 ml water.
6. Elute and collect the purified fluorescent dye-labelled oligonucleotide dropwise with 1 ml 20% acetonitrile in water.

Fractions containing fluorescent dye-labelled oligonucleotide will be easily distinguished by an absorbance maximum for the bases at 260 nm and the fluorescent dye maximum at 490–540 nm. Fractions showing only absorbance around 260 nm are unlabelled oligonucleotides and those showing a maxima at 490–540 nm are excess dye. Pure fluorescent dye labelled oligonucleotides will have a ratio of $A_{490-540}/A_{260} = 0.2–0.4$ (*Figure 4*).

## 4. Enzyme labelling of oligonucleotides

Amino- and thiol-linked oligonucleotides can be conjugated to active enzymes to function in multi-step, multi-component assay systems. Oligonucleotide conjugates of horseradish peroxidase (HRP), alkaline phosphatase (AP), and glucose-6-phosphate dehydrogenase (G6PD) are used to probe sequences and sometimes PCR products (20, 21). DNA-probe based diagnostic assays that employ enzyme-labelled oligonucleotides with chemiluminescent or bioluminescent detection are extremely sensitive. Typically the amino- or thiol-linked oligonucleotide is reacted with a bifunctional linker molecule

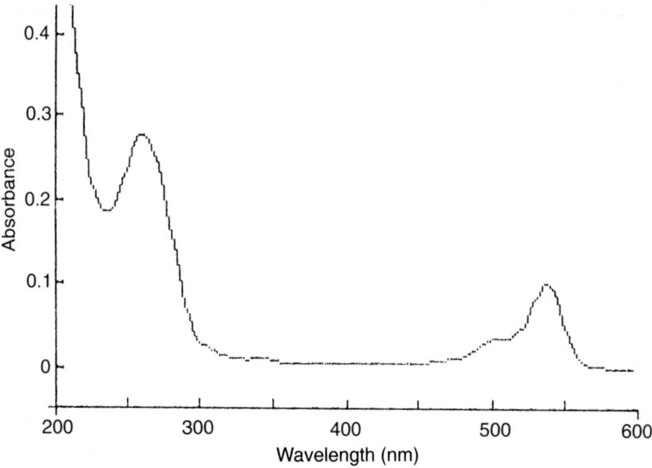


**Figure 4.** Absorbance scan of hexachloro-6-carboxyfluorescein (HEX) 20-mer 5′ (HEX) ACA TCT CCC CTA CCG CTA TA 3′ synthesized at 40 nmol scale and purified by OPC. Yield 5.5 $A_{260}$ units; 180 μg.

followed by coupling to a thiol derivatized enzyme. Protocols have been described elsewhere (4, 7).

## 5. Biotin-labelled oligonucleotides

Certainly the most popular non-isotopic oligonucleotide label is biotin. The first methods for biotinylating oligonucleotides pre-date PCR and involve biotin NHS ester coupling to aminolink–oligonucleotides (22). Biotin-labelled primers and probes have been used in many different PCR experiments. An example is amplification with biotin-labelled primers and solid-support bound streptavidin isolation of the amplified products (23). The availability now of biotin phosphoramidite reagents facilitates synthesis and purification. The preferred molecule, reported by Pon (24, 25), has DMT protection of the biotin nitrogen, which allows trityl-specific purification, minimizes side reactions, and ensures high coupling efficiency. *Protocol 6* describes this approach.

### **Protocol 6.** Biotin phosphoramidite coupling during synthesis of biotin-labelled oligonucleotides

*Reagents*

- Biotin amidite (85 mg, P/N 401395, 250 mg, P/N 401396, Applied Biosystems)
- Concentrated ammonium hydroxide (see *Protocol 2*)
- Dry acetonitrile
- 20% acetonitrile in water
- Oligonucleotide purification cartridge (OPC, Applied Biosystems P/N 400771)
- 2% trifluoroacetic acid (TFA)
- 2 M triethylammonium acetate (TEAA)

A. *Synthesis, cleavage, and deprotection*

1. Dissolve the biotin phosphoramidite in dry acetonitrile (<100 p.p.m. $H_2O$) to make a 0.1 M solution (0.05 M for 40 nmol scale synthesis).
2. Install the diluted reagent on DNA synthesizer at any monomer position.
3. Define the biotin-labelled oligonucleotide sequence with the monomer position of the biotin phosphoramidite as the 5′-terminus base.
4. Conduct the synthesis 'Trityl ON'.
5. Follow the manufacturer's recommended procedures for synthesis, which may include a longer coupling time for the biotin phosphoramidite, e.g. 120 sec.
6. Cleave and deprotect the biotin-labelled oligonucleotide in concentrated ammonium hydroxide under the recommended conditions. See *Figure 2* for an example of a crude biotin-labelled oligonucleotide.

B. *Purification of Trityl ON biotin-labelled oligonucleotides*

1. Pass 5 ml acetonitrile through the OPC to waste, then 5 ml 2 M TEAA.
2. Dilute the biotin oligonucleotide/ammonia deprotection solution (from step 6 above) with an equal volume of water, e.g. 1 ml each.
3. Pass this solution through the OPC at a rate of about one drop per sec.
4. Collect the eluate and pass it through a second time.
5. Pass 10 ml water through the OPC to waste.
6. Pass a portion of 5 ml 2% trifluoroacetic acid (TFA) through the OPC to waste. Let the OPC stand 10 min before passing the remainder through to waste.
7. Pass 10 ml water through the OPC to waste.
8. Elute and collect the purified biotin labelled oligonucleotide dropwise with 1 ml 20% acetonitrile in water.

## 6. Digoxigenin-labelled oligonucleotides

Digoxigenin is a steroid molecule that functions as a hapten label on oligonucleotides, allowing capture by a high-affinity, specific antibody. A number of different nucleic acid detection systems using digoxigenin-labelled probes, some including PCR, have been described (20). Labelling is achieved either with digoxigenin-dNTP incorporation during PCR or by labelling aminolink oligonucleotides with digoxigenin NHS (27). *Protocol 7* describes the latter approach.

**Protocol 7.** Digoxigenin NHS and aminolink–oligonucleotide coupling reaction

*Equipment and reagents*

- Vacuum centrifuge
- Digoxigenin NHS (Boehringer Mannheim Biochemica P/N 1333 054, 5 mg)
- Desalted or purified aminolink-oligonucleotide (see Sections 1 and 2)
- 100 mM sodium borate pH 8.5
- Dimethylformamide (Burdick & Jackson, 4 litres, P/N 076–4)

A. *Coupling digoxigenin NHS to aminolink–oligonucleotide*

1. Dry 100–200 nmol (20–40 $A_{260}$ units of a 20-mer) desalted or purified aminolink–oligonucleotide in a microcentrifuge tube and dissolve it in 100 μl of 100 mM sodium borate.
2. Dissolve 1.3 mg digoxigenin NHS (2 μmol) in 50 μl dimethylformamide and add this to the aminolink–oligonucleotide solution. Mix and let stand 12–24 h at room temperature.
3. Add 500 μl ethanol, mix, chill on ice for 30 min, and spin for 5 min in a microcentrifuge. Remove the supernatant.
4. Repeat step 3.
5. Dry the pellet in a vacuum centrifuge. Dissolve the crude digoxigenin-labelled oligonucleotide in aqueous media for quantification, analysis, and purification.

B. *Purification of the digoxigenin-labelled oligonucleotide*

1. Pass 5 ml acetonitrile through an OPC to waste, then 5 ml 2 M TEAA.
2. Dissolve the crude digoxigenin-labelled oligonucleotide in 1 ml 0.1 M TEAA.
3. Pass this solution through the OPC at a rate of about one drop per sec.
4. Collect the eluate and pass it through a second time.
5. Pass 10 ml water through the OPC to waste.
6. Elute and collect the purified digoxigenin-labelled oligonucleotide dropwise with 1 ml 50% acetonitrile in water.
7. Analyse the digoxigenin-labelled oligonucleotides by gel CE, HPLC, or PAGE. *Figure 5* shows a typical analysis of digoxigenin-labelled oligonucleotides by PAGE.

# Acknowledgements

The author thanks Will Bloch, Lynn Wuischpard, Bill Giusti, Jenny Andrus, and Elaine Heron for their assistance.

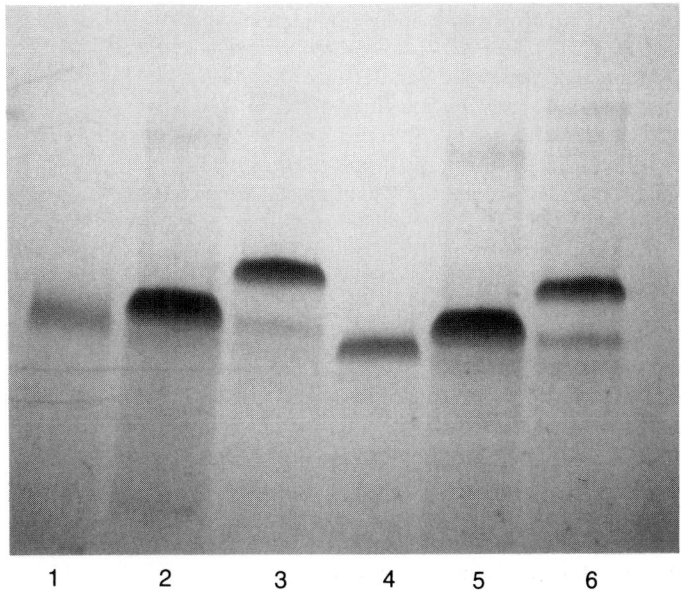


**Figure 5.** Analysis of digoxigenin-labelled oligonucleotides by polyacrylamide gel electrophoresis (15% acrylamide, 25 μamps, 500 volts, 0.75 mm gel thickness, UV shadow analysis) Lane 1, 5′ AAT CTG GGC GAC AAG AGT GA 3′ 20-mer; lane 2, 5′-aminolink–AAT CTG GGC GAC AAG AGT GA 3′; lane 3, 5′-digoxigenin–AAT CTG GGC GAC AAG AGT GA 3′; lane 4, 5′ ACA TCT CCC CTA CCG CTA TA 3′ 20-mer; lane 5, 5′-aminolink–ACA TCT CCC CTA CCG CTA TA 3′; lane 6, 5′ digoxigenin–ACA TCT CCC CTA CCG CTA TA 3′.

# References

1. The Survey Committee of the Association of Biomolecular Resource Facilities (1992). *ABRF News*, **3** (4), 3.
2. Goodchild, J. (1990). *Bioconjugate Chem.*, **1**, 165.
3. Sauvaigo, S., Fouque, B., Roget, A., Livache, T., Bazin, H., Chypre, C., and Teoule, R. (1990). *Nucleic Acids Res.*, **18**, 3175.
4. Keller, G. H. and Manak, M. M. (1989). *DNA probes*. Stockton Press, New York.
5. Applied Biosystems (1988). *Aminolink 2*, User Bulletin No. 49, Applied Biosystems.
6. Connolly, B. A. and Rider, P. (1985). *Nucleic Acids Res.*, **13**, 4485.
7. Levenson, C. and Chang, C. (1990). In *PCR protocols* (ed. M. A. Innis, D. H. Gelfand, J. J. Sninsky, and T. J. White), p. 99. Academic Press, San Diego.
8. Applied Biosystems (1992). *Evaluating and isolating synthetic oligonucleotides*. (available upon request, 850 Lincoln Centre Dr., Foster City, CA, 94404).
9. Zon, G. and Thompson, J. A. (1986). *Biochromatography*, **1**, 22.

10. Efcavitch, J. W. (1990). In *Gel electrophoresis of nucleic acids: a practical approach* (ed. D. Rickwood, and B. D. Hames), p. 125. IRL Press, Oxford.
11. Andrus, A. (1993). In *Protocols for oligonucleotide conjugates* (ed. S. Agrawal), p. 277. Humana Press Inc., New Jersey.
12. Andrus, A. (1992). *Methods: A Companion to Methods in Enzymology*, **4**, 213.
13. Applied Biosystems (1991). *New applications for the oligonucletide purification cartridge*, User Bulletin No. 59, Applied Biosystems.
14. McBride, L. J., McCollum, C., Davidson, S., Efcavitch, J. W., Andrus, A., and Lombardi, S. J. (1988). *Biotechniques*, **6**, 362.
15. Lundeberg, J., Wahlberg, J., and Uhlén, M. (1991). *Biotechniques*, **10**, 68.
16. Chehub, F. F. and Kan, Y. W. (1989). *Proc. Natl Acad. Sci. USA*, **86**, 9178.
17. Mayrand, P. E., Corcoran, K. P., Ziegle, J. S., Robertson, J. M., Hoff, L. B., and Kronick, M. N. (1992). *Appl. Theor. Electrophoresis*, **3**, 1.
18. Giusti, W. G. and Adriano, T. (1993). *PCR Methods and Applications*, **2**, 223.
19. Theisen, P., McCollum, C., and Andrus, A. (1992). *Nucleic Acids Symp. Ser.*, **27** p. 99. IRL Press, Oxford.
20. Kessler, C. (1992). In *Nonisotopic DNA probe techniques* (ed. L. J. Kricka), p. 30. Academic Press, San Diego.
21. McInnes, J. L. and Symons, R. H. (1989). In *Nucleic acid probes* (ed. R. H. Symons), p. 33. CRC Press, Inc., Boca Raton, Florida.
22. Chollet, A. and Kawashima, E. H. (1985). *Nucleic Acids Res.*, **13**, 1529.
23. Saiki, R. K., Walsh, P. S., Levenson, C. H., and Erlich, H. A. (1989). *Proc. Natl Acad. Sci. USA*, **86**, 6230.
24. Pon, R. (1991). *Tetrahedron Lett.*, **32**, 1715.
25. Applied Biosystems (1992). *Biotin labeling of oligonucleotides on a DNA/RNA synthesizer*, User Bulletin No. 70. Applied Biosystems.
26. Kessler, C. (1992). In *Nonradioactive labeling and detection of biomolecules* (ed. C. Kessler), p. 27. Springer Verlag, Berlin.
27. Epplen, J. T. *et al.* (1989). *Human Genetics*, **82**, 223.

# 4

# Solid phase PCR

STEFAN STAMM and JÜRGEN BROSIUS

## 1. Introduction

The method of solid-anchored PCR uses specific oligonucleotides coupled to a solid phase as primers for cDNA synthesis (1) and results in cDNA that is covalently linked to the solid phase. Typical solid phase matrices are agarose (1), acrylamide (1), magnetic (2–4), or latex beads (5). A solid phase with cDNA attached, generated using oligo(dT) as a primer, contains sequence information similar to a cDNA library, that is, it represents a 'solid phase library' (1, 3, 4). The cDNA that is attached to the solid phase can be used directly as a template for PCR reactions or can be modified enzymatically prior to the PCR reaction. Oligonucleotides that are attached to a solid phase can also serve for affinity purification of RNA (2). RNA isolated in this way can be directly reverse transcribed, using the primer that is coupled to the solid phase. Subsequent PCR reactions can employ this primer with or without additional internal primers. Since the cDNA is coupled to a solid phase, changing buffer conditions or primer composition is conveniently achieved by washing the solid phase and resuspending it in a different PCR reaction mixture. The general principle of solid anchored PCR is shown in *Figure 1*.

PCR products immobilized on a solid phase have already been used in a wide variety of applications:

- for automated single-strand sequencing of PCR products (6, 7)
- for detection of PCR products in automated clinical assays (8–10)
- to generate single-stranded DNA probes with high specific activity (11)
- to construct enriched genomic and cDNA libraries (12–14)
- for hybrid selection of RNA (5).

Finally, nucleic acids that have been immobilized on nylon membranes have been successfully amplified (15).

This chapter describes the attachment of specific oligonucleotides to a solid phase, the use of these oligonucleotides as primers for cDNA synthesis, the modification of this cDNA by homopolymer tailing, and the handling of this cDNA in PCR experiments.

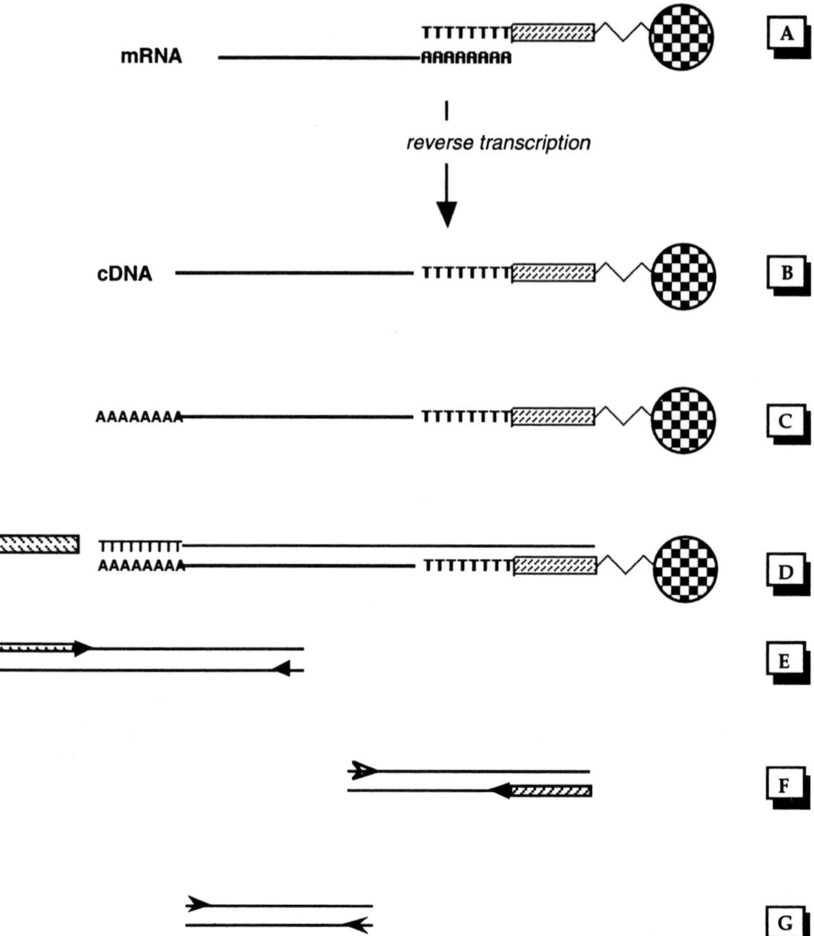


**Figure 1.** The principle of solid phase PCR. The solid phase matrix is represented as a chequered circle on which the oligonucleotide is attached. In this example oligo(dT) is attached to an adaptor primer shown as a hatched rectangle (A). Reverse transcription results in cDNA that is covalently linked to the solid phase, (B). This cDNA can be tailed (C) and amplified using the adaptor primers without (D) or in combination with internal primers (E,F) or nested internal primers (G).

## 2. Coupling of oligonucleotides to a solid phase

### 2.1 Introduction

In order to couple an oligonucleotide to a solid phase, the oligonucleotide has to contain a chemical side group that can be activated. As no strong nucleophiles are present in ordinary oligonucleotides, the introduction of a strong nucleophile, such as a primary amine, allows coupling of the oligo-

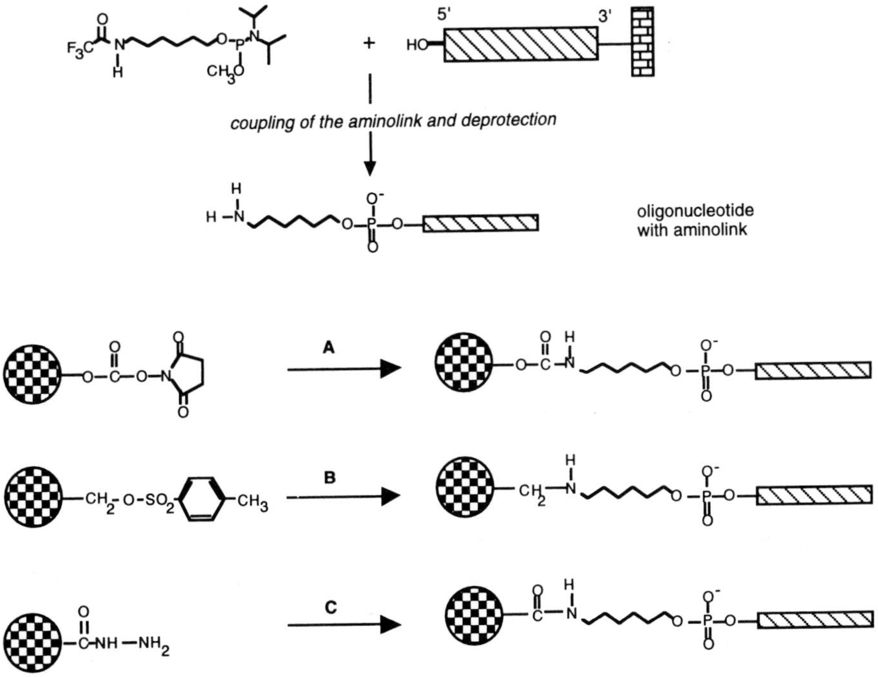

**Figure 2.** Coupling of an oligonucleotide using an aminolink. The aminolink is introduced to the 5′-end of the oligonucleotide (shown as a hatched rectangle) prior to cleavage of the oligonucleotide from the synthesizer support. The synthesizer support is indicated as a box with bricks. The $NH_2$ group of the aminolink can be coupled to *N*-hydroxysuccinimide (A), tosyl (B), and hydrazine groups (C).

nucleotide. A primary amine can be introduced into the oligonucleotide using an aminolink during automated synthesis (*Figure 2*) (16). The aminolink can be coupled to biotin (17; Chapter 2) which leads to immobilization of the oligonucleotide on streptavidin-coated solid phases (7, 8), or it can be coupled directly to an activated solid phase. It is important that the buffer in which the coupling is done does not contain any amino groups such as Tris (tris(hydroxymethyl) aminomethane), for example in TE buffer. We couple the aminolink group directly to a solid phase, because this leads to covalent binding between the oligonucleotide and the solid phase. Coupling of the oligonucleotide via the aminolink leaves the 3′-end of the oligonucleotide accessible for enzymatic reactions such as priming cDNA synthesis.

## 2.2 Choice of the solid phase

Several different solid matrices for performing solid phase PCR are commercially available. Aminolink activated oligonucleotides can be coupled to *N*-hydroxysuccinimidyl activated agarose, polyacrylamide hydrazide particles, or tosyl activated magnetic beads. The active groups that are present on these matrices can also be chemically introduced into other solid phases by standard methods (18). We have found that agarose is best suited for re-amplification experiments of coupled cDNA and that polyacrylamide or magnetic beads are better suited for RNA affinity isolation from more viscous solutions. Magnetic beads are easier to see in a polypropylene tube, but we have found that agarose beads stick better to the tube. Choosing the solid phase depends on the specific application and preference of the investigator.

## 2.3 The coupling reaction

The coupling conditions for several matrices are given in *Protocols 1–3* and the structure of the reaction products is schematically shown in *Figure 2*.

**Protocol 1.** Coupling of oligonucleotides to tosyl activated beads

*Reagents*

- Oligonucleotide with aminolink (100 nmol)
- Tosyl activated Dynabead suspension (Dynal, M450)
- 500 mM sodium borate buffer pH 9.5. Dissolve 1.55 g boric acid in 50 ml of water and bring to pH 9.5 by adding approximately 2 g NaOH (the alkaline pH is important for coupling).

*Method*

1. Incubate 100 nmol oligonucleotide (modified with the aminolink) with 200 µl Dynabead suspension.
2. Add 1/10 vol. of 500 mM sodium borate solution, pH 9.5.
3. Incubate this mixture on a shaking platform or rotating wheel overnight at room temperature.
4. Wash the solid phase twice with water using 500 µl water each time and resuspend the beads in 100 µl water.
5. Store the coupled oligonucleotides at 4°C in the aqueous slurry. This has an approximate concentration of 0.5 nmol oligonucleotide/µl. Long term storage is recommended at −70°C in aliquots.

**Protocol 2.** Coupling of oligonucleotides to *N*-hydroxysuccinimide groups

*Reagents*

- Oligonucleotide with aminolink (100 nmol)
- *N*-Hydroxysuccinimide-coupled agarose (Sigma H3512)
- 2 M $Na_2CO_3$/$NaHCO_3$ pH 9.5. Dissolve 16.8 g $NaHCO_3$ in water, adjust pH with 2 M NaOH and bring volume up to 100 ml.
- Cold saturated (3.3 M) glycine solution

*Method*

1. Wash a 10-fold excess (1 mmol matrix for 100 nmol oligonucleotide) of matrix-bound coupling groups (as specified by the manufacturer) twice by resuspending the gel particles containing the *N*-hydroxy-succinimide groups in water, followed by a brief centrifugation to pellet the matrix.
2. Add 50 μl 2 M $Na_2CO_3/NaHCO_3$ buffer to 100 nmol oligonucleotide in a 500 μl reaction volume. Mix this solution with the washed matrix.
3. Perform the coupling overnight at room temperature with agitation.
4. Stop the reaction by adding 60 μl of a 3.3 M glycine solution to the reaction mixture. Incubate for 2 h at room temperature.
5. Wash the solid phase twice with water and resuspend the beads in 100 μl water.
6. Store the coupled oligonucleotides at 4°C in the aqueous slurry. This has an approximate concentration of 0.5 nmol oligonucleotide/μl. Long term storage is recommended at −70°C in aliquots.

**Protocol 3.** Coupling of oligonucleotides to hydrazide groups

*Reagents*

- Oligonucleotide with aminolink (100 nmol)
- Hydrazide coupled to polyacrylamide powder (Sigma P8885)
- 0.1 M amidosulphonic acid ($H_3NO_3S$)
- 0.1 M HCl
- 0.1 M $NaNO_2$
- 5 M $NH_4(OH)$

*Method*

1. Dissolve the dry polyacrylamide powder containing approximately 1–10 μmol hydrazide groups (typically 20 mg) in 250 μl water.
2. Activate the hydrazine groups by adding 250 μl 0.1 M HCl and 250 μl 0.1 M $NaNO_2$. Place the reaction mixture on ice for 20 min. Wash twice with 0.1 M HCl and once with 0.1 M amidosulphonic acid, to remove NO.
3. Resuspend the activated matrix in 500 μl 0.2 M $Na_2CO_3/NaHCO_3$ buffer, pH 9.5, that contains the coupling oligonucleotide at a concentration of 100 nmol/500 μl.
4. Perform the coupling reaction at room temperature overnight.
5. Stop the reaction by adding 10 μl of 5 M $NH_4(OH)$.
6. After 1 h at room temperature, wash the matrix twice with water and resuspend it in 100 μl of water.

**Protocol 3.** *Continued*

7. Store the coupled oligonucleotides at 4°C in the aqueous slurry. This has an approximate concentration of 0.5 nmol oligonucleotide/μl. Long-term storage is recommended at −70°C in aliquots.

# 3. Synthesis of solid phase coupled cDNA

## 3.1 Reverse transcription of RNA using solid phase coupled oligonucleotides

Reverse transcription is performed in 10 μl reactions containing the solid phase with the coupled oligonucleotides as described in *Protocol 4*. The reverse transcription reaction is performed in the same tube in which the PCR reaction will be performed, as it is difficult to transfer small amounts of solid phase matrix. In order to optimize reaction conditions such as the $Mg^{2+}$ concentration, we usually prepare six tubes with identical reaction mixtures. One of the reactions is spiked with [$\alpha$-$^{32}$P]dCTP to monitor cDNA synthesis (see Section 3.2).

**Protocol 4.** Reverse transcription reaction using solid phase oligonucleotides coupled to a solid phase

*Reagents*

- RT mixture (7 μl of the RT mixture contain: 2 μl 5 × reverse transcription buffer (250 mM Tris–HCl pH 8.3; 15 mM $MgCl_2$, 375 mM KCl), 1 μl 100 mM DTT, and 1 μl of 10 mM of each dNTP solution). The RT mixture is made up in a large volume and stored in 70 μl aliquots at −20°C.
- RNase inhibitor (Boehringer 799025)
- MMLV H$^-$ reverse transcriptase (Superscript, BRL 8053SA)
- Solid phase coupled oligonucleotide resuspended in water (contains usually about 0.5 nmol of coupled oligonucleotide; see *Protocols 1–3*)

*Method*

1. Into a 500 μl polypropylene PCR reaction tube, pipette the following:
   - RT mixture — 7 μl
   - solid phase coupled oligonucleotide — 1 μl
   - total RNA (1 μg/μl) — 1 μl
   - RNase Inhibitor (20 U) — 0.5 μl
   - MMLV (Moloney murine leukaemia virus) H$^-$ reverse transcriptase (100 U) — 0.5 μl
2. Incubate the reaction at 37°C for 1 h.
3. Wash the cDNA coupled to the solid phase twice with water.
4. Store the solid phase cDNA at −20°C or use it in a PCR experiment (see *Protocol 7*). The PCR experiment is performed in the same tube in which the cDNA was synthesized.

## 3.2. Monitoring of cDNA synthesis

The amount of cDNA created by reverse transcription can be estimated by the incorporation of [$\alpha$-$^{32}$P]dCTP into cDNA (*Protocol 5*). The final dCTP concentration in the reverse transcription reaction is 1 mM dCTP. Since the reaction volume is 10 μl, this equals 10 nmol dCTP. The amount of cDNA synthesized can be estimated from the dCTP incorporation. The average formula weight of one nucleotide is 330 g/mol. Since DNA is composed of four nucleotides, for each mole of dCTP one mol/l of nucleic acid equalling $330 \times 4 = 1320$ g is incorporated. The amount of cDNA synthesized is then given by the formula in step 7 of *Protocol 5*. A typical reverse transcription reaction spiked with [$\alpha$-$^{32}$P]dCTP containing 1 μg of total RNA emits about $2.1 \times 10^6$ c.p.m. After three wash steps, about 800 c.p.m. are usually still bound to the solid phase, representing an incorporation of about 0.04% equivalent to 5 ng cDNA. As we use total RNA instead of poly(A)$^+$RNA, this first strand synthesis is lower than the cDNA synthesis usually observed in making cDNAs in solution. A typical washing pattern of solid phase coupled cDNA is shown in *Figure 3*. As shown in this figure, most of the unincorporated nucleotides are removed after two washes, indicating the high washing efficiency of this method.

**Protocol 5.** Estimation of amount of cDNA synthesized by [$\alpha$-$^{32}$P]dCTP incorporation

*Reagents*

- [$\alpha$-$^{32}$P]dCTP (3000 Ci/mmol)
- all other reagents indicated in *Protocol 4*

*Method*

1. Perform a reverse transcription reaction as in *Protocol 4*, but, in addition, add 0.5 μl [$\alpha$-$^{32}$P]dCTP, (3000 Ci/mmol) to the reaction mixture.
2. Count the radioactivity in the reaction tube.
3. Add 100 μl water to the tube, vortex, and spin for 5 min in a microcentrifuge at maximum speed.
4. Remove all the liquid from the tube.
5. Repeat the washing procedure twice more.
6. Count the tube after the three wash steps.
7. Calculate the approximate amount of cDNA synthesized (in g) by:

$$\text{cDNA(g)} = \frac{\text{counts after wash}}{\text{counts before wash}} \times 10 \times 10^{-9}\ (\text{mol}) \times 1320\ (\text{g/mol})$$

**Protocol 5.** *Continued*

As an example, with the typical incorporation data given in the text, we obtain about:

$$\frac{800}{2.1 \times 10^6} \times 10 \times 10^{-9} \times 1320 = 5.03 \times 10^{-12} \text{ g cDNA}$$

## 3.3 Addition of homopolymers to the coupled oligonucleotides

Using terminal deoxynucleotidyl transferase, nucleotides can be added to the free 3′-end of the coupled oligonucleotides and to coupled cDNA (4). This introduces a homopolymer tail, which is attached to the cDNA or to the oligonucleotide (*Protocol 6*). When added to oligonucleotides, the homopolymer tail is useful to determine whether the oligonucleotides have coupled to the solid phase. By using radioactive nucleotides in the terminal transferase reaction, radioactively labelled homopolymers are generated. As the exact length of the homopolymer is hard to control, the amount of incorporation is only a rough indication of the coupling efficiency. However, the absence of incorporation indicates that the oligonucleotides were not coupled to the

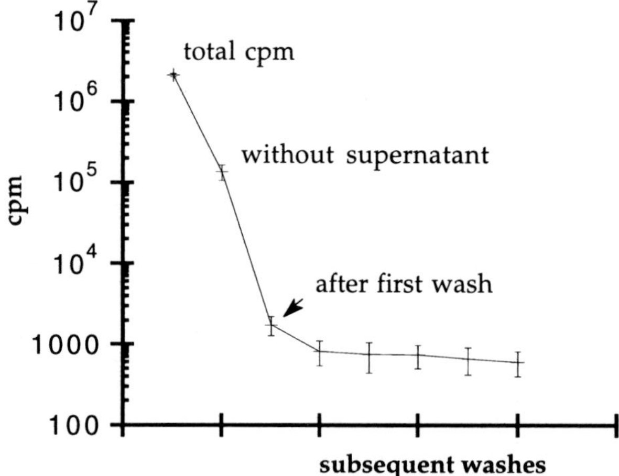


**Figure 3.** Washing behaviour of cDNA coupled to agarose beads. 1 μg of total rat brain RNA was reverse transcribed in the presence of [α-$^{32}$P]dCTP with primer SS020 (see *Table 1*, footnote *a*) attached to agarose. The beads were spun down and counted after removal of the supernatant ('without supernatant' in the figure). The beads were subsequently resuspended in 500 μl water and repelleted ('after first wash' in the figure). After one wash, virtually all of the unincorporated label was removed. Standard errors from five experiments are indicated. Note that the scale is logarithmic.

solid support. Furthermore, homopolymer tails attached to cDNA can be used for RACE experiments (19).

**Protocol 6.** Terminal transferase reaction with oligonucleotides coupled to a solid phase

*Reagents*

- Solid phase with attached oligonucleotide resuspended in water
- 5 × Terminal transferase buffer (1 M potassium cacodylate, 125 mM Tris–HCl, pH 7.2)
- [α-$^{32}$P]dATP (3000 Ci/mmol)[a]
- Terminal deoxynucleotidyl transferase (50 U, Boehringer)

*Method*

1. Add the following reagents in order into a 500 μl PCR tube:

| | |
|---|---|
| solid phase with attached oligonucleotide | 1 μl |
| 5 × terminal transferase buffer | 2 μl |
| [α-$^{32}$P]dATP 3000 Ci/mmol | 1 μl |
| water | 4 μl |
| terminal deoxynucleotidyl transferase | 1 μl |

2. Incubate for 30 min at 37 °C.
3. Remove the unincorporated label by resuspending the solid phase in 100 μl of water followed by a 5 min centrifugation step.
4. Repeat the wash twice more.

[a] For the addition of non-labelled homopolymer tails, add 2.5 mM dNTP instead of labelled dATP. The dNTP used depends on which tail is to be synthesized.

## 4. Use of solid phase coupled cDNA in PCR experiments

The cDNA that is created with oligonucleotides coupled to a solid phase can be repeatedly used in PCR experiments using different primer combinations. *Protocol 7* describes the procedures used. An example of such an experiment is shown in *Figure 4*. In this experiment, 1 μg of total rat brain RNA was reverse transcribed using primer SS020 (see *Table 1*, footnote *a*) coupled to agarose beads.

**Protocol 7.** Use of solid phase coupled cDNA in PCR experiments

*Reagents*

- cDNA coupled to a solid phase from *Protocol 4*
- PCR mixture (16.6 mM $(NH_4)_2SO_4$, 170 μg/ml BSA, (fraction V, Sigma A7906) 67 mM Tris–HCl pH 8.8, at 20 °C, 1.5 mM $MgCl_2$, 200 μM of each dNTP and 7.8 μg/ml of each primer[a])
- Mineral oil

**Protocol 7.** *Continued*

A. *Removal of the reverse transcription reaction components*

1. After reverse transcription (*Protocol 4*), add 100 μl of water to the entire reaction, vortex and centrifuge for 10 min.
2. Remove and discard the aqueous phase.
3. Repeat the washing step in step 1 above twice more.

B. *Use of the cDNA in a PCR reaction*

1. Remove all the fluid from the tube and overlay the solid phase with 50 μl PCR mixture.
2. Overlay the aqueous phase with mineral oil, centrifuge the tubes briefly, and perform the PCR reaction.
3. After the PCR reaction, centrifuge for 10 min.
4. Remove the oil. Recover the aqueous phase and analyse the PCR products in the aqueous phase as usual. Do not disturb the pellets; leave about 5 μl fluid in the tubes.

C. *Reamplification of the cDNA*

1. After the PCR reaction and the removal of the product, add 100 μl of water to the tube, vortex, and centrifuge for 10 min to wash the solid phase.
2. Repeat the washing step twice more.
3. Go back to step B, using a new primer combination.

[a] The exact primer and $MgCl_2$ concentration have to be optimized: the concentrations given here are only a starting point.

The fact that the same beads can be re-used several times shows that:

(a) the cDNA is covalently linked to the beads;

(b) the cDNA is chemically stable enough to be intact after being exposed to heat and alkaline conditions in numerous PCR cycles.

## 5. Isolation and reverse transcription of RNA using solid phase coupled primers

Oligonucleotides coupled to a solid phase can be used for direct affinity isolation of RNA (2). Using the solid phase coupled primer, this RNA can be directly reverse transcribed and amplified. As any oligonucleotide can be

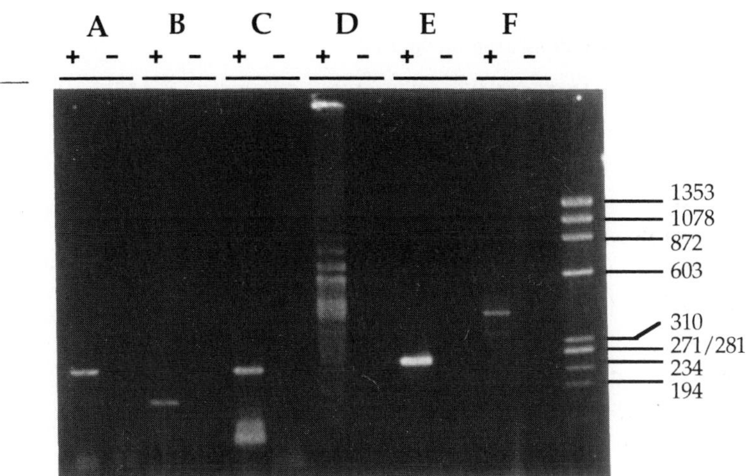


**Figure 4.** Re-use of cDNA attached to agarose beads in PCR experiments. 1 μg total liver RNA was reverse transcribed with primer SS020 (see *Table 1*, footnote *a*) attached to agarose beads and amplified with the PCR mixes and reaction conditions specified in *Table 1*. Denaturation was for 30 min at 94°C and 30 cycles were used. The buffer contained 16.6 mM $(NH_4)_2SO_4$, 67 mM Tris–HCl, pH 8.8, and the $MgCl_2$ concentration indicated. The accuracy of the thermocycler was controlled with an external thermocouple (22). For analysis, 1/5th of each reaction was loaded on a 2% TBE agarose gel. It is interesting that the 700 bp artefact band showing up in lane A disappears in lane C due to different PCR conditions. This further indicates that the washing effectively removes products of prior reactions. The formation of a product using the amplification primer SS015 in lanes D and F indicates that the amplification primer has access to the anchor primer close to agarose bead structures. Lanes indicated '+' represent complete reactions, lanes indicated '−' indicate control reactions performed using beads and RNA but without the reverse transcription step. Size markers were electrophoresed in the lane on the right hand side of lane F; the corresponding sizes (bp) are indicated alongside this lane.

coupled to a solid phase using the aminolink, special RNA sequences can be enriched by this isolation procedure. By coupling specific oligonucleotides to a solid phase, one is no longer limited to the use of commercially available oligo(dT) containing matrices. A common modification is, for example, the introduction of adaptor sequences (20) next to the oligo(dT) that can be used in the PCR amplification. The use of adaptor primers during the amplification step eliminates the need to employ oligo(dT) primer which may be a source of mispriming (*Figure 1F*). Adaptor primers are furthermore useful in introducing restriction sites for cloning purposes.

A typical application of this direct isolation and amplification method (*Protocol 8*) is shown in *Figure 5*.

**Table 1.** Reaction conditions used in *Figure 4*

| Lane | Oligonucleotide[a] | T[b] (°C) | $t_1$[c] (min) | $t_2$[d] (min) | $MgCl_2$ (mM) | Gene/target sequence | Ref. | Size (bp) |
|---|---|---|---|---|---|---|---|---|
| A | SRS015<br>SRS016 | 60 | 1 | 2 | 1.50 | NILE[e] | 23 | 238 |
| B | SS003<br>SS004 | 55 | 1 | 2 | 2.00 | Clathrin light chain B | 24 | 152 |
| C | SRS015<br>SRS016 | 60 | 10 | 2 | 2.00 | NILE | 23 | 238 |
| D | SS015<br>SS029 | 55 | 1 | 2 | 1.50 | anchor primer<br>ID elements | 25 | various |
| E | SS032<br>SS033 | 55 | 1 | 2 | 1.50 | Clathrin light chain A | 24 | 256 |
| F | SS003<br>SS015 | 55 | 1 | 3 | 1.75 | Clathrin light chain B<br>anchor primer | 24 | >500 |

[a] Oligonucleotides (5′→3′):
SRS015: GCATCCGAATTCGAGGACACTGAGGTAGATTCCGAGGCCCGG
SRS016: CTCGAGAAGCTTGCCGATGAAAGAGCCATCCTCATTGAACTG
SS003: TGCCTCGAAGGTGAACCGAAC
SS004: GGTCTCCTCCTTGGATTCTTTC
SS015: GCCTTCGAATTCAGCACC
SS021: GGGGTTGGGGATTTAGC
SS020: AL-GCCTTCGAATTCAGCACCTTTTTTTTTTTTT (where AL = Aminolink)
SS029: AL-GGGGTTGGGGATTTAGC (where AL = Aminolink)
SS032: TGGTACGCAAGGCAGGATGAGC
SS033: AGATCGGAGACGTAGTGTTTCCA
[b] annealing temperature (°C).
[c] annealing time (min).
[d] extension time (min); the extension temperature was 72°C.
[e] NILE: nerve growth factor-inducible large external glycoprotein.

## Protocol 8. One-step procedure to isolate RNA and reverse transcribe it on a solid phase

### *Equipment and reagents*

- PBS (0.14 M NaCl, 2.7 mM KCl, 10 mM $Na_2HPO_4$, 1.8 mM $KH_2PO_4$ pH 7.0)
- 18 cm cell scraper (Falcon 3085)
- Lysis buffer: 10 mM Tris–HCl, 0.14 M NaCl, 5 mM KCl and 1% NP40 (Nonidet P-40, Sigma N0896)
- 2 × Binding buffer: 20 mM Tris–HCl pH 8.0, 1 M LiCl, 2 mM EDTA, 1% SDS, 10 mM DTT
- Oligonucleotide coupled to a solid phase
- Washing buffer: 10 mM Tris–HCl pH 8.0, 0.15 M LiCl, 1 mM EDTA, 0.3 % SDS

### *Method*

1. Wash the cells twice with ice-cold PBS.[a]
2. Harvest the cells by scraping them off the plate using a disposable 18 cm cell scraper (Falcon 3085).
3. Pellet the cells in a PCR microtube for 1 min and resuspend the pellet in 50 μl lysis buffer. Keep on ice for 1 min.

4. Add 50 μl 2 × binding buffer and 10 μl solid phase coupled oligonucleotide and allow to bind for 5 min at room temperature with gentle agitation on a shaking platform.
5. Pellet the solid phase at maximum speed in a microcentrifuge for 5 min.
6. Add 100 μl ice-cold washing buffer, vortex, and spin again for 5 min.
7. Repeat the washing step (step 6) twice more.
8. Remove all the liquid and reverse transcribe the solid phase bound RNA as in *Protocol 4*.
9. Perform the PCR reaction as in *Protocol 7*.

[a] This step is optional for cells in small culture volumes or serum-free grown cells

## 6. Inclusion of an oligonucleotide with an aminolink does not interfere with the PCR reaction

Various primer combinations with an aminolink at the 5′-end of an oligonucleotide used for PCR were found not to interfere with the amplification efficiency or the electrophoretic properties of the product. The DNA generated using PCR with a primer containing an aminolink can be coupled in the same manner as an oligonucleotide to a solid phase. Oligonucleotides can be chemically synthesized only to a limited size of about 100 bases. In contrast, generating a PCR product in which one primer contains an aminolink has virtually no size limitations. This allows, for example, the synthesis of defined DNA segments that can be coupled to solid phases to generate affinity purification matrices (21).

## 7. Precautions and troubleshooting

As with other PCR applications, contamination is the most serious problem in working with immobilized cDNAs. We therefore routinely use aerosol filter tips, aliquot all our solutions (including small aliquots of water used for washing) and use, whenever possible, different locations for RNA isolation and cDNA synthesis. Possible problems and their solutions are listed in *Table 2*.

## 8. Discussion

The major advantage of solid phase PCR is that, by binding any kind of DNA to a solid phase, the DNA is concentrated in a small volume and can be easily

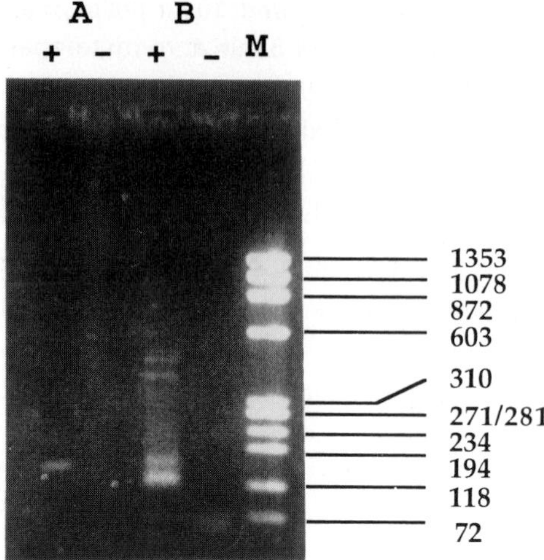


**Figure 5.** Direct RNA isolation using solid phase coupled oligonucleotides. RNA from $10^5$ αT3–1 cells was isolated and reverse transcribed using primer SS020. The cDNA was then used in a PCR reaction containing primers SS003/SS004 (A) and subsequently re-used with primers SS015/SS029 (B), under the conditions given in *Figure 4*. Details of the primers are given in *Table 1*, footnote *a*. Lanes indicated '+' represent complete reactions; lanes indicated '–' represent negative control reactions using SS020 containing solid phase without RNA in the PCR reaction. Lane M contained size markers.

**Table 2.** Troubleshooting in solid phase PCR experiments

| Problem | Possible reason | Solution |
|---|---|---|
| No coupling of oligonucleotide | Activated matrix might be too old | Check reference date on matrix, retry with new matrix; |
| | Wrong pH of coupling buffer | Use only freshly made buffers for coupling; try different activated groups |
| No cDNA synthesis | Oligonucleotide might not have been coupled | Check oligonucleotide coupling with terminal transferase reaction (*Protocol 6*) |
| | RNA degraded or absent | When using one step RNA isolation procedures, use conventionally made RNA as control |
| No PCR product | PCR reaction conditions suboptimal | Titrate $Mg^{2+}$ concentration in 0.25 mM increments from 1.5 mM to 5 mM in either subsequent reamplification or parallel experiments |
| | Loss of solid phase | Label cDNA (*Protocol 4*) and monitor PCR tubes for loss; use 10 min centrifugation steps before changing the PCR products |

manipulated in different buffer systems. Furthermore, this DNA can be reused in a whole series of PCR experiments. A portion of cDNA attached to a solid phase has the same information as a cDNA library. By using different solid phase coupled oligonucleotides first in an affinity step during RNA isolation, and subsequently as primers for cDNA synthesis, specific solid phase 'libraries' can be constructed. The re-use of these solid phase 'libraries', in combination with the option to isolate RNA from small amounts of tissue or cells, makes solid phase PCR a useful tool for identifying new genes and analysing gene expression.

## Acknowledgements

This work was supported by a fellowship of the Gottlieb Daimler- and Karl Benz-Stiftung 2.88.9 to S.S. and NIMH grant MH38819 to J.B.

## References

1. Stamm, S. and Brosius, J. (1991). *Nucleic Acids Res.*, **19**, 1350.
2. Jacobsen, K. S., Breivold, E., and Horns, E. (1990). *Nucleic Acids Res.*, **18**, 3669.
3. Lambert, K. N. and Williamson, V. M. (1993). *Nucleic Acids Res.*, **21**, 775.
4. Raineri, I., Moroni, C., and Senn, H. P. (1991). *Nucleic Acids Res.*, **19**, 4010.
5. Kuribayashi-Ohta, K., Tamatsukuri, S., Hikata, M., Miyamoto, C., and Furuichi, Y. (1993). *Biochim. Biophys. Acta*, **1156**, 204.
6. Hultman, T., Bergh, S., Moks, T., and Uhlén, M. (1991). *Biotechniques*. **10**, 84.
7. Hultman, T., Stahl, S., Hornes, E., and Uhlén, M. (1989). *Nucleic Acids Res.*, **17**, 4937.
8. Holmberg, M., Wahlberg, J., Lundeberg, J., Pettersson, U., and Uhlén, M. (1992). *Mol. Cell. Probes*, **6**, 201.
9. Wahlberg, J., Lundeberg, J., Hultman, T., and Uhlén, M. (1990). *Proc. Natl Acad. Sci. USA*, **87**, 6569.
10. Olson, J. D., Panfili, P. R., Zuk, R. F. and Sheldon, E. L. (1991). *Mol. Cell. Probes*, **5**, 351.
11. Espelund, M., Prentice Stacy, R. A., and Jackobsen, K. S. (1990). *Nucleic Acids Res.*, **18**, 6157.
12. Morgan, J. G., Dolganov, G. M., Robbins, S. E., Hinton, L. M., and Lovett, M. (1992). *Nucleic Acids Res.*, **20**, 5173.
13. Tsurui, H., Hara, E., Oda, K., Suyama, A., Nakada, S., and Wada, A. (1990). *Gene*, **88**, 233.
14. Tagle, D. A., Swaroop, M., Lovett, M., and Collins, F. (1993). *Nature*, **361**, 751.
15. Skryabin, B. V., Khalchitsky, S. E., Kuzjmin, A. I., Kaboev, O. K., Kalinin, V. N., and Schwartz, E. I. (1990). *Nucleic Acids Res.*, **18**, 4289.
16. Applied Biosystems (1989). *Synthesis of fluorescent labelled dye oligonucleotides for use as primers in fluorescent-based DNA sequencing*, User Bulletin No. 11. Applied Biosystems.
17. Keller, G. H. and Manak, M. M. (1989). *DNA probes*. Stockton Press, New York.

18. Dean, P. D. G., Johnson, W. S., and Middle, F. A. (1985). *Affinity chromatography: a practical approach*. IRL Press, Oxford.
19. Frohman, M. A., Dush, M. K., and Martin, G. R. (1988). *Proc. Natl Acad. Sci. USA*, **85**, 8998.
20. Frohman, M. A. (1990). In: *PCR protocols* (ed. M. A. Innis, D. H. Gelfand, J. J. Sninsky, and T. J. White) pp. 28. Academic Press, New York.
21. Eckstein, F. (1991). *Oligonucleotides and analogues: a practical approach*. IRL Press, Oxford.
22. Stamm, S., Gillo, B., and Brosius, J. (1991) *Biotechniques*, **10**, 430.
23. Prince, J. T., Milona, N., and Stallcup, W. B. (1989). *J. Neuroscience*, **9**, 876.
24. Kirchhausen, T., Scarmato, P., Harrison, S. C., Monroe, J. J., Chow, P. E., Mattaliano, R. J., Ramachandran, K. L., Smart, J. E., Ahn, A. H., and Brosius, J. (1987). *Science*, **236**, 320.
25. DeChiara, T. M. and Brosius J. (1987). *Proc. Natl Acad. Sci. USA*, **84**, 2624.

# 5

# Solid phase sequencing of PCR products

JOHAN WAHLBERG, THOMAS HULTMAN, and MATHIAS UHLÉN

## 1. Introduction

Solid phase methods have proven to be extremely useful for the separation and synthesis of biomolecules. The advantages of solid phase approaches are a combination of good yields and reproducible results. Furthermore, the facilitated separation of the solid phase from the reaction solution allows routine methods suitable for automation to be designed. This chapter describes protocols for the use of magnetic beads in combination with the biotin–streptavidin system for solid phase DNA sequencing of PCR products. The use of a solid support for the preparation of sequencing templates enables efficient production of single-stranded DNA with simultaneous removal of PCR buffers, dNTPs, and PCR primers. The immobilization of the PCR product also makes it possible to avoid competition of the sequencing primer with the complementary strand of the template obtained when using the double-stranded DNA fragments directly.

## 2. Different methods for DNA sequencing of PCR products

It is possible to sequence PCR products directly after alkali denaturation followed by precipitation (1) or after heat denaturation followed by 'snap-cooling' (rapid freezing in dry-ice/ethanol) (2). However, it is difficult to maintain sufficient dissociation of the two rapidly re-annealing complementary strands while allowing annealing and extension of the sequencing primer. Successful routines have commonly involved reagents or procedures designed to maximize duplex dissociation. Many of these protocols were first employed for sequencing double-stranded plasmids where an alkali or heat denaturation step is sufficient to denaturate the two strands (3, 4). However, in general these plasmid-sequencing methods require further manipulation to allow sequencing of linear PCR products (5). Many reagents have been used for

Cycle sequencing

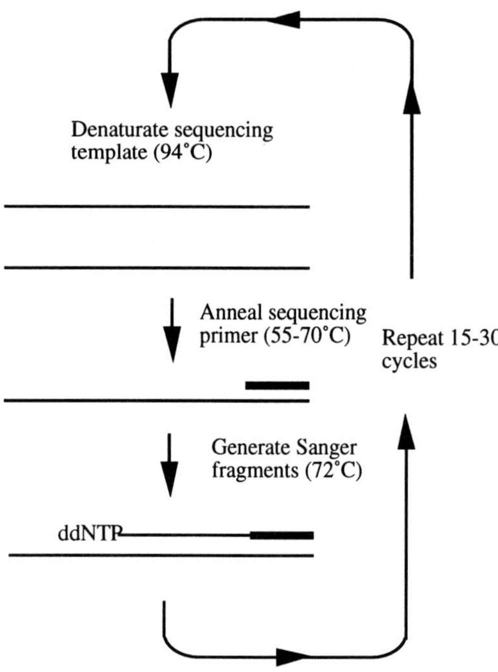


**Figure 1.** A schematic drawing of cycle sequencing; see the text for details.

maintaining the single-stranded state, during the elongation reaction, including detergents (6), DMSO (7, 8), and single-strand binding protein (9).

An alternative approach to sequencing double-stranded PCR products is to take advantage of a thermostable DNA polymerase, such as *Taq* DNA polymerase, and cycle the temperatures of annealing, extension, and denaturation as with PCR (*Figure 1*). In this method, called cycle sequencing (10–12), the amount of specific extension products accumulates linearly with the number of cycles when the right reaction conditions are chosen (dNTP and ddNTP). A reporter molecule for detection of the generated fragments is introduced. The dNTP concentration is particularly important because *Taq* DNA polymerase incorporates ddNTPs less efficiently than dNTPs (13). However, the uneven incorporation of ddNTPs and dNTP by *Taq* DNA polymerase therefore results in variation in the relative signal intensity of the extension products. This may make interpretation by the software algorithm used in automated fluorescent electrophoresis units more difficult (14).

DNA sequencing using double-stranded PCR products is associated with these problems, so various methods have been developed for converting double-stranded PCR fragments to single-stranded DNA (*Figure 2*). The first

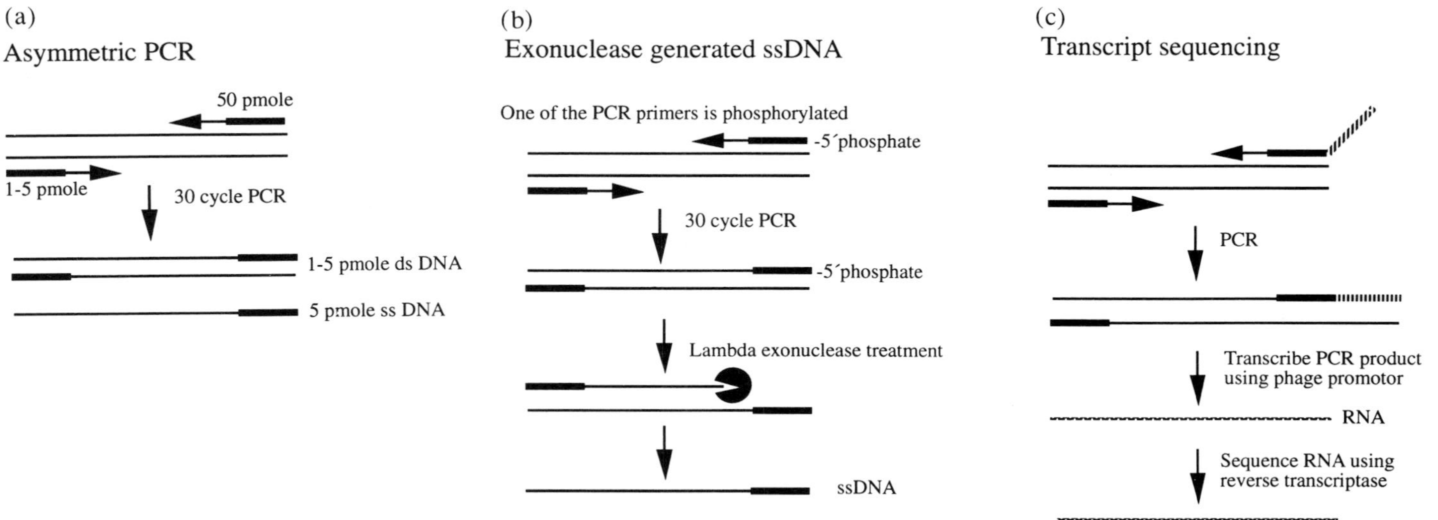


**Figure 2.** Procedures for generation of single-stranded DNA from double-stranded PCR products. (a) Asymmetric PCR using unequal amounts of the two amplification primers, (b) exonuclease generated ssDNA technique in which one PCR primer is phosphorylated so that the strand can later be digested by lambda exonuclease, (c) transcript sequencing which relies on the addition of a phage promoter sequence to one PCR primer, followed by transcription of this strand after PCR and sequencing of the RNA using reverse transcriptase. See the text for more details.

method developed (*Figure 2a*) was asymmetric PCR (15) in which unequal amounts of the two amplification primers are used. During the first 20–25 cycles, dsDNA is exponentially generated and, when the limiting primer is exhausted, ssDNA is accumulated linearly for the next 5–10 cycles. The disadvantage is that the PCR must be run with suboptimal conditions of primers and thus the result may be non-reproducible (16, 17). Obviously, the overall efficiency of amplification is lower when unequal amounts of primers are used as compared to conditions where both primers are present in the same amount and in vast excess. Titration is therefore usually necessary to find the optimal primer ratio for each new template and for each strand of that template. An alternative method involves lambda exonuclease (*Figure 2b*), which can be used to digest one strand of the amplification product to generate ssDNA (18). Prior to PCR-amplification, one of the amplification primers is phosphorylated. When the double-stranded PCR product is exposed to lambda exonuclease, the strand containing the 5′-phosphate is digested resulting in ssDNA. This method depends largely on the efficiency of the phosphorylation and the lambda exonuclease treatment. The results of a typical reaction appear to be very similar to that of an asymmetric PCR, with only a fraction of the templates reduced to ssDNA. However, compared to an asymmetric PCR, no titration of primer ratios is needed. Another relatively simple method of generating a single-stranded RNA from double-stranded DNA amplification products is the addition of a phage promoter sequence (e.g. T7 promoter) on one of the amplification primers (*Figure 2c*). Following amplification, T7 RNA polymerase is used in a second reaction to create single-stranded RNA (19, 20). The RNA can then be used as sequencing template using reverse transcriptase. This approach is, however, susceptible to background signal generation caused by misprimed amplification products (21).

Amplification by PCR utilizes oligonucleotide primers and deoxynucleotides in a large excess and the remaining components have to be removed prior to DNA sequencing in all protocols. The PCR primers will otherwise compete with the sequencing primer during the annealing step, thus giving rise to false priming. Thus an optimal deoxy/dideoxynucleotide ratio is hard to obtain due to the large excess of deoxynucleotides used in the amplification. Various methods have been developed to remove PCR buffer, nucleotides, and primers before proceeding with DNA sequencing: different spin columns (22), ultrafiltration using Centricon apparatus (23), or HPLC (24). Another method is to cut out an agarose slice containing the PCR product, recover the DNA, and then perform the sequencing reactions. When using low melting point agarose, sequencing can be performed directly in the agarose plug (25). However, all these methods are relatively cumbersome and difficult to scale-up to handle a large through-put of samples. To allow direct sequencing of PCR products, limited amounts of primer and dNTPs have been used, resulting in a very small additional contribution of oligonucleotides and dNTPs from the amplification (26). The obvious disadvantage from such a protocol is that the small amounts of dNTPs and primer that have to be used make it difficult to amplify different target DNA sequences with reproducible results.

## 3. The principle of solid phase sequencing

The solid-phase DNA sequencing scheme (27, 28) is outlined in *Figure 3*. First, biotin is introduced into one of the strands of the DNA. This is preferably performed during the amplification using one of the primers 5′ end biotinylated. The PCR product is then immobilized through the extremely stable interaction between biotin and streptavidin ($K_d = 10^{-15}$ M) on to a solid support (preferably monodispersed paramagnetic beads) with covalently-coupled streptavidin on the surface. After immobilization the beads, with the immobilized DNA, are washed extensively to remove all the reaction components resulting from the amplification. The immobilized double-stranded DNA is then denatured by incubation with 0.10 M NaOH for a few minutes. Note that both the immobilized single strand and the recovered eluted strand may be used for standard Sanger DNA sequencing (29) after neutralization by HCl.

The following sections describe protocols for the amplification of plasmid inserts, immobilization of the PCR product and strand separation. The amplification of target genes in genomic DNA is also covered. Finally, several alternative solid phase sequencing protocols are given.

## 4. Amplification of plasmid vector inserts for solid phase sequencing

Universal primers can be used to amplify inserts in plasmid vectors such as pUC, pBluescript, pEMBL, pGEM etc. Simplified protocols have therefore been designed for direct colony lysis (*Protocol 1*) and amplification with general primer sets containing biotinylated primers (*Protocol 2*). The biotinylated PCR products are directly immobilized to streptavidin covered paramagnetic beads (*Protocol 3*). After denaturation of the immobilized double-stranded DNA bound to the beads and elution of the non-biotinylated strand, single-stranded DNA suitable for sequencing is obtained (*Protocol 4*). By using primer set A (*Protocol 2*) the immobilized strand can be sequenced with the universal sequencing primer. Alternatively the immobilized strand can be sequenced with the T3 or the SP6 promoter primers. The eluted (supernatant) strand can be sequenced using the reverse sequencing primer or the T7 promoter primer. By using primer set B (*Protocol 2*) the immobilized strand is sequenced with the reverse sequencing primer or the T7 promoter primer. The supernatant strand can be sequenced using the universal sequencing primer, T3 or the SP6 promoter primers.

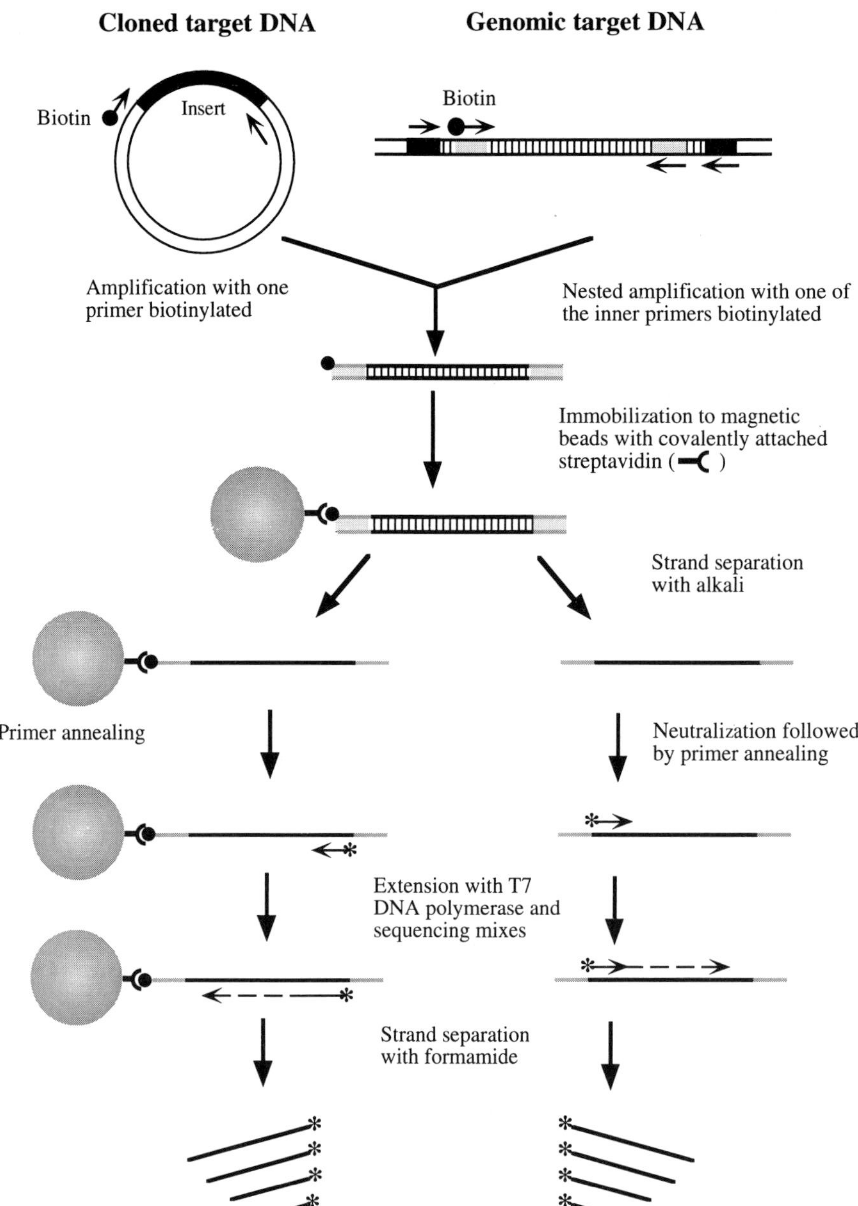


**Figure 3.** Scheme for solid-phase DNA sequencing.

## Protocol 1. Colony lysis

*Equipment and reagents*

- Thermocycler
- Sterile toothpicks
- PCR tubes
- Lysis buffer: 100 mM Tris–HCl pH 8.3 (20°C), 20 mM $MgCl_2$, 500 mM KCl, 1.0% Tween 20

*Method*

1. Dispense 10 μl of lysis buffer into each tube.
2. Take a part of one colony from the cultivation plate using a toothpick and place it in the PCR tube.
3. Incubate in the thermocycler at 99 °C for 5 min and then place the tube directly on ice.

## Protocol 2. Amplification of plasmid vector insert for solid phase sequencing

*Equipment and reagents*

- Thermocycler
- 10 × PCR buffer: 100 mM Tris–HCl pH 8.3 (20°C); 20 mM $MgCl_2$; 500 mM KCl; 1% Tween 20
- AmpliTaq polymerase (Perkin–Elmer)
- Template DNA or colony lysate (from *Protocol 1*)
- Nucleotide solution (2 mM of each dNTP)
- PCR primer solution (containing two primers each having a concentration of 2.5 μM) (obtained from Dynal AS)
  Primer set A (Dynal AS):
  5′-Biotin-GCTTCCGGCTCGTATGTTGTGTG-3′
  3′-GCGGAACGTCGTGTAGGGGGAAA-5′
  Primer set B (Dynal AS):
  5′-GCTTCCGGCTCGTATGTTGTGTG-3′
  3′-GCGGAACGTCGTGTAGGGGGAAA-Biotin-5′

*Method*

1. Set up the PCR master mix in a microcentrifuge tube:
   - 10 × PCR buffer 5 μl
   - primers 2 μl
   - dNTP solution 5 μl
   - AmpliTaq 1 U
   - sterile water (−volume of DNA, step 2) to 50 μl
2. Add 1–2 pmol of template DNA (or 5 μl of the colony lysate).
3. Overlay with 25 μl of light mineral oil, if necessary, and cycle as follows:
   96°C 30 sec
   72°C 0.5–3.0 min (depending on the length of target sequence) for 30 cycles,
   72°C 10 min
4. Check the PCR product on an agarose gel.

## **Protocol 3.** Preparation of magnetic beads

*Equipment and reagents*

- Dynabeads M-280 Streptavidin (10 mg/ml) (Dynal AS)
- A strong magnet (e.g. MPC-E from Dynal AS or equivalent)
- Washing/binding solution: 10 mM Tris–HCl pH 7.5 (20 °C), 1 mM EDTA, 2 M NaCl
- 1.5 ml microcentrifuge tubes

*Method*

1. Prepare a bulk wash of beads, depending upon the number of templates which need to be purified. To do this, resuspend the beads by pipetting and, using 20 μl of resuspended beads per PCR template, pipette the suspended beads into a clean 1.5 ml microcentrifuge tube.
2. Place the tube in the magnetic holder and allow for beads to adhere to side of the wall. Remove the supernatant using a pipette (*do not remove the tube from magnetic holder*).
3. *Now* remove the tube from the magnetic holder. Add an equal volume of washing/binding solution and gently pipette to suspend.
4. Once again, using the magnetic holder, allow the beads to adhere to the side of the tube and remove the supernatant.
5. Resuspend the beads in washing/binding solution using twice the original volume. The bead concentration is now 5 μg/μl.

## **Protocol 4.** Immobilization of the PCR product and strand separation

*Equipment and reagents*

- Dynabeads M-280 Streptavidin (10 mg/ml) (Dynal AS)
- A strong magnet (e.g. MPC-E from Dynal AS or equivalent)
- 1.5 ml microcentrifuge tubes
- Washing/binding solution: 10 mM Tris–HCl pH 7.5 (20 °C), 1 mM EDTA, 2 M NaCl
- 1 × TE buffer: 10 mM Tris–HCl pH 7.5 (20°C), 1 mM EDTA
- 0.1 M NaOH (freshly made) [a]
- 0.333 M HCl (freshly made) [b]
- T buffer: 10 mM Tris–HCl pH 7.5 (20°C)

*Method*

1. Transfer 45 μl of the total 50 μl PCR amplification reaction to a fresh 1.5 ml microcentrifuge tube and add 40 μl of the pre-washed Dynabeads (from *Protocol 3*).
2. Incubate at room temperature for 15 min. Mix once or twice during the immobilization reaction by carefully tapping on the outside of vial.

3. Collect the beads, by placing the vials in the magnetic holder, and remove the supernatant with a pipette.
4. Wash the beads once with 50 μl washing/binding solution.[c]
5. Wash once with 50 μl 1 × TE buffer. Remove the 1 × TE buffer carefully, avoid droplets on the walls and the bottom of the tube.
6. Resuspend the beads in *exactly* 10 μl 0.1 M NaOH.[a]
7. Incubate at room temperature for 10 min.
8. Collect the beads (now with only ssDNA attached) by placing the tube in the magnetic holder and transfer the 10 μl of NaOH supernatant (containing the non-biotinylated strand) to a clean tube. Neutralize the NaOH supernatant with 3 μl 0.333 M HCl[b] and mix *immediately*. Add 2 μl of T buffer and save the supernatant strand.
9. Wash the beads once with 50 μl 0.1 M NaOH, once with 50 μl washing/binding solution and once with 50 μl 1 × TE buffer. Carefully remove the 1 × TE without leaving any droplets.
10. Resuspend the beads in 13 μl of sterile water (or the appropriate buffer for the sequencing protocol to follow).

[a] Important: use a 1.000 ± 0.005 M volumetric solution of NaOH and dilute this to 0.1 M. Aliquot this and store at −20°C.

[b] Important: use a 1.000 ± 0.005 M volumetric solution of HCl and dilute this to 0.333 M. Aliquot this and store at −20 °C.

[c] The immobilized DNA can be stored at 4 °C for several weeks.

# 5. Amplification of genomic DNA for solid phase sequencing

## 5.1 Biotinylated primers

Specific primers for the target gene must be designed and synthesized. A nested primer procedure is often recommended to yield a clean product (30). It is possible to use a 'handle' sequence in one of the primers, hereby introducing a 'universal' sequencing into the PCR product as described by Wahlberg *et al.* (31). A general biotinylated primer can subsequently be used thus avoiding the synthesis of new biotinylated primers for each target gene.

Many protocols now exist for biotinylation of oligonucleotides to be used for the amplification of the target gene. Two of these are described in *Protocols 5* and *6*. In addition, biotinylated oligonucleotides can be obtained from several commercial sources offering oligonucleotide synthesis services. It is of great importance that the biotinylated oligonucleotide is purified from unbound biotin, preferably by reverse phase FPLC or HPLC, as free biotin will occupy binding sites on the beads and reduce the binding capacity of biotinylated PCR products.

**Protocol 5.** 5′-end biotinylation of oligonucleotide using biotin phosphoramidites

*Equipment and reagents*

- Automated DNA synthesizer
- Reverse phase FPLC or HPLC
- A biotin phosphoramidite containing a spacer arm, e.g. DMT-BIOTIN-C6-PA (Cambridge Research Biochemicals Ltd) or Biotin-ON™ Phosphoramidites (Clontech Laboratories Inc.) or Biodite (Pharmacia LKB Biotechnology AB)

*Method*

1. Incorporate the biotin phosphoramidite containing a spacer arm directly on to the 5′-end of the oligonucleotide during synthesis using an automated DNA synthesizer. Some of the biotin phosphoramidites can be incorporated at any position (single or multiply) in the oligonucleotide whereas other reagents contain a protecting group and are limited to 5′-end biotinylation only (see Chapter 3).
2. Purify the biotinylated oligonucleotide by a reverse phase HPLC or FPLC.

**Protocol 6.** 5′-end biotinylation of oligonucleotide using 5′-amino-modified phosphoramidites

*Equipment and reagents*

- Automated DNA synthesizer
- Reverse phase FPLC or HPLC
- An amino-linker, e.g.
  Aminolinker 2™ (Applied Biosystems Inc.), or
  THF-linker (Pharmacia LKB Biotechnology AB)
- A biotin ester, e.g.
  Biotin-X-NHS-ester (Clontech Laboratories Inc.), or
  D-Biotin-*N*-hydroxysuccinimide ester (Boehringer Mannheim GmbH), or
  NHS-LC-Biotin (Pierce Europe B.V.), or
  Biotinaminocaproate *N*-hydroxysuccinimide ester (Sigma)
- *N,N*-dimethylformamide
- NAP™-10 Column (Pharmacia LKB Biotechnology AB)

*Method*

1. Incorporate the amino-linker phosphoramidite directly into 5′-end of the oligonucleotide during synthesis using an automated DNA synthesizer.
2. Dissolve the 5′-end amino-modified oligonucleotide in 0.7 ml of sterile distilled water.
3. Add 0.1 ml of 1.0 M $NaHCO_3/NaCO_3$ buffer, pH 9.0 (20°C)
4. Prepare a fresh 10 mg/ml solution of the biotin ester in *N,N*-

dimethylformamide and add 0.2 ml of this solution to the reaction mixture.

5. Leave at room temperature for at least 2 h and preferably overnight.
6. Remove traces of unincorporated biotin ester by using the NAP$^{TM}$-10 spin column.
7. Purify the biotinylated oligonucleotide by reverse phase FPLC or HPLC according to the manufacturer's instructions.

### 5.2 Amplification and isolation of single-stranded target DNA

Amplification of the target gene sequence in the genomic DNA is carried out using the biotinylated primers according to the procedure described in *Protocol 2*. However, obviously the choice of primers influence the conditions used in the amplification reaction. Thus *Protocol 2* must be adapted to fit the primer pair chosen for the amplification. After amplification, *Protocols 3* and *4* are used to prepare the magnetic beads and then use these to isolate single-stranded target DNA for sequencing.

## 6. Solid phase sequencing

Three alternative solid phase sequencing protocols are described below. *Protocol 7* sequences using a labelled primer (either a radiolabelled primer or a primer containing a fluorescent dye label), *Protocol 8* describes sequencing using fluorescent labelled dATP, and *Protocol 9* uses four fluorescent dye labelled primers. The sequencing primer used in *Protocols 7–9* can be:

(a) a custom-designed primer complementary to a sequence inside the target DNA being amplified (27);

(b) a universal primer complementary to a sequence introduced by a 'handle' sequence in one of the PCR primers (32);

(c) one of the primers used in the PCR (33).

The choice as to which solid phase sequencing protocol to use (*Protocols 7–9*) is determined by the sequencing techniques available to the reader. *Protocol 7* is suitable for manual sequencing using radioactively labelled primer or for automated DNA sequencing using the Pharmacia A.L.F. system. *Protocol 8* again requires access to a Pharmacia A.L.F. system and *Protocol 9* needs access to an automated DNA sequencer from Applied Biosystems Inc.

## Protocol 7. Solid phase DNA sequencing using T7 DNA polymerase and labelled primer

### *Equipment and reagents*

- Water baths or incubators at 37°C and 65°C
- A Pharmacia A.L.F. system or a polyacrylamide sequencing gel and electrophoresis equipment
- Annealing buffer: 280 mM Tris–HCl pH 7.5 (20°C), 100 mM $MgCl_2$
- 0.5 μM labelled sequencing primer. The label could be either $^{32}P$ for radioactive sequencing or one fluorescent dye label such as fluorescein isothiocyanate, FITC (Pharmacia LKB Biotechnology AB)
- Extension buffer: 300 mM citric acid pH 7.0 (20°C), 318 mM DTT, 40 mM $MnCl_2$
- T7 DNA polymerase with enzyme dilution buffer (Pharmacia LKB Biotechnology AB)
- Four nucleotide mixes each containing 40 mM Tris–HCl pH 7.5 (20°C), 50 mM NaCl, 1.0 mM of each dNTP,[a] and 5.0 μM of one specific ddNTP. Thus the 'A' mix contains 5 μM ddATP, the 'C' mix contains 5 μM ddCTP, the 'G' mix contains 5 μM ddGTP, and the 'T' mix contains 5 μM ddTTP.
- Stop solution (shake 100 ml formamide with 5 g Amberlite MB-1 resin and 300 mg dextran blue for 30 min. Filter through 0.45 μM pore-size filter)

### *Method*

1. Add single-stranded template DNA (from *Protocol 4*) to a fresh tube.[b] Adjust the volume to 15 μl with sterile water.
2. Add 2 μl (1 pmol) labelled primer.
3. Add 2 μl of annealing buffer and mix gently with a pipette. Incubate at 65°C for 10 min. Mix gently and leave to cool at room temperature for at least 10 min, mix two or three times during cooling.
4. Add 1 μl of extension buffer and mix gently.
5. Dilute the T7 DNA polymerase to 1.5 U/μl using cold dilution buffer; 2 μl of this diluted 'stock solution' will be required for each template (do not remove the T7 DNA polymerase stock solution from the −20°C freezer for more than a few seconds and keep the tube with diluted 'stock solution' (1.5 U/μl) on ice).
6. Label four new tubes[b] 'A', 'C', 'G' and 'T'. Pipette 2.5 μl of the corresponding dNTP/ddNTP sequencing mixes into the tubes.
7. Warm the dispensed nucleotide sequencing mixes at 37°C for at least 1 min.
8. Add 2 μl of the T7 polymerase diluted 'stock solution' (from step 5) to the template mixture (from step 4) and mix gently. Immediately add 4.5 μl of this mixture to each of the preincubated nucleotide sequencing mixes.
9. Incubate at 37°C for 5 min.
10. Add 5 μl of stop solution to each reaction and mix gently.
11. Incubate at 85–90°C for 2–3 min and then put the tubes on ice.

**12.** Load the samples on to the polyacrylamide sequencing gel.[c]

[a] c7dGTP is preferred instead of dGTP for resolving band compression during electrophoresis.
[b] If many samples are to be analysed, a microtitre plate might be more convenient.
[c] The samples may be stored at −20°C if not loaded immediately.

## Protocol 8. Solid phase DNA sequencing using fluorescent labelled dATP

*Equipment and reagents*

- Water baths or incubators at 37°C and 65°C
- A Pharmacia A.L.F. system (Pharmacia LKB Biotechnology AB)
- Annealing buffer: 280 mM Tris–HCl pH 7.5 (20°C), 100 mM $MgCl_2$
- 5 μM sequencing primer
- Extension buffer: 300 mM citric acid pH 7.0 (20°C), 318 mM DTT, 40 mM $MnCl_2$
- T7 DNA polymerase with enzyme dilution buffer (Pharmacia LKB Biotechnology AB)
- Labelling mixture: 10 μM fluorescein-15-dATP, 1 μM of each dCTP, dGTP, and dTTP
- Four nucleotide mixes each containing 40 mM Tris–HCl pH 7.5 (20°C), 50 mM NaCl, 1.0 mM of each dNTP,[a] and 5.0 μM of one specific ddNTP. Thus the 'A' mix contains 5 μM ddATP, the 'C' mix contains 5 μM ddCTP, the 'G' mix contains 5 μM ddGTP, and the 'T' mix contains 5 μM ddTTP
- Stop solution (shake 100 ml formamide with 5 g Amberlite MB-1 resin and 300 mg dextran blue for 30 min. Filter through 0.45 μM pore-size filter)

*Method*

1. Add single-stranded DNA template (*Protocol 4*) to a fresh microcentrifuge tube.[b] Adjust the volume (with sterile water) to 15 μl.
2. Add 2 μl (10 pmol) sequencing primer.
3. Add 2 μl of annealing buffer and mix gently with a pipette. Incubate at 65°C for 10 min. Mix gently and leave to cool at room temperature for at least 10 min, mixing two or three times during cooling.
4. Add 1 μl of labelling mixture to each sample and mix gently.
5. Dilute the T7 DNA polymerase to 3 U/μl using cold dilution buffer; 2 μl of this diluted 'stock solution' will be required for each template (do not remove the T7 DNA polymerase stock solution from the −20°C freezer for more than a few seconds and keep the tube with diluted 'stock solution' (3 U/μl) on ice).
6. Add 2 μl of the T7 polymerase diluted 'stock solution' (from step 5) to the template mixture (from step 4) and mix gently.
7. Incubate at 37°C for 10 min. Then place the tube on ice.
8. Add 1 μl of extension buffer and mix gently.
9. Label four new tubes[b] 'A', 'C', 'G' and 'T'. Pipette 2.5 μl of the corresponding dNTP/ddNTP sequencing mixes into the tubes.

**Protocol 8.** *Continued*

**10.** Warm the dispensed nucleotide sequencing mixes at 37 °C for at least 1 min.

**11.** Add 4.5 μl of this annealing–labelling mixture to each of the pre-heated dNTP/ddNTP sequencing mixes.

**12.** Incubate at 37 °C for 5 min.

**13.** Add 5 μl of stop solution to each reaction and mix gently.

**14.** Incubate at 85–90 °C for 2–3 min and then put the tubes on ice.

**15.** Load the samples on to the sequencing gel in the A.L.F. apparatus.[c]

[a] c7dGTP is preferred instead of dGTP for resolving band compression during electrophoresis.
[b] If many samples are to be analysed, a microtitre plate might be more convenient.
[c] The samples may be stored at −20 °C if not loaded immediately.

## **Protocol 9.** Solid-phase DNA sequencing using T7 DNA polymerase and four dye fluorescent primers[a]

### *Equipment and reagents*

- Incubators or water baths at 37 °C and 65 °C
- A strong magnet (MPC-E from Dynal AS or equivalent)
- An ABI automated DNA sequencer
- 1 × TE buffer: 10 mM Tris–HCl pH 7.5 (20 °C), 1 mM EDTA
- Annealing buffer: 11.2 mM Tris–HCl pH 7.5 (20 °C), 0.02% Tween 20, 4.0 mM $MgCl_2$
- Extension buffer: 300 mM citric acid pH 7.0 (20 °C), 318 mM DTT, 40 mM $MnCl_2$
- Four nucleotide mixes each containing 40 mM Tris–HCl pH 7.5 (20 °C), 50 mM NaCl, 1.0 mM of each dNTP,[b] and 5.0 μM of one specific ddNTP. Thus the 'A' mix contains 5 μM ddATP, the 'C' mix contains 5 μM ddCTP, the 'G' mix contains 5 μM ddGTP, and the 'T' mix contains 5 μM ddTTP
- T7 DNA polymerase with enzyme dilution buffer (Pharmacia LKB Biotechnology AB)
- 10 × TE: 100 mM Tris–HCl pH 7.5 (20 °C), 10 mM EDTA
- Loading solution (shake 100 ml formamide with 5 g Amberlite MB-1 resin and 300 mg dextran blue for 30 min. Filter through 0.45 μM pore-size filter)

### *Method*

1. Dissolve the immobilized single stranded template (*Protocol 4*) in 19 μl of annealing buffer in a microcentrifuge tube.[c]
2. Dilute the T7 DNA polymerase to 0.7 U/μl using cold dilution buffer; 6 μl of this diluted 'stock solution' will be required for each template (do not remove the T7 DNA polymerase stock solution from the −20 °C freezer for more than a few seconds and keep the tube with diluted 'stock solution' (0.7 U/μl) on ice).
3. Divide the paramagnetic beads bearing the immobilized template DNA into four tubes labelled 'A', 'C', 'G' and 'T'. Add primer as indicated

below. Mix gently and heat to 65°C for 10 min. Mix gently and leave to cool at room temperature for at least 10 min, mixing two or three times during cooling.

| | Reaction tube | | | |
|---|---|---|---|---|
| | *A* | *C* | *G* | *T* |
| Beads in annealing buffer (μl) | 3 | 3 | 6 | 6 |
| Primer (μl) | 1 | 1 | 2 | 2 |
| Extension buffer (μ) | 0.5 | 0.5 | 1 | 1 |
| Nucleotide mixes (μl) | 1.5 | 1.5 | 3 | 3 |
| T7 DNA polymerase diluted 'stock solution' (μl) | 1 | 1 | 2 | 2 |

4. Add an appropriate volume of the relevant dNTP/ddNTP sequencing mixes (A, C, G, or T) to each tube as indicated in the table above. Keep the samples on ice while doing this.
5. Add the T7 DNA polymerase diluted 'stock solution' as shown above. Incubate at 37°C for 5 min.
6. Stop the reaction by adding 40 μl *ice-cold* 10 × TE to each tube and place the tubes on ice.
7. Pool the A, C, G and T reactions for each sample and collect the beads using the magnet. Discard the supernatant. Wash the beads once with 1 × TE buffer and discard the supernatant. Resuspend the beads in 5 μl loading buffer.[d]
8. Just before loading the samples on to the ABI automated DNA sequencer, heat the samples to 95°C for 2 min and place on ice.

[a] The eluted strand from *Protocol 2* cannot be used with this protocol.
[b] c7dGTP is preferred instead of dGTP for resolving band compression during electrophoresis.
[c] If many samples are to be analysed, a microtitre plate might be more convenient.
[d] The samples may be stored at −20°C if not loaded immediately.

## 7. Discussion

Solid-phase purification of PCR products provides many advantages over conventional methods that are used to prepare PCR products for sequencing, such as gel purification or ultrafiltration. With the solid-phase approach, it becomes possible to separate the complementary strands and sequence them independently. It is also possible to sequence the biotinylated strand repeatedly. Here, we have described solid phase sequencing using T7 DNA polymerase (Sequenase) which yields uniform band intensities due to the high processivity of this enzyme. However, the immobilized DNA template can also be used for cycle sequencing by standard protocols using *Taq* DNA polymerase (11).

A limitation of the current solid phase method is that DNA fragments larger than approximately 3.0 kb do not bind efficiently to the solid support. Although the biotin-binding capacity of the beads is approximately 3.0 pmol/μl, the observed binding of PCR-amplified DNA is much less. Recently, Yura *et al.* have obtained improved binding of larger PCR fragments by introducing two biotin molecules in the PCR primer (34).

The manual sequencing protocols described in this chapter have also been used as the basis for developing semi-automated sequencing systems (28, 35). *Protocols 3, 4* and *7* for template preparation and performing the sequencing reactions have been implemented on a Beckman Biomek 1000 workstation, while *Protocol 9* preceded by *Protocols 3* and *4* have been implemented on an ABI Catalyst workstation (36).

## Acknowledgement

The authors thank Dr Eric Hornes for considerable help with the protocol on biotinylation of primers and comments on the rest of the manuscript.

## References

1. Wrischnik, L. A., Higuchi, R. G., Stoneking, M., Erlich, H., Arnheim, N., and Wilson, A. C. (1987). *Nucleic Acids Res.*, **15**, 529.
2. Kusukawa, N., Uemori, T., Asada, K., and Kato, I. (1990). *Biotechniques*, **9**, 66.
3. Vizard, D., Hadman, M., Bianca, D., and Manzer, D. (1990). *Biotechniques*, **8**, 430.
4. Saunders, S. E. and Burke, J. F. (1990). *Nucleic Acids Res.*, **18**, 4948.
5. Casanova, J. L., Pannetier, C., Jaulin, C., and Kourilsky, P. (1990). *Nucleic Acids Res.*, **18**, 4028.
6. Bachmann, B., Lüke, W., and Hunsmann, G. (1990). *Nucleic Acids Res.*, **18**, 1309.
7. Seto, D. (1990). *Nucleic Acids Res.*, **18**, 5905.
8. Winship, P. R. (1989). *Nucleic Acids Res.*, **17**, 1266.
9. Kaspar, P., Zadrazil, S., and Fabry, M. (1989). *Nucleic Acids Res.*, **17**, 3616.
10. Carothers, A. M., Urlaub, G., Mucha, J., Grunberger, D., and Chasin, L. A. (1989). *Biotechniques*, **7**, 494.
11. Craxton, M. (1991). *Methods: A Companion to Methods in Enzymology*, **3**, 20.
12. Murray, V. (1989). *Nucleic Acids Res.*, **17**, 8889.
13. Innis, M. A., Myambo, K. B., Gelfand, D. H., and Brow, M. A. D. (1988). *Proc. Natl Acad. Sci. USA*, **85**, 9436.
14. Khurshid, F. and Beck, S. (1993). *Anal. Biochem.*, **208**, 138.
15. Gyllensten, U. B. and Erlich, H. A. (1988). *Proc. Natl Acad. Sci. USA*, **85**, 7652.
16. Hopgood, R., Sullivan, K. M., and Gill, P. (1992). *Biotechniques*, **13**, 82.
17. Hunkapiller, T., Kaiser, R. J., Koop, B. F., and Hood, L. (1991). *Current Opinion in Biotechnology*, **2**, 92.
18. Higuchi, R. G. and Ochman, H. (1989). *Nucleic Acids Res.*, **17**, 5865.
19. Sarkar, G. and Sommer, S. S. (1988). *Nucleic Acids Res.*, **16**, 5197.
20. Stoflet, E. S., Koeberl, D. D., Sarkar, G., and Sommer, S. S. (1988). *Science*, **239**, 491.

21. Bevan, I. S., Rapley, R., and Walker, M. R. (1992). *PCR Methods and Applications*, **1**, 222.
22. DuBose, R. F. and Hartl, D. L. (1990). *Biotechniques*, **8**, 271.
23. Mihovvilovic, M. and Lee, J. E. (1989). *Biotechniques*, **7**, 14.
24. Katz, E. D. and Dong, M. W. (1990). *Biotechniques*, **8**, 546.
25. Kretz, K. A., Carson, G. S., and O'Brien, J. S. (1989). *Nucleic Acids Res.*, **17**, 5864.
26. Meltzer, S. J., Mane, S. M., Wood, P. K., Johnson, L., and Needleman, S. W. (1990). *Biotechniques*, **8**, 142.
27. Hultman, T., Ståhl, S., Hornes, E., and Uhlén, M. (1989). *Nucleic Acids Res.*, **17**, 4937.
28. Hultman, T., Bergh, S., Moks, T., and Uhlén, M. (1991). *Biotechniques*, **10**, 84.
29. Sanger, F., Nicklen, S., and Coulson, A. R. (1977). *Proc. Natl Acad. Sci. USA*, **74**, 5463.
30. Mullis, K. B. and Faloona, F. A. (1987). In *Methods in enzymology* (ed. R. Wu), Vol. 155, p. 335. Academic Press, London.
31. Wahlberg, J., Lundeberg, J., Hultman, T., and Uhlén, M. (1990). *Molecular and Cellular Probes*, **4**, 285.
32. Wahlberg, J., Albert, J., Lundeberg, J., von Gegerfelt, A., Broliden, K., Utter, G., Fenyö, E.-M., and Uhlén, M. (1991). *AIDS Res. Hum. Retrov.*, **7**, 983.
33. Wahlberg, J., Albert, J., Lundeberg, J., Cox, S., Wahren, B., and Uhlén, M. (1992). *The FASEB Journal*, **6**, 2843.
34. Yura, T., Mori, H., Nagai, H., Nagata, T., Ishihama, A., Fujita, N., Isono, K., Mizobuchi, K., and Nakata, A. (1992). *Nucleic Acids Res.*, **20**, 3305.
35. Wahlberg, J., Holmberg, A., Bergh, S., Hultman, T., and Uhlén, M. (1992). *Electrophoreses*, **13**, 547.
36. Holmberg, A., Fry, G., and Uhlén, M. (1993). In *Automated DNA sequencing and analysis techniques* (ed. C. Ventor). Academic Press, London.

# 6

# cDNA cloning by RT–PCR

JEAN BAPTISTE DUMAS MILNE EDWARDS, PHILIPPE RAVASSARD, CHRISTINE ICARD-LIEPKALNS, and JACQUES MALLET

## 1. Introduction

The methods available for the characterization of mRNAs have progressed enormously since the mid 1970s. This includes the isolation of RNAs, cloning and bacterial amplification of specific mRNAs, Northern blotting studies, *in situ* hybridization, and nuclease protection assays that allow the description of mRNA metabolism in cells.

PCR (1) in conjunction with reverse transcription (RT–PCR) can be used to study mRNA almost at the level of a single cell thus allowing investigations that were not previously possible. The result is a revolution in the scale of the nucleic acid manipulations and in the number of samples that can easily be handled. This chapter focuses on RT–PCR and the cloning of the PCR products.

Cloning by RT–PCR requires sequence data on stretches of the target mRNA and various strategies have been devised (*Figure 1*). Their use depends on the information available and, they can be split broadly into two groups:

(a) *At least the two ends of the sequence to be amplified are known.* In this case, specific primers can be used (*Figure 1A*). This is the most common situation and does not present real technical difficulties.

(b) *Only one stretch of the sequence is available.* In this situation, a second sequence can be added to the 3′-end or to the 5′-end of the mRNA-derived cDNA. The addition at the 3′-end can be performed by taking advantage of the poly(A)$^+$ tail (*Figure 1B*). The addition to the 5′-end can be performed by terminal transferase or T4 RNA ligase (SLIC strategy) (*Figure 1C*). Note that, in this chapter, all references to positions within a given nucleic acid sequence refer to the sequence of the mRNA unless otherwise stated.

In the case where only the amino acid sequence is available, a mixture of oligonucleotides derived from the degeneracy of the genetic code can be used.

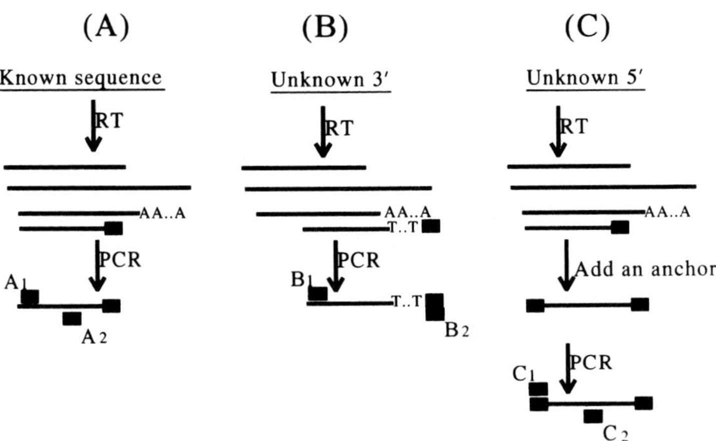


**Figure 1.** RT–PCR strategies. (*a*) *RT–PCR for a known sequence of mRNA.* Retrotranscription can be primed either with an oligodeoxyribonucleotide specifically designed to copy the sequence of the mRNA (primer extension), an oligo $dT_{[12-14]}$ (priming at the poly(A) tail) or random hexamers (random priming). PCR is then carried out with the primer pair ($A_1$, $A_2$) with sequences specific to the mRNA to be amplified. $A_1$ is located in the 5′-part of the fragment to be amplified and is a sense oligonucleotide. $A_2$ is located in the 3′-part of the fragment to be amplified and is an antisense oligodeoxyribonucleotide. (*b*) *RT–PCR of an unknown 3′-end.* Priming retrotranscription with a chimeric oligodeoxyribonucleotide composed of a homopolymeric $dT_{[14]}$ tail at its 3′-end and a known sequence at its 5′-end allows the synthesis of a single strand cDNA with a tagged 5′-end. PCR is then carried out using oligonucleotide $B_1$ (sense oligonucleotide), the sequence of which is chosen from the known sequence of the mRNA, and $B_2$ (anti-sense oligonucleotide) which is targeted to the now defined 5′-end of the ss-cDNA. (*c*) *RT-PCR of an unknown 5′-end.* It is currently impossible to add a known sequence to the 5′-ends of mRNA. Sequences are thus added to the 3′-end of the corresponding ss-cDNA. PCR is then carried out with the primer $C_1$, the sequence of which is the anti-complementary sequence of the sequence anchored at the 3′-end of the ss-cDNA and the primer $C_2$ (anti-sense oligonucleotide). The strategies B and C together can be used to amplify total mRNA from an RNA preparation and lead to the generation of cDNA libraries. This allows investigation of the transcription within cell populations.

Once amplified cDNA is available, it can be cloned. This can be facilitated by the use of PCR primers containing (within their 5′-ends) sequences recognized by appropriate restriction enzymes.

# 2. Reverse transcription

## 2.1 General considerations

Reverse transcription is the critical step in RT–PCR experiments. The success of the manipulation relies on the quality of the RNA preparation and on the characteristics of the enzyme used for the reverse transcription. Reverse

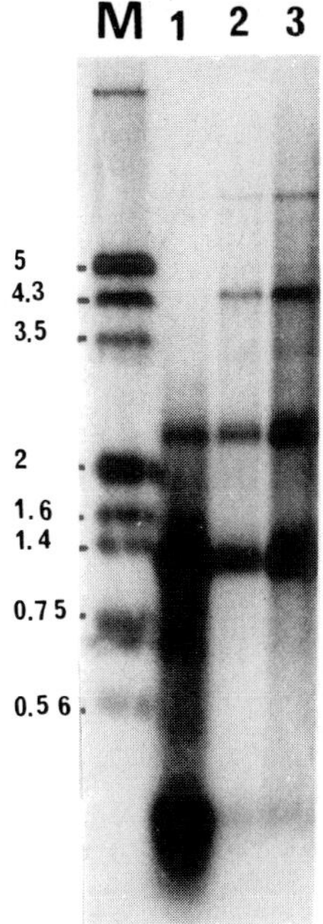


**Figure 2.** Efficiency of retrotranscription. Retrotranscription was carried out as described in *Protocol 2* (except that the concentrations of dATP, dTTP, dGTP, and dCTP were 2 mM, 2 mM, 2 mM, and 0.1 mM, respectively) using 1 μg of a poly(A)$^+$ mRNA ladder from GIBCO–BRL and an oligo(dT) primer. The reaction was carried out with the same material and reagents on two different days. Note that retrotranscription has stopped at 2.5 kb in one sample (lane one) and goes up to 7.5 kb in the other samples (lanes 2 and 3). M is [$^{32}$P]DNA marker III (Boehringer Mannheim; lambda DNA cut with *Hin*dIII and *Eco*RI).

transcription remains a delicate manipulation and is, in some cases, difficult to reproduce especially for long retrotranscripts (*Figure 2*).

## 2.2 Isolation of RNAs

RNAs can be isolated from cell cultures, organs, or samples of organs from living or dead specimens (prokaryote, plant, animal, or human). Recent reports also describe the isolation of RNA and successful RT–PCR experiments from faecal (2) or dried blood (3) specimens.

For many applications, total RNAs can be used for reverse transcription. However, as mRNA represents only 1–5% of the total RNA, random priming strategies must be avoided on such preparations to avoid copying rRNA and tRNA. We have examined a variety of protocols to extract RNAs from cell cultures and organs. The quickest is the guanidine/phenol (RNAzol™B, Biotecx Laboratories Inc.) extraction procedure based on the method described in reference 4.

To isolate the poly(A)$^+$ fraction from a total RNA preparation, we routinely use the Dynabeads kit (610–01 available from Dynal) and follow the supplier's recommendations.

## 2.3 Quality control

If the amount of RNA available is more than 100 ng, a good and fast optical method for verification of the quality of the RNA is to run an agarose minigel in RNase-free conditions (*Protocol 1*). If the RNA samples are intact, clear 28S and 18S rRNA bands will be visible. The 28S band should be about twice as intense as the 18S band (*Figure 3*). The smear spanning these two bands represents the mRNA. The presence of an intense band or smear of lower size than the 18S band indicates that the mRNA has been degraded (*Figure 3*). Bands near the sample well of the gel indicate contamination by genomic DNA which can be eliminated by the use of RNase-free DNase.

### **Protocol 1.** Agarose minigel electrophoresis of RNAs

*Precautions*

- Tips, tubes and all the material used in this manipulation must be very clean and sterilized.
- Wear gloves during all the RNA manipulation to avoid contamination with RNases.
- Use double-distilled water.
- If you are not **absolutely sure** that an aliquot of reagent is RNase free, discard it.
- These precautions represent the minimum acceptable protection against RNase. It is advisable to read ref. 5 for further advice.

**CAUTION: Ethidium bromide** (used to stain nucleic acids after gel electrophoresis) **is a mitogen.** Therefore avoid contact with skin or mucous membranes and also avoid inhalation of ethidium bromide vapours.

*Equipment and reagents*

- Minigel apparatus (Horizon, GIBCO–BRL)
- 0.1 M NaOH
- TBE (A 10 × stock contains 55 g Tris base, 55 g boric acid, 40 ml 0.5 M EDTA, water to 1 litre)
- Loading buffer (0.05% w/v bromophenol blue, 60% w/v glycerol)
- 20% SDS
- 1 mg/μl ethidium bromide
- Agarose

*Method*

1. Immerse the minigel apparatus and the combs in 0.1 M NaOH (to destroy RNA and RNase). Wash at least twice with double-distilled water.
2. Prepare a sufficient volume of 0.8% agarose solution and pour the solution into the clean and dry minigel gel apparatus. Avoid bubbles. The gel must be at least 5 mm thick.
3. When the gel is completely set, remove the comb and fill the tank with TBE buffer.

4. Dilute the RNA sample (1–2 μg) in 7 μl double-distilled water. Add 1 μl 20% SDS and incubate for 2 min at 75°C to dissociate the RNA. Add 3 μl loading buffer. Mix well and load the samples on to the gel.
5. Electrophorese the samples using a constant voltage of 50 V until the bromophenol blue has migrated two-thirds of the length of the gel.
6. After electrophoresis, add ethidium bromide to the electrophoresis buffer to 1 μg/ml. Stain the gel for 10–15 min and visualize nucleic acids under UV light (wavelength 260 nm).

If less than 10 ng of RNA is available, the quality of the mRNA preparation cannot be checked by ethidium bromide staining of gels. Two others strategies can be used:

- Synthesize radioactive single-stranded cDNA (ss-cDNA) primed with and without oligo(dT) (*Protocol 2*). Determine the radioactivity incorporated into each sample. This can be determined either by using ethanol precipitations or by using a spin column (*Protocol 3*). The radioactivity incorporated in the sample containing the primer should be at least 20-fold higher than for the unprimed sample.

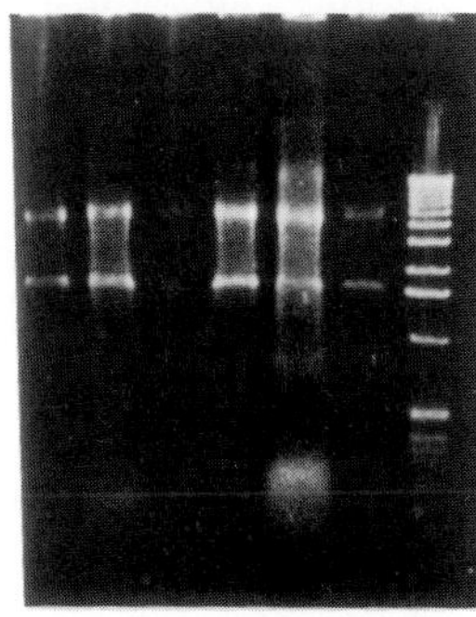

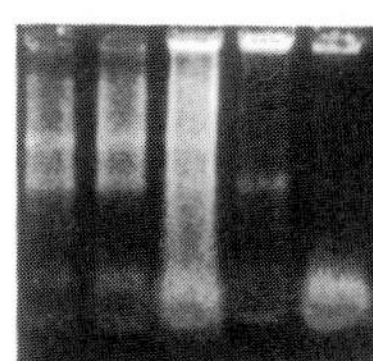


**Figure 3.** RNA quality. Total RNAs were extracted from rat (panel A, lanes 1–4, 6) or human (panel A, lane 5; panel B, all lanes) brain samples with $RNA_{ZOL}{}^{TM}$ according to the supplier's recommendations and separated by electrophoresis as described in *Protocol 1*. (A) Undegraded (lanes 1, 4, 6) or partly degraded RNA (lanes 2, 5). The two bands clearly visible in each lane represent the rRNA: the upper band is 28S rRNA and the lower one 18 S rRNA. The smear between these two bands represents the mRNAs. A faint contamination with genomic DNA can be seen in lanes 4 and 5 (faint band above the 28S band). Partly degraded RNAs are in lanes 2 and 5. Note the low molecular weight material and the clear smear under the 18S band. If RNA samples are difficult to obtain (in particular post mortem samples), it is possible to use partly degraded RNAs such as these in RT–PCR experiments. (B) Banding pattern of various samples of degraded RNA which are obviously unusable for RT–PCR experiments.

- Synthesize unlabelled ss-cDNA (*Protocol 4*) and check by PCR (*Protocol 5*) using primers chosen from the sequence of an ubiquitous mRNA.

To avoid degradation, RNAs must be extracted as soon as possible after collection of the tissue samples, or samples must be frozen in liquid nitrogen and stored at −70 °C until use. Working with post-mortem human samples is very difficult because it is impossible to avoid degradation of mRNAs (*Figure 3*) due to the activity of intracellular RNases which are liberated on cell death. In general, degradation increases with the post-mortem delay.

## 2.4 cDNA synthesis

Many kits are now available for cDNA synthesis and seem to be satisfactory. Special attention must be paid when full-length cDNAs are expected. Secondary structures, processivity of the enzyme used, and sometimes uncontrollable parameters (see *Figure 2*) limit the length of the ss-cDNA obtained.

In most RT–PCR experiments it is not necessary to synthesize radioactive ss-cDNA (*Protocol 2*). Moreover, when the amount of starting mRNA is very low, labelling ss-cDNA to a level which is detectable introduces an imbalance between dCTP (the technique requires a high ratio between radioactive and unlabelled dCTP) and the other dNTPs. This greatly reduces the efficiency of reverse transcriptase. Thus, it is preferable that synthesis of unlabelled ss-cDNA should be performed (*Protocol 3*) and, if necessary, the length of the ss-cDNA can be assessed by PCR using sets of primers allowing amplification of reference cDNAs.

### **Protocol 2.** Synthesis of a radioactive ss-cDNA[a]

*Precautions* (see *Protocol 1*)

*Equipment and reagents*

- Water baths (90°C and 42°C)
- Powdered dry ice
- 10 mg/ml Acetylated bovine serum albumin (New England Biolabs)
- 5 units/μl RNasin (Promega Biotech). RNasin is provided at the concentration of 40 U/μl; dilute it to 5 U/μl in RT buffer.
- 350 mM β-Mercaptoethanol (Sigma). The commercial β-mercaptoethanol is provided at the concentration of 14.4 M; to make a 350 mM solution, mix 50 μl with 1950 μl of double distilled water.
- 10 X RT buffer: 100 mM Tris–HCl pH 8.3 at 42°C, 80 mM KCl, 16 mM $MgCl_2$
- 100 mM solutions of each dNTP (Pharmacia or Boehringer)
- 1 mM dCTP
- 20 mM sodium pyrophosphate
- [$\alpha$-$^{32}$P]dCTP 400 Ci/mmol (Amersham PB 10165)
- AMV reverse transcriptase (Stratagene or Promega Biotech)
- The oligonucleotide used to prime the reverse transcriptase in solution (100 ng/μl in $ddH_2O$). This will be oligo $(dT)_{15-17}$ (Boehringer), BM-3′ or A-3′ NV, random hexamers, at a specific primer.

*Method*

1. Dilute the RNA[b] sample in 18 μl of water.
2. Denature the RNA by heating the tubes for 5 min at 90°C. Briefly

centrifuge, freeze the tubes in dry ice powder, then gently warm the tube in ice.

**3.** To anneal primers to the RNA, add 2.5 μl of 10 × RT-buffer and 3 pmol of primer. Adjust the volume to 25 μl.
   (a) If the retrotranscription is primed by oligo(dT) incubate for 15 min at 4°C.
   (b) If the retrotranscription is primed with random hexamers, incubate for 15 min at room temperature (approx. 18°C).
   (c) If the retrotranscription is primed with a specific primer, incubate for 15 min at the $T_m$ of the oligonucleotide (see *Protocol 5*, footnote *c*).

**4.** Mix the following components in a microcentrifuge tube:

| | |
|---|---|
| 10 × RT-buffer | 2.5 μl |
| 5 μg/μl BSA | 1 μl |
| 5 U/μ RNAsin | 1 μl |
| 350 mM β-mercaptoethanol | 2 μl |
| 100 mM dATP | 0.5 μl |
| 100 mM dTTP | 0.5 μl |
| 100 mM dGTP | 0.5 μl |
| 1 mM dCTP | 2.5 μl |
| 200 mM Sodium pyrophosphate [c] | 1 μl |
| Water to | 22 μl |

Then add 2 μl [α-$^{32}$P]dCTP (400 Ci/mmol) and 1 μl (approx. 15 U) of AMV reverse transcriptase. Mix with the annealed RNAs mixture. Incubate for 45 min at 42°C.

**5.** To favour extension of the ss-cDNA, add dCTP to 1 mM (e.g. 0.5 μl 100 mM dCTP).

**6.** Incubate for 30 min at 42°C. Store at −20°C until required.

[a] This protocol has been tested with RNA isolated from 10, 100, and 1000 cells.
[b] RNA quantities as low as 1 ng–5 μg of poly(A)$^+$ RNA can be used.
[c] Sodium pyrophosphate is added to prevent the formation of hairpin loop structures at the end of the ss-cDNA (6).

## **Protocol 3.** Determination of the radioactivity incorporated into radioactive ss-cDNA using spin column chromatography [a]

*Equipment and reagents*

- 0.5 M Trisacryl GF (Sepracor 259112)
- Glass beads (3 mm diameter, Ateliers Cloup 00 065.03)
- 0.1 × TE (1 mM Tris–HCl pH 7.6, 0.1 mM EDTA pH 8.0)

**Protocol 3.** *Continued*

A. *Preparation of the resin*

1. Suspend 1 vol. 0.5 M Trisacryl GF (Sepracor 259112) in 1 vol. 0.1 × TE.
2. Decant the resin (this can be speeded up by a low speed centrifugation at 50 *g* for 7 min)
3. Remove the supernatant and repeat steps 1 and 2 a further 3 times.
4. Resuspend the resin in a small volume of 0.1 × TE and store at 4°C (for up to 5 months).

B. *Assay of the radioactive cDNAs*

1. Pierce the bottom of a 0.5 ml microcentrifuge tube with a needle and place a glass bead in the bottom of the tube.
2. Place the 0.5 ml tube in a 2 ml microcentrifuge tube and then fill the 0.5 ml microcentrifuge tube with resin.
3. Centrifuge in a swing-out rotor at 700 *g* for exactly 2 min. If the liquid (suspension buffer of the column) which collects in the bottom of the 2 ml tube touches the bottom of the column, discard it and spin the column again.
   The resin should fill the 0.5 ml tube to within 3–5 mm of the top. If this is not the case, the addition of further resin and another centrifugation will be required.
4. Transfer the micro-column to a 1.5 ml microcentrifuge tube. Load the micro-column with the sample to be desalted, in a volume less than 30 μl.
5. Centrifuge in a swing-out rotor at 700 *g* for exactly 2 min.
6. Count the radioactivity in the column ($T$) and the eluate ($I$).
7. Calculate the mass ($M$) of ss-cDNA using the formula:
   $$M = (I/T) \times (\text{total mass of dCTP in the mixture reaction}) \times 4.$$
   For the conditions used in *Protocol 2*,
   $$M(\text{ng}) = (I/T) \times 330.^{b}$$

[a] If a column is used for desalting, proceed as described in steps 1 to 4, then discard the column and store the eluate until use. Under the conditions described here, these columns retain salts and small oligonucleotides (up to an approximate length of 10 nt).

[b] Typically, 10–30% of starting mRNAs are retrotranscribed into ss-cDNA.

**Protocol 4.** Synthesis of non-radioactive ss-cDNA

*Precautions* (see *Protocol 1*)

*Equipment and reagents* (see *Protocol 2*)

*Method*

**1–3.** Proceed as in *Protocol 2*, steps 1–3.

4. Mix the following components in a microcentrifuge tube

| | |
|---|---|
| 10 × RT buffer | 2.5 μl |
| 5 μg/μl BSA | 1 μl |
| 5 U/μl RNAsin | 1 μl |
| 350 mM β-Mercaptoethanol | 2 μl |
| 100 mM dATP | 0.5 μl |
| 100 mM dTTP | 0.5 μl |
| 100 mM dGTP | 0.5 μl |
| 100 mM dCTP | 0.5 μl |
| 200 mM Sodium pyrophosphate[c] | 1 μl |
| Water to | 24 μl |

Mix with the annealed RNAs/primers mixture and add 1 μl AMV reverse transcriptase. Incubate for 45 min at 42°C.

5. Store at −20°C until required.

## 2.5 PCR on ss-cDNA

Synthesized ss-cDNA can be used for PCR without further purification (*Protocol 5*). Typically an aliquot of 1/10th to 1/100th of the ss-cDNA is used. Many protocols have been devised for reverse transcription and PCR in the same tube (ref. 7, or Gene Amp PCR kit from Cetus). However it has recently been reported that large amounts of reverse transcriptase may inhibit *Taq* polymerase activity (8).

### **Protocol 5.** PCR using ss-cDNA as substrate

*Precautions*

- In all PCR protocols, contamination by exogenous DNA *must* be avoided. Since the amplification is exponential, even minute amounts of DNA contaminants can give rise to a significant amplification signal.
- There are various guidelines to avoid contamination. Use aliquots of buffer, dNTP, and all the reagents involved in the manipulation. If the PCR buffer has been prepared in the laboratory, sonication for 20–30 min will eliminate any contaminant DNA. Wear disposable plastic gloves at all times. For micropipetting, use tips containing cotton plugs and use a set of pipettes dedicated to the preparation of RT and PCR experiments. If possible, use positive displacement pipettes. Avoid the formation of aerosols by limiting the opening and closure of microcentrifuge tubes. When preparing a PCR reaction, keep all the tubes on ice with their caps open. Close the caps just before transfer to the PCR machine. Treat each tube as a separate entity and do not use multipettes.
- To verify the absence of contamination, a variety of controls are required (see step 2 of this protocol).

*Reagents*

- *Taq* polymerase (Boehringer)
- 10 × PCR buffer (Boehringer)
- 100 mM dATP; 100 mM dGTP; 100 mM dCTP; 100 mM dTTP (Boehringer). Mixing equal volumes of each of these dNTP solutions leads to a 25 mM dNTP stock solution.
- Pairs of primer for PCR. In each pair, the 5′-

**Protocol 5.** *Continued*

oligonucleotide[a] (oligonucleotide $A_1$ *Figure 1A*) is a sense sequence and the 3′-oligonucleotide (oligonucleotide $A_2$, *Figure 1A*) is an anti-sense sequence.

- A programmable thermal cycler (Perkin-Elmer Cetus or Hybaid) or three water baths
- Mineral oil (eg. Sigma M3516)

*Method*

1. For each of the reaction tubes (see step 2) mix on ice:
   - 10 × PCR buffer — 5 μl
   - 25 mM dNTP — 0.5 μl
   - Primer 1 (100 ng/μl in water) — x[b] μl
   - Primer 2 (100 ng/μl in water) — x[b] μl
   - *Taq* polymerase — 1 to 2 U
   - *Add* water to — 45 μl[c]
2. Prepare a series of 0.5 ml microcentrifuge tubes (a–e) on ice containing the following:
   (a) 5 μl of water
   (b) 5 μl of RT mix
   (c) in 5 μl the non-retrotranscribed RNAs
   (d) 5 μl of the reverse transcriptase product (*Protocol 4*, step 3)
   (e) if possible, 5 μl containing 100 ng–1 μg of genomic DNA
3. Add 45 μl of the mix prepared in step 1 to each tube (a–e), changing the pipette tip for each addition.
4. Overlay each reaction mixture with 50 μl of mineral oil.
5. Perform PCR. Typically, start with a denaturation step of 3 min at 93°C then 30–40 cycles of amplification. Each cycle is typically (i) denaturation: 94°C for 30 sec; (ii) annealing: close to the $T_m$ of the oligonucleotide[d] for 30 sec; (iii) extension: 72°C for 1 min/kb.[e]
6. Analyse 5–10 μl of each PCR product by agarose gel electrophoresis.

[a] We routinely use crude oligonucleotide preparations (i.e. unpurified) for all PCR experiments, since the oligonucleotides used are shorter than 30 nt. Each oligonucleotide is diluted in $ddH_2O$ to 100 ng/μl (working solution) and stored at −20°C.

[b] In these PCR experiments, the oligonucleotides are used at 0.1 μM final concentration. The volume of the 100 ng/μl solution to be used to give 1 μM in a final volume of 50 μl is given by the formula:

$$V_{\mu l} = 0.165\ n_T$$

where $V$ is the volume in μl and $n_T$ is the length of the oligonucleotide in residues.

[c] We routinely perform PCR in a final volume of 50 μl but a volume of 20 μl can also be used.

[d] When we started to use PCR in collaboration with Jacques Delort, in 1988, there was no software available to help choose PCR primers or to indicate an appropriate annealing temperature. Thus the rule of thumb that we have adopted to calculate the $T_m$ uses the formula:

$$T_m = 2\ (A+T) + 4\ (G+C)$$

which is equivalent to $T_m = 2\ (n_T + (G+C))$ where $A$, $T$, $G$, $C$, represent the number of corresponding nucleotides and $n_T$ is the length (in nucleotides) of the oligonucleotide. Using annealing temperatures close to the calculated $T_m$ always yields good results without (or with a low amount of) non-specific amplification.

[e] *Taq* polymerase is usually considered to synthesize DNA at a rate of 1 kb/min. However, we routinely amplify 1.15 kb of sequence using an extension time of 45 sec and for fragments shorter than 150 bp we generally omit the extension step.

The choice of the primers is of great importance in this technique; the now well-known rules of $T_m$ compatibility, avoiding self-hybridization and/or inter-hybridization have to be followed (see Chapter 1). The particular characteristics of mRNA introduce other criteria in the choice of primers:

(a) To ensure that mRNA sequences are amplified and not contaminating genomic DNA, it is possible to prime ss-cDNA synthesis with an oligonucleotide composed of the three or four most 3′-bases of the mRNA and a homopolymeric dT-tail. This directs the synthesis of a ss-cDNA close to the 3′-end of the mRNA studied (9).

(b) To target processed mRNA and not hnRNA and/or genomic DNA sequences, it is possible to use a pair of primers in which at least one of the primers spans the junction of two exons.

(c) To detect and/or prevent the amplification of genomic DNA, pairs of primers spanning introns are widely used.

The product of RT–PCR, using mRNA as template, should give a single band on agarose gel electrophoresis, corresponding to the length of the amplified sequence. The presence of additional bands is usually due to contamination. Such contaminating bands can be eliminated by changing primers, increasing the annealing temperature, changing the $Mg^{2+}$ concentration, performing hot starts, and/or digesting genomic DNA, which could be in the RNA preparation, with an RNase-free DNase. However, alternative splicing of mRNA can also result in several related mRNA molecules and hence multiple PCR products. This is best verified by isolating, cloning, and sequencing each of the amplification products.

## 3. Anchored PCR

### 3.1 General considerations

Anchoring a defined sequence to the 3′-end of an mRNA is not a real difficulty as the priming of the ss-cDNA synthesis offers the opportunity to add a known tag to the 5′-end of the ss-cDNA corresponding to the 3′-end of the mRNA (*Figures 1B* and *5A* and *C*). When the 5′-end of an mRNA is unknown, it is necessary to add a sequence to target the amplification. There is currently no reproducible means to add a defined sequence to the 5′-end of mRNAs; instead, these sequences have to be added to the 3′-end of the corresponding cDNAs.

This chapter will not deal with strategies devised to anchor defined sequences to double-stranded cDNA. Such strategies add a step, namely the synthesis of the second strand, which can result in loss of material and can constitutively truncate the 5′-end of the corresponding mRNA.

There are two ways to add a defined sequence to the 3′-end of ss-cDNAs (*Figure 4*). The first is the addition of an homopolymeric tail (*Figure 4A*; ref.

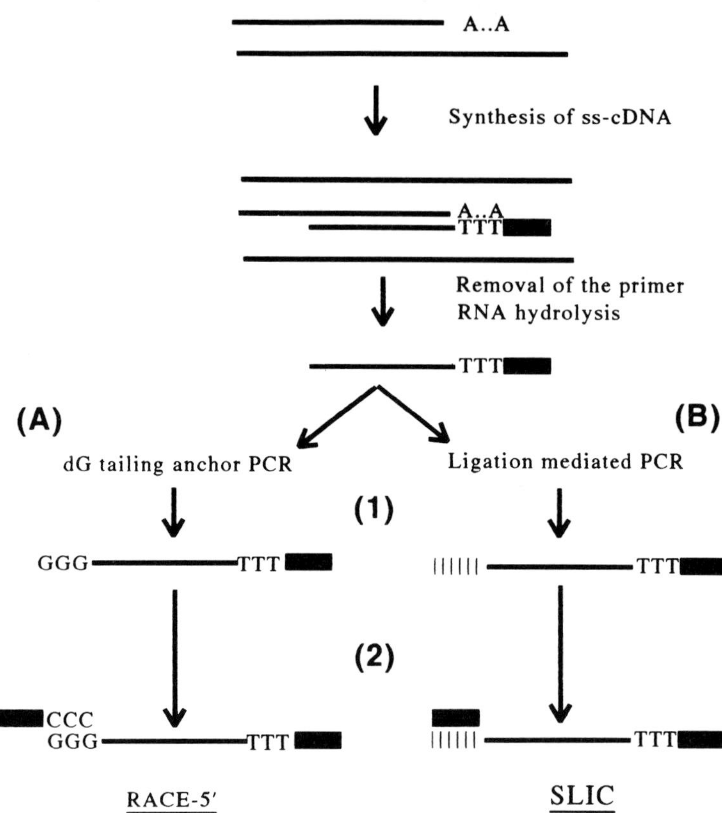


**Figure 4.** Anchor-PCR strategies. Prior to adding a known sequence to the 3′-end of ss-cDNA, it is necessary to remove the oligonucleotides used to prime ss-cDNA synthesis and to eliminate all RNA. (A) Tailing-mediated anchor-PCR. (1) Terminal deoxynucleotidyl transferase is used to add a homopolymeric tail to the 3′-end of the ss-cDNA. (2) Then the synthesis of the second strand and the further PCR experiments are carried out with an anchor primer containing at its 3′-end an anti-complementary homopolymeric tail. (B) Ligation-mediated anchor PCR. (1) An oligodeoxyribonucleotide is added to the 3′-end of the ss-cDNA using T4 RNA ligase. (2) The ss-cDNA can be amplified by PCR: it has two defined extremities. Further PCR experiments can be carried out with pairs of primers chosen according to the normal rules.

10). The second is the direct addition of a defined sequence oligodeoxyribonucleotide (*Figure 4B*; ref. 11).

## 3.2 Removal of the primer and RNA hydrolysis

In both of these 3′-addition strategies, the oligonucleotide used to prime reverse transcription must be removed to avoid competition with the ss-cDNA for the addition of the known sequence. Similarly, remaining RNA must also be removed. If the length of the ss-cDNA is greater than 200 nt the Prep-A-Gene DNA purification matrix (Bio-Rad 732–6010) provides a fast and efficient means of removing the primers and RNA (*Protocol 6*). If the reverse transcribed product is shorter than 200 nt, the ss-cDNA may be purified by alkaline hydrolysis and fractionated precipitation (*Protocol 7*).

**Protocol 6.** Purification of ss-cDNA longer than 200 nt

*Reagents*

- Prep-A-Gene DNA purification kit (Bio-Rad 732–6010). The silica matrix used in the kit does not bind RNA or small DNA fragments in oxidizing conditions.

*Method*

1. Adjust the volume of the reverse transcriptase product (*Protocol 4,* step 3) to 50 μl with double distilled water.
2. Heat for 5 min at 90–95°C to denature the RNA:DNA heteroduplexes. Add 150 μl of binding buffer, mix, then add 5 μl of resuspended matrix. Mix well. Incubate for 10 min at room temperature.
3. Briefly spin (30 sec) in a benchtop centrifuge. Discard the supernatant. Wash the matrix once with 150 μl of binding buffer (to remove RNA and protein weakly bound to the matrix) and twice with 150–200 μl of prepared wash buffer (i.e. after addition of ethanol to the wash buffer).
4. Carefully remove all the wash buffer after the last wash. Traces of ethanol can be removed by drying the tubes for 3 min in a Speedvac or equivalent rotary vacuum desiccator.
5. Add 5 μl [a] of double-distilled water to the resin and, heat for 5 min at 65°C. Spin for 30 sec and collect the supernatant.

[a] If the ss-cDNA is purified for the SLIC protocol (see Section 3.4), only 1 μl of the ss-cDNA will be added to the ligation mixture. The elution volume should therefore be as small as possible; 5 μl is suitable.

**Protocol 7.** Purification of ss-cDNA shorter than 200 nt [a]

*Equipment and reagents*

- Water bath at 50°C
- Speedvac
- 2 M NaOH
- 2.2 M acetic acid
- 4 M Ammonium acetate
- 2-propanol
- 75% ethanol

*Method*

1. To the reverse transcription product (*Protocol 4,* step 3), add 0.15 vol. 2 M NaOH. Incubate for 30 min at 50°C (alkaline hydrolysis). Then add 0.15 vol. 2.2 M acetic acid to neutralize.

**Protocol 7.** *Continued*

2. After alkaline hydrolysis (step 1) add 1*V* (*V* is the final volume of the alkaline hydrolysis solution) 4 M ammonium acetate and 2*V*[b] 2-propanol to the sample and mix. Incubate at room temperature for at least 10 min.
3. Centrifuge at 10 000 rpm (7000 g) for 10 min at room temperature. Carefully discard the supernatant. Generally the pellets are hardly visible.
4. Add 5*V*[b] 75% of ethanol. Resuspend the pellets. Spin at 10 000 rpm (7000 g) for 10 min at room temperature. Discard the supernatant.
5. Dry the pellets under vacuum. Avoid drying times longer than 3 min to avoid the ss-cDNA sticking to the wall of the tube.
6. Resuspend the pellets in 5 μl of double distilled water (see *Protocol 6*).

[a] This protocol is suitable for primers even shorter than 30 nt.
[b] *V* is the final volume of the alkaline hydrolysis solution in step 1.

## 3.3 Tailing-mediated anchored PCR

This was the first anchored PCR strategy to be reported (10) and the first approach that we used (12). A homopolymeric tail is added to the 3′-end of the ss-cDNA using terminal deoxyribonucleotide transferase. Four different homopolymeric tails can be produced. Only two are used in practice:

- *homopolymeric dA tails*. These mimic the poly(A) tail of mRNA and allow the use of the same primer at both ends of the ss-cDNA.
- *homopolymeric dG tails*. This procedure is described in *Protocol 8*. The extension of a dG tail creates secondary structures which limit the length of the tail to between 16–20 residues. For tails generated using the other three nucleotides (A, T, or C) the length of the tail is more difficult to control.

**Protocol 8.** Addition of a homopolymeric dG-tail to ss-cDNA and further PCR experiments

*Reagents*

- 10 × Tailing buffer (100 mM Tris–HCl pH 8.0, 100 mM $MgCl_2$)
- [α-$^{32}$P]dGTP 3000 Ci/mmol (Amersham PB 10206)
- Terminal deoxyribonucleotide transferase (TdT, Boehringer)
- 5 mg/ml BSA
- PCR primers. The tail-specific primer in particular must be chosen with care. In general use a chimeric primer containing 6 dC residues (12) at its 3′-end and a known sequence at its 5′-end.

*Method*

1. Assemble in a microcentrifuge tube, in a final volume of 20 μl:

| | |
|---|---|
| purified ss-cDNA | 2 μl |
| 10 × tailing buffer | 2 μl |
| 5 mg/ml BSA | 2 μl |
| [$\alpha$-$^{32}$P]dGTP 3000 Ci/mmol | 2 μl |
| TdT | 5 U |

2. Prepare:
   (a) a 'non-tailed' sample with the same mix, as in step 1, minus the enzyme
   (b) a control tube containing all the components of the tailing reaction except the ss-cDNA.
3. Incubate all tubes for 30 min at 37°C. Then heat for 5 min at 72°C.
4. Eliminate free nucleotides (see *Protocol 3,* steps 1 to 4) to check tailing.
5. Store at −20°C until required.
6. PCR experiments can be carried out as described in *Protocol 5* with a slight modification in the first annealing step to allow priming of the synthesis of the double strand cDNA. With a tail-specific primer containing 6 dCs, the first annealing step should be at 30°C for 30 sec and then incubation at 45°C for 5 min prior to the subsequent normal PCR cycles.

This strategy presents two difficulties:

- the length of the homopolymeric tail added to the 3′-end of the ss-cDNA is difficult to control.
- the anti-complementary homopolymeric tail used with the tail-specific primer (*Figure 1C* primer $C_1$ and *Figure 4A*) can hybridize with and thus prime the synthesis of DNA from short homopolymeric stretches within the target sequence (9, 10).

To circumvent these drawbacks, we have devised a new strategy to tag the 3′-end of ss-cDNA by ligation of an oligodeoxyribonucleotide to the ss-cDNA. This is described in the following section.

## 3.4 Ligation-mediated anchored PCR: SLIC strategy

This strategy (referred to as SLIC for single strand ligation of cDNA) uses T4 RNA ligase to join ss-cDNA to an oligodeoxyribonucleotide. As the enzyme is at least 200-fold slower in the ligation of DNA than in the ligation of RNAs (13), all RNAs must be removed prior to the single strand ligation step.

To avoid circularization or concatemerization of the ss-cDNA, the primer

**A**

| Name | 5' Sequence 3' |
|---|---|
| BM 3' | CGAATACGACTCACTATAGGAAGCTGCGGCCGCTGCAGTACTTTTTTTTTTTTTT |
| BM 1-3' | CGAATACGACTCACTATAGG |
| BM 2-3' | ACTCACTATAGGAAGCTGCG |
| BM 3-3' | AGGAAGCTGCGGCCGCT |

**B**

| Name | 5' Sequence 3' |
|---|---|
| BM 5' | GCATTGCATCATGATCGATCGAATTCTTTAGTGAGGGTTAATTGCC |
| BM 1-5' | GGCAATTAACCCTCACTAAAG |
| BM 2-5' | TCACTAAAGAATTCGATCGATC |
| BM 3-5' | CGATCGATCATGATGCAATGC |

**C**

| Name | 5' Sequence 3' |
|---|---|
| A 3' NV | ATCGTTGAGACTCGTACCAGCAGAGTCACGAGAGAGACTACACGGTACTGGTTTTTTTTTTTTTT |
| A 3'_1 | ATCGTTGAGACTCGTACCAGCAGAG |
| A 3'_2 | TCGTACCAGCAGAGTCACGAGAGAG |
| A 3'_3 | CACGAGAGAGACTACACGGTACTGG |

**D**

| Name | 5' Sequence 3' |
|---|---|
| A 5' NV | CTGCATCTATCTAATGCTCCTCTCGCTACCTGCTCACTCTGCGTGACATC |
| A 5'_1 | GATGTCACGCAGAGTGAGCAGGTAG |
| A 5'_2 | AGAGTGAGCAGGTAGCGAGAGGAG |
| A 5'_3 | CGAGAGGAGCATTAGATAGATGCAG |

used to prime the reverse transcription must not contain a phosphate group at its 5′-end. To target the ligation of the oligodeoxyribonucleotide to the 3′-end of the ss-cDNA and avoid concatemerization, the 3′-end of the oligodeoxyribonucleotide is blocked (no OH group) and the 5′-end must have a phosphate group. In our original report (11) we blocked the 3′-end by adding

**Figure 5.** Primers used in the anchored PCR strategies. (A) Oligonucleotides used to anchor a defined sequence to the 3′-end of mRNA (e.g. BM 3′ series) (B). Oligonucleotides used to anchor a defined sequence to the 5′ end of mRNAs (i.e. BM 5′ series). The oligodeoxyribonucleotides BM 3′ and BM 5′ have been designed to allow the utilization of pairs of overlapping oligonucleotides useful in nested-PCR experiments. The oligonucleotides described in these tables can be used in PCR experiments with annealing temperatures of 50–55 °C. (C) Oligonucleotides used to anchor a defined sequence to the 3′-end of mRNA (i.e. A3′ NV series). (D) Oligonucleotides used to anchor a defined sequence to the 5′-end of mRNA (i.e. A5′ NV series). These oligonucleotides can be used in PCR experiments with annealing temperatures ranging from 60–65 °C. Higher temperatures are recommended to avoid formation of primer-dimers. These oligonucleotides have been chosen using the following criteria: First, the sequence of A5′ NV and A3′ NV are random sequences with a G + C content of 60%. Second, the oligonucleotides are studied in pairs (A5′-2/A3′-2) using the classical rules of choice of primers.

---

a dideoxyribonucleotide. However the protection of the 3′-end was weak. Thus, we now use oligonucleotides purchased from Genset which are 3′-$NH_2$ blocked during synthesis. This protection is very efficient provided repeated freeze–thaw of the oligonucleotide is avoided: it is therefore recommended to aliquot these reagents before use.

Since publication of the SLIC strategy we have found that the elimination of the non-ligated oligodeoxyribonucleotide greatly enhances the yield of PCR products. In fact, it is obvious that since the tail-specific PCR primers (*Figure 1C* primer $C_1$ and *Figure 4B*) can hybridize to the ligated oligodeoxyribonucleotide, the removal of the non-ligated oligodeoxyribonucleotide will reduce competition for targets.

The sequences of the oligodeoxyribonucleotides that we have used are given in *Figure 5*. The sequences of the BM 5′ series and BM 3′ series oligonucleotides exhibit some special features:

- The 5′-end of BM 3′ has the target sequence of the T7 RNA polymerase and a sequence recognized by the restriction enzyme *Not*I is close to the poly(dT) tail.
- The 5′-end of BM 5′ is the sequence of the T3 RNA polymerase promoter and in the 3′-region it contains sequences recognized by restriction enzymes *Pvu*I and *Bsp*HI.

However we observed that the use of T3 and/or T7 promoters during the first PCR reaction can introduce PCR contaminants originating from clones carrying these sequences close to the polylinker (e.g. pBluescript). Furthermore it proved very difficult to obtain a satisfactory yield of clones using the restriction sites that we chose. Thus a second set of oligonucleotides has been designed to overcome these difficulties (*Figure 5C* and *D*). Step by step instructions for the SLIC protocol are given in *Protocol 9*.

In cloning 5′-ends of mRNAs it is best to avoid the use of the primer that primed the reverse transcription in the PCR experiments. This eliminates the

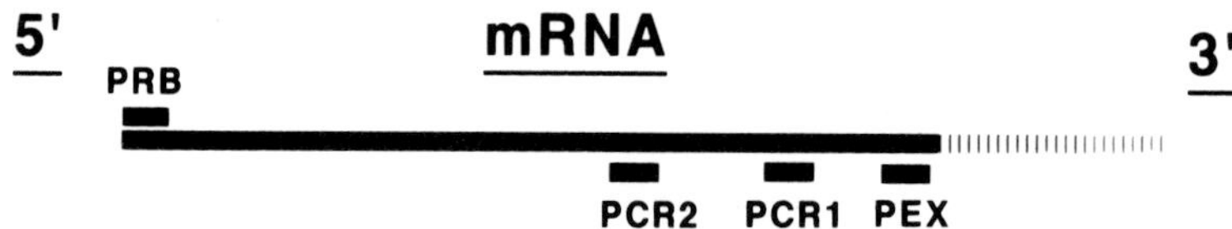


**Figure 6.** Location and use of oligonucleotides for cloning 5′-ends. The thick line represents the sequence of the mRNA. Sense and antisense oligonucleotides are represented above and below respectively. PEX is used to prime reverse transcription. PCR1 and PCR2 are used in nested PCR after the ligation step. PRB is used as probe in Southern blotting to monitor the PCR products and screen the clones. Note that a PCR carried out with the product of the reverse transcription and oligonucleotides PRB/PCR1 is a good way to check if the primer extension product is long enough to be ligated and then cloned. In this control experiment PCR2 can be used as probe.

amplification of mis-primed ss-cDNA. It is possible to derive three oligodeoxyribonucleotides from the sequence of the ligated oligodeoxyribonucleotide to allow nested-PCR (*Figure 5*). A usable signal is obtained after at least the second PCR for all the 5′-ends that we have cloned. Thus, in cloning 5′-ends of mRNAs it is useful to plan the use of four oligodeoxyribonucleotides derived from the sequence of the target mRNA studied (*Figure 6*).

### **Protocol 9.** Ligation of an oligodeoxyribonucleotide to the 3′-end of ss-cDNA (SLIC strategy)

*Reagents*

- 10 × RNA ligase buffer: 500 mM Tris–HCl pH 8.0, 100 mM $MgCl_2$, 100 μg/ml BSA, 200 μM ATP, 10 mM hexamine cobalt chloride. Filter through a 0.22 μm membrane. Store at −20°C
- 40% PEG 6000 (Appligène) made in double distilled water and stored at −20°C
- Functionalized 3′-$NH_2$ oligodeoxyribonucleotide diluted in double distilled water at 50–100 ng/μl
- 20 U/μl T4 RNA Ligase (New England Biolabs or Amersham)
- Prep-A-Gene DNA purification matrix (Bio-Rad 732–6010)

*Method*

1. Prepare three tubes at room temperature.
   (a) The 'ligated sample tube' contains:
   - purified ss-cDNA (*Protocol 6* or *7*) — 1 μl
   - 10 × RNA ligase buffer — 1 μl
   - protected oligonucleotide — 0.5 μl
   - 40% PEG 6000 — 6.25 μl
   - T4 RNA ligase — 1 μl

   (b) The same mix as in (a) except omit the T4 RNA ligase. This is the 'non-ligated sample tube' and will allow the detection of non-specific amplification of the ss-cDNA.

(c) The same mix as in (a) except omit the ss-cDNA. This is the control to test for contaminants in the ligation mix.

2. Incubate for 24 h at 22°C.
3. Remove 5 μl of each sample and store at −20°C until required.
4. Incubate the remaining mixtures for another 24 h at 22°C.
5. Store the samples at −20°C until required.
6. Adjust the volume of the samples to 50 μl with double distilled water. Add 150 μl of Prep-A-Gene binding buffer. Mix.[a] Add 5 μl of resuspended matrix. Incubate for 10 min at room temperature.
7. Proceed as described in *Protocol 6*, steps 2–4, except resuspend the sample in a volume of 10 μl. Store at −20°C until required.
8. PCR experiments can be carried out as described in *Protocol 5*. Add to the control for the SLIC procedure a PCR control (the PCR mix without enzyme).

[a] PEG precipitates in the binding buffer but it does not interfere with the binding of the ss-cDNA to the matrix.

## 3.5 Other uses of the SLIC strategy

Undoubtedly, appropriate strategies to obtain full length cDNAs and to clone 5′-ends of mRNA will evolve. Strategies more efficient than SLIC in the cloning of 5′-ends of mRNA will probably be developed (see Section 3.6). Nevertheless, the SLIC strategy will remain helpful in the cloning of:

- promoters in genomic DNA
- extremities of long clones (YAC, cosmids)
- insertion sites of transgenes or retrovirus

These uses are displayed in *Figure 7*.

## 3.6 Further improvement in cloning 5′-ends of mRNA

Since publication of the SLIC strategy, we have cloned many 5′-ends of mRNA. The limiting factor in these experiments is the efficiency of the reverse transcriptase. There is now a need to improve the synthesis of full length cDNAs. We have tried to improve this step by various means. Thus Chien *et al.* (14) claimed that the *Taq* DNA polymerase they have isolated could be valuable for the synthesis of full length cDNA. We have tried it unsuccessfully. Similarly, we have tried many commercially available thermostable DNA polymerases possessing reverse transcriptase activity. All have been unsuccessful.

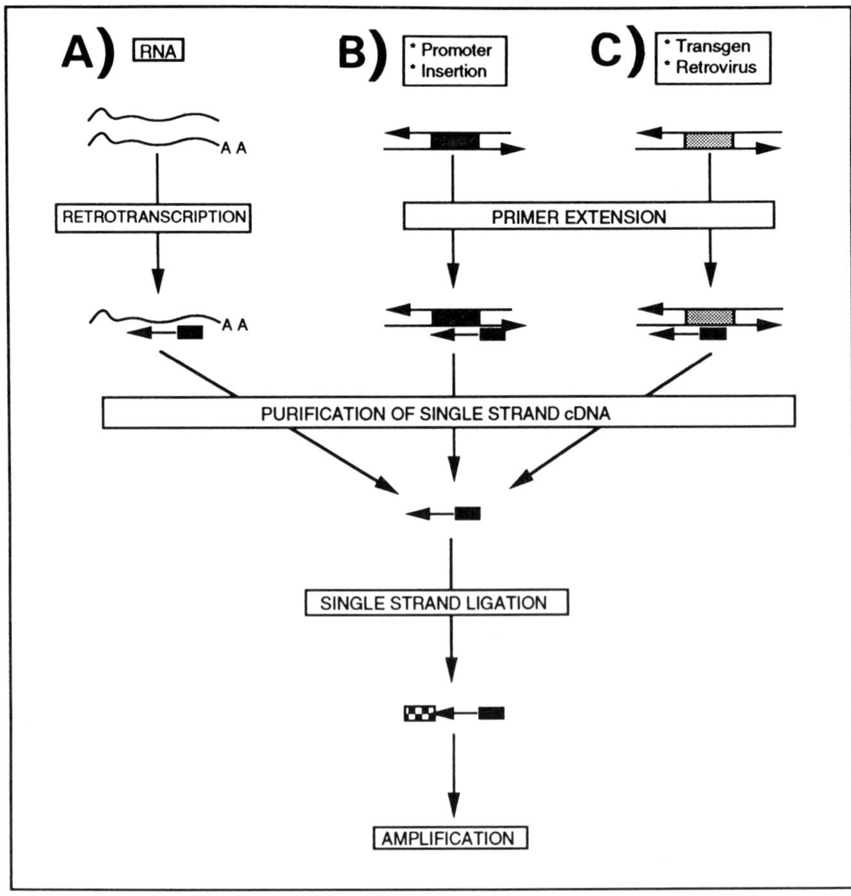


**Figure 7.** Various uses of the SLIC strategy. (A) Cloning 5′-ends of mRNAs. (B) Cloning promoters or extremities of long clones. In this case, the unknown sequence is included in the known one and an oligonucleotide can be designed to clone the proximal parts of the unknown sequences. (C) Cloning insertion sites of transgenes or of retroviruses. In this case, the known sequence is included in the unknown. PEX and PCR primers can be designed on the two strands to clone the insertion site. In (B) and (C), the use of a 5′-biotinylated PEX primer will greatly facilitate the purification of the single-stranded neo-synthesized DNA prior to carrying out the ligation step.

# 4. Cloning PCR products

## 4.1 General considerations

In our hands, blunt-ending PCR products and then ligating them to a blunt-ended, phosphatased plasmid is the best way to clone 5′-ends of mRNA. To clone PCR products resulting from the amplification of total mRNA, blunt-ended products are ligated to *Eco*RI linkers, and then inserted into lambda ZAP (Stratagene). Otherwise the main difficulty that we have encountered in these experiments is cleaving the ends of the PCR products where restriction enzyme sites are incorporated into PCR primers; in particular *Not*I, *Bsp*H1, *Pvu*I and *Bam*HI sites are resistant to cutting. However recent experiments with *Eco*RI have been successful.

Prior to cloning the PCR products it is advisable to assess the cDNA. To do this quickly we have developed a fast and accurate method called 'Flying Southern' which is based on the fixation of DNA to positively-charged nylon membrane in alkaline blotting.

## 4.2 Analysis of PCR products by 'Flying Southern'

This method has been previously published in ref. 15; *Protocol 10* is an updated version.

### **Protocol 10.** 'Flying Southern' and hybridization procedures

*Material*

- A plastic box to hold the agarose gel
- Microtitre plate gondolas and centrifuge
- Filter paper sheets cut to the size of the gel (1 cm total thickness)
- Nylon membrane e.g. Hybond $N^+$ (Amersham) or other positively charged nylon membrane cut to the size of the gel
- Four sheets of Whatman 3 MM paper cut to the size of the gel
- 0.4 M NaOH
- 6 × SSC: 1 × SSC is 0.15 M NaCl, 0.015 M sodium citrate
- γ-ATP (3000 Ci/mmol, Amersham PB10168)
- Polynucleotide kinase (Boehringer)
- Kinase buffer: 500 mM Tris–HCl pH 7.5, 100 mM $MgCl_2$, 1 mM spermidine, 1 mM EDTA
- 100 mM DTT
- Hybol buffer: 1 × Denhardt's (0.02% BSA, 0.02% PVP, 0.02% Ficoll), 6 × SSC, 25 mM sodium phosphate buffer pH 7, 25 mM EDTA, 250 μg/ml denatured herring sperm, 1% SDS

*Method*

1. After UV visualization of the nucleic acids in an agarose gel, soak the gel in 0.4 M NaOH for 10 min.
2. Soak the Hybond-$N^+$ membrane in water and then in 6 × SSC for 1 min.
3. Stack in a plastic box, from the bottom to the top:
   - filter paper sheets (1 cm total thickness)
   - Nylon membrane (Hybond $N^+$)
   - agarose gel (avoiding bubbles between the membrane and the gel)
   - four sheets of Whatman 3 MM paper soaked in 0.4 M NaOH
4. Spin the plastic box in the microtitre plate gondola at 1000 *g* for 20–30 min (time depends on the agarose gel; limits are for agarose gels from 0.8 to 2%). This centrifugation step causes DNA transfer.
5. Kinase the oligodeoxyribonucleotide probe during the DNA transfer (centrifugation). To do this, mix in a tube:

**Protocol 10.** *Continued*

| | |
|---|---|
| • oligodeoxyribonucleotide | 100 ng |
| • 10 × kinase buffer | 1 μl |
| • 100 mM DTT | 0.5 μl |
| • [$\gamma$-$^{32}$P] ATP (3000 Ci/mmol) | 2.5 μl |
| • polynucleotide kinase | 1 U |
| • water to | 10 μl |

Incubate for at least 30 min at 37 °C. Remove unincorporated nucleotides using the columns as described in *Protocol 3*. About 50% of the radioactivity should be incorporated into the oligodeoxyribonucleotide.

6. Remove the Whatman paper. Mark the position of the wells on the membrane. Remove the membrane from the gel and rinse twice for 5 min each time in 6 × SSC.
7. Prehybridize in 1 ml (for a 7.5 × 10.5 cm membrane) of Hybol buffer. To do this, place the buffer in a Petri dish, then place the membrane (DNA face down) on the buffer. Avoid air bubbles.
8. Close the Petri dish.
9. Incubate for 20 min in an incubator at 42 °C.
10. Remove the prehybridization buffer and the membrane. Add 300 μl of Hybol buffer containing 2–3 × $10^6$ c.p.m. of $^{32}$P-kinased oligodeoxyribonucleotide. Replace the membrane as described above.
11. Incubate for 15–30 min in an incubator at 42 °C.
12. Wash consecutively in wash solutions from 6 × SSC to 1 × SSC containing 1% SDS at 42 °C. Monitor the radioactivity and thus hybridization.

## 5. Subtractive hybridization (general considerations)

Classically in subtraction experiments (probes or libraries), the aim is to isolate sequences specifically expressed in a cell population (e.g. sequences specific to particular tissues, or to physiological or pathological conditions). It is now straightforward to generate cDNA libraries with defined ends and it is possible to address the subtraction strategies to cell populations characterized by a low number of cells.

In a typical experiment, the mRNAs of the subtracted population are isolated and retrotranscribed to give ss-cDNAs. The ss-cDNAs are then hybridized to the driver mRNAs. When the two populations studied yield low amounts of mRNA, this strategy is problematic. Fortunately, the SLIC strategy can be used to amplify low amounts of mRNA and is thus potentially useful in subtraction experiments. The first step is to obtain two populations of cDNAs tagged with two sets of primers. Then asymmetric-PCR is used to produce single strands of the two populations. Judicious choice of the primers leads to the production of the + (sense) strand of one population and of the − (anti-sense) strand of the other population. The use of a 5′-biotinylated primer to amplify the driver population will facilitate the removal of driver and common sequences.

We have applied existing protocols of hybridization and capture with streptavidin magnetic beads. In all cases, up to 20% of the driver population remains in the collected subtracted population.

# 6. Cloning cDNAs of multigene families

## 6.1 General considerations

Multigene families are characterized by several conserved amino acid domains. PCR technology provides a powerful tool for cloning new members of a family by amplifying the DNA sequence between two conserved domains. The specific amplified DNA can then be sequenced and used to screen a cDNA library.

The information available is usually an amino acid sequence. As the genetic code is degenerate, PCR has to be performed with degenerate primers. Designing a PCR protocol with degenerate oligodeoxyribonucleotides involves some strategic choices:

(a) the identification of two conserved domains to design the PCR primers;

(b) the choice of degenerate primers giving the best possible representation of the chosen domains;

(c) the choice of the optimal cDNA substrate (in particular, the experimenter has to choose the tissue from which the RNAs will be extracted and to choose between the three ways of priming the cDNA synthesis: oligo-dT, random hexamers, or specific priming)

(d) the design of a PCR generating a highly specific amplification even though the degeneracy of the primers may result in a non-specific signal.

The difficulty is finding an appropriate compromise between the experimental conditions giving the optimal hybridization of the primers on their respective targets and the amplification itself. This is considered in detail in Sections 6.2–6.4.

Before screening the cDNA library, the PCR products have to be cloned and sequenced in order to identify new members of the multigene family. This could yield a large amount of clone analysis and sequencing work. We have therefore developed a fast and easy method to identify recombinant clones and this is described in Section 6.5.

## 6.2 The choice of degenerate primers

### 6.2.1 Identification of two conserved domains for the design of the PCR reaction

(a) Compile all the members of the family and identify the conserved amino acid domains.

(b) For the PCR reaction choose domains as close as possible to one another. This will minimize sequencing work.

(c) If possible, try to choose a region with conserved residues between the two domains to help identify the new clones as members of the family. In any case, find domains that will lead to primers with a length of about 25–30 nt.

### 6.2.2 Designing primers for PCR from an amino acid sequence

A few simple rules have to be followed. The most important is to choose primers with the smallest possible degeneracy. This can be done as follows:

(a) Back-translate the conserved domains and choose the areas of minimum ambiguity. Some domains are very small and therefore choice will be limited. In this situation, to reduce the ambiguity, the family should be divided into subclasses and primers chosen for each subclass.

(b) Reduce the degeneracy by using a codon usage data table (16).

(c) Use an intercodon dinucleotide frequencies table. Studies have shown that intercodon dinucleotide frequencies are not random (17).

(d) It is possible to further decrease the degeneracy by using deoxyriboinosine (18).

(e) The 3′-end of the primer should not be degenerate.

The final choice is made by combining these rules with the general rules for designing PCR primers, such as $T_m$ compatibility and avoiding self-hybridization and/or inter-hybridization.

For determining $T_m$ compatibility, calculate the average $T_m$, defined here as $T_{m_{av}}$ of each primer. Various software packages (e.g. OLIGO4) are available for such calculations (see *Protocol 5*). Also calculate the $T_m$ of the oligodeoxyribonucleotide containing the highest percentage of AT, defined as $T_{m_{ATmax}}$.

### 6.2.3 An example of primer choice: the cloning of new members of the class III POU-domain proteins

POU-domain proteins are characterized by two conserved domains; the POU-specific-domain and the POU-homeo-domain (19). Sequence analysis has led to the identification of five subclasses in the family. Alignment of the amino acid sequences from the five class III proteins clearly shows the conserved residues. Suitable degenerate primers can be chosen as shown in *Figure 8*.

## 6.3 cDNA synthesis

In our hands, the synthesis of a randomly primed cDNA (*Protocol 11*) provides the best substrate for PCR with degenerate oligodeoxyribonucleotides particularly compared to oligo(dT) priming.

**Protocol 11.** Randomly-primed cDNA synthesis

Perform the reverse transcription according to *Protocol 2* with the following modifications.

*Reagents*

- poly(A)$^+$ mRNA (see Section 2.2)
- 10 × PCR buffer (Boehringer)
- 25 mM dNTP solution (Boehringer; see *Protocol 5*)
- 0.4 μM random hexamers
- 28 mM $MgCl_2$
- 40 U/μl RNasin (Promega Biotech)
- 7.5 U/μl AMV RTase

*Method*

1. Denature 2 μl[a] of each sample of poly(A)$^+$ mRNA for 2 min at 72°C. Spin, freeze in powdered dry ice and allow to thaw slowly on ice.
2. The mixture for 10 reactions is as follows:

| | |
|---|---|
| 10 × PCR buffer | 20 μl |
| 28 mM $MgCl_2$ | 25 μl |
| 100 mM DTT | 2 μl |
| 25 mM dNTP | 4 μl |
| 0.4 μM random hexamer | 5 μl |
| double distilled water | 117 μl |
| 40 U/μl RNasin | 2 μl |

3. For each sample of poly(A)$^+$ mRNA, prepare a reaction as follows. Add 17.5 μl of reaction mixture (step 2) and 0.5 μl of AMV RTase (7.5 U/μl) to the 2 μl of denatured poly(A)$^+$ mRNA.
4. Prepare the following controls:
   - one reaction without AMV RTase
   - one reaction without poly(A)$^+$ mRNA
5. Incubate for 1 h at 42°C.
6. Store at −20°C until required.

[a] The volume of the poly(A)$^+$ mRNA should be one-tenth of the total reaction volume, which should not exceed 100 μl.

## 6.4 PCR with degenerate primers

The sequence heterogeneity of the primers necessitates the use of large concentrations of oligonucleotides for the PCR reaction. We usually use

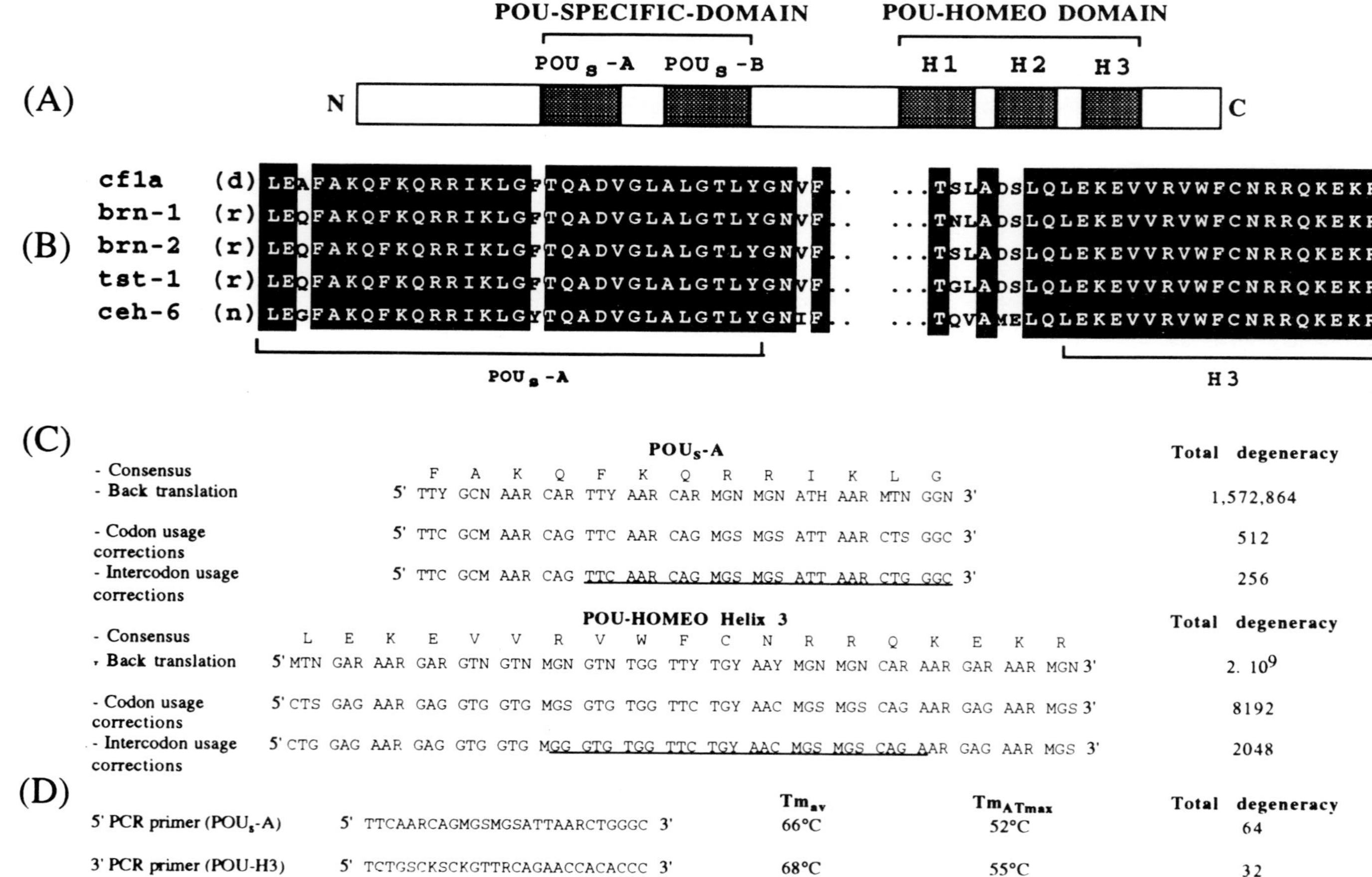

(A)
POU-SPECIFIC-DOMAIN
POU-HOMEO DOMAIN
POUs-A POUs-B
H1 H2 H3
N
C
(B)
cf1a (d) LEAFAKQFKQRRIKLGFTQADVGLALGTLYGNVF.. ...TSLADSLQLEKEVVRVWFCNRRQKEKR
brn-1 (r) LEQFAKQFKQRRIKLGFTQADVGLALGTLYGNVF.. ...TNLADSLQLEKEVVRVWFCNRRQKEKR
brn-2 (r) LEQFAKQFKQRRIKLGFTQADVGLALGTLYGNVF.. ...TSLADSLQLEKEVVRVWFCNRRQKEKR
tst-1 (r) LEQFAKQFKQRRIKLGFTQADVGLALGTLYGNVF.. ...TGLADSLQLEKEVVRVWFCNRRQKEKR
ceh-6 (n) LEGFAKQFKQRRIKLGYTQADVGLALGTLYGNIF.. ...TQVAMELQLEKEVVRVWFCNRRQKEKR
POUs-A
H3
(C)
POUs-A
Total degeneracy
- Consensus F A K Q F K Q R R I K L G
- Back translation 5' TTY GCN AAR CAR TTY AAR CAR MGN MGN ATH AAR MTN GGN 3' 1,572,864
- Codon usage corrections 5' TTC GCM AAR CAG TTC AAR CAG MGS MGS ATT AAR CTS GGC 3' 512
- Intercodon usage corrections 5' TTC GCM AAR CAG TTC AAR CAG MGS MGS ATT AAR CTG GGC 3' 256
POU-HOMEO Helix 3
Total degeneracy
- Consensus L E K E V V R V W F C N R R Q K E K R
- Back translation 5' MTN GAR AAR GAR GTN GTN MGN GTN TGG TTY TGY AAY MGN MGN CAR AAR GAR AAR MGN 3' 2. 10^9
- Codon usage corrections 5' CTS GAG AAR GAG GTG GTG MGS GTG TGG TTC TGY AAC MGS MGS CAG AAR GAG AAR MGS 3' 8192
- Intercodon usage corrections 5' CTG GAG AAR GAG GTG GTG MGG GTG TGG TTC TGY AAC MGS MGS CAG AAR GAG AAR MGS 3' 2048
(D)
Tm av
Tm ATmax
Total degeneracy
5' PCR primer (POUs-A) 5' TTCAARCAGMGSMGSATTAARCTGGGC 3' 66°C 52°C 64
3' PCR primer (POU-H3) 5' TCTGSCKSCKGTTRCAGAACCACACCC 3' 68°C 55°C 32

**Figure 8.** Choice of degenerate primers for PCR with class III POU-domain proteins. (A) General structure of a POU-domain protein. The three helical domains are indicated as H1, H2, and H3. (B) Partial comparison of the POU-domain sequence of members of the class III POU-domain proteins (cf1-a, brn-1, brn-2, tst-1, ceh-6). Identity is indicated by black highlighting. Species from which the factors were cloned are rat (r), *Caenorhabditis elegans* (n), *Drosophila melanogaster* (d). (C) Back-translation of 16 amino acids of the POUs-A domain and the Helix 3 domain. The degenerate nucleotide symbols are K (G,T), M (A,C), R (A,G), S (C,G), Y (C,T), H (A,C,T), N (A,C,G,T). The codon and intercodon usage data are rat specific. Two 27 nucleotide primer sequences obtained after $T_m$ compatibility analysis are underlined. (D) The 3′ PCR primer is the complementary sequence of the underlined POU-HOMEO Helix 3 sequence.

10-fold more primer than for standard PCR conditions. A greater excess of primers may increase non-specific amplification.

To optimize the PCR conditions with degenerate primers, we have designed a PCR protocol (*Protocol 12*) using two different annealing temperatures; a low temperature for the first few cycles and a much higher one for the remaining cycles. The first temperature we use is between $T_{m_{ATmax}}$ − 15 °C and $T_{m_{ATmax}}$. The second temperature is the calculated $T_{m_{av}}$ of the primers.

## Protocol 12. PCR conditions for amplification with degenerate primers and rapid cloning of the PCR product

*Precautions* (see *Protocols 1* and *5*)

### *Equipment and reagents*

- A programmable thermal cycler (e.g. Perkin Elmer Cetus or Hybaid) or three water baths
- 5 U/μl *Taq* polymerase (Boehringer)
- PCR buffer (Boehringer)
- 2.5 mM each dNTP (Boehringer)
- Randomly primed cDNA (see *Protocol 11*) and the various controls for the reverse transcription reaction
- 5′ and 3′ primers each at 125 μM
- Mineral oil (Sigma M3516)
- Nusieve agarose (FMC)
- 4 U/μl T4 DNA polymerase (Amersham)

### *Method*

1. For a single PCR run, mix[a]
   - 10 × PCR buffer 5 μl
   - 2.5 mM dNTP 4 μl
   - 125 μM 5′ primer 1 μl
   - 125 μM 3′ primer 1 μl
   - double distilled water 37.5 μl
2. Prepare on ice, four 0.5 ml tubes containing
   - 1 μl cDNA[b] (from *Protocol 11*)

**Protocol 12.** *Continued*

- 1 μl control without RT (from *Protocol 11*)
- 1 μl control without poly(A)$^+$ mRNA (from *Protocol 11*)
- 1 μl double distilled water respectively

3. Add 48.5 μl of PCR mixture (from step 1) to each tube.
4. Place the tubes in the thermal cycler. Heat to 75°C, then add 0.5 μl of *Taq* DNA polymerase (5 U/μl), and immediately add 100 μl of mineral oil.
5. Run the following program:
   - 3 min denaturation at 94°C
   - 2–5 cycles (94°C for 30 sec, $T_{m_{ATmax}}$ −15°C to $T_{m_{ATmax}}$ 45 sec, 72°C for one min/kb)
   - 38 or 35 cycles (94°C for 30 sec, $T_{m_{av}}$ 45 sec, 72°C for one min/kb)
6. Cool the PCR products on ice.
7. Fill-in reaction: add 1μl of T4 DNA polymerase (4 U/μl) to the PCR products and incubate for 20 min at 16°C.
8. Run the end-filled PCR products on a low melting agarose gel and cut out the band of desired size.
9. Isolate the DNA from the agarose by phenol extraction and phosphorylate the 5′-end of the fragment with T4 polynucleotide kinase (20).
10. Clone into a dephosphorylated linearized, blunt-ended vector[c] with a 2 to 5 ratio of insert to vector.

[a] Optimize the $MgCl_2$ concentration according to the primer pair chosen.
[b] This represents one-twentieth of the previously synthesized cDNA (see *Protocol 11*). Do not exceed this ratio.
[c] We use pBluescript II or pUC19 digested with *Sma*I and dephosphorylated.

PCR using cDNA will usually give several discrete bands. The three PCR controls (*Protocol 12*, step 2) should not give any amplification products. Should this not be the case, one should check for the presence of non-specific amplification products by performing two extra PCR controls on the cDNA (each containing one primer alone). In some instances, the same primer has been found to anneal at both ends of the PCR product.

We use vectors such as pUC or pBluescript to allow simple detection of the recombinant clones by the white and blue test. The recombinant colonies can be directly tested by PCR as follows.

## 6.5 Analysis of recombinants

First, identify the white clones on XGal–IPTG medium (20). Before sequencing the cloned PCR products, the recombinant clones have to be analysed to determine the sizes of the inserted fragments. To do this, and to rule out false positive recombinant clones, we have developed a direct PCR analysis procedure (*Protocol 13*).

**Protocol 13.** Direct PCR on colonies

*Precautions* (see *Protocol 1*)

*Equipment and reagents*

- Inoculating needles (Nunc)
- Thermal cycler (e.g. Techne, Hybaid, Cetus)
- 5 U/μl *Taq* polymerase (Boehringer)
- PCR buffer (Boehringer)
- 2.5 mM each dNTP (Boehringer)
- PCR primers[a] (50 ng/μl each)
- A programmable thermal cycler (e.g. Perkin-Elmer Cetus or Hybaid) or three water baths
- Mineral oil (Sigma M3516)

*Method*

1. Pick a colony with an inoculating needle.
2. Touch the bottom of a 0.5 ml microcentrifuge tube (or a well in a microtitre plate) with the needle.
3. Inoculate, with the same needle, 3 ml of liquid bacterial growth medium in a 15 ml tube, and incubate at 37°C for 18 h.
4. Repeat steps 1–3 for all the colonies to be analysed.
5. Prepare the PCR mixture on ice as follows. The volumes given are sufficient for one reaction.

| | |
|---|---|
| • 10 × PCR buffer | 2.5 μl |
| • 2.5 mM dNTP | 2 μl |
| • primer 1[b] (50 ng/μl) | 1 μl |
| • primer 2[b] (50 ng/μl) | 1 μl |
| • *Taq* DNA polymerase (5 U/μl) | 0.1 U |
| • double distilled water to | 25 μl. |

6. Distribute the PCR mixture into every tube on ice and add 100 μl of mineral oil.
7. Place the tubes or the microtitre plate in the thermal cycler and run the following program:
   - 3 min denaturation at 94°C
   - 35 cycles (94°C for 30 sec, 50°C for 45 sec, 72°C for 1 min/kb)
8. Analyse the PCR products on an agarose gel.
9. Prepare plasmid DNA of the positive clones from the cultures (prepared in step 3). Use this plasmid DNA for sequencing.

[a] With pBluescript vectors, use T7 and T3 sequencing primers or M13 universal sequencing primer and M13 reverse sequence primer. With pUC vectors, use M13 universal sequencing primer and M13 reverse sequence primer.

[b] Any pair of primers may be used. Do not forget to adjust the annealing temperature accordingly.

Sequence all the positive clones which arise from *Protocol 13*. This can be done by any available double-strand sequencing system.

## 6.6 Isolation of full length cDNA clones

Analyse the various sequences and identify the clones corresponding to putative new members of the family. Use those clones as probes for screening a full-length cDNA library. Another route to obtain full-length clones is to follow strategies outlined in *Figure 1B* and *1C* and described in *Section 3*.

# References

1. Mullis, K. B. and Faloona, F. (1987). In *Methods in enzymology* (ed R. Wu), Vol. 155, p. 335. Academic Press, New York.
2. Eiden J. J., Wilde J., Foroozmand F., and Yolken, R. (1991). *J. Clinical Microbiol.*, **29**, 539.
3. Zhang, Y.-H. and McCabe, E. R. B. (1992). *Human Genet.*, **89**, 311.
4. Chomczynky, P. and Saachi, N. (1987). *Anal. Biochem.*, **162**, 156.
5. Blumberg, D. D. (1987). In *Methods in enzymology* (ed. S. L. Berger and A. R. Kimmel), Vol. 152, p. 20. Academic Press, New York.
6. Rhyner, T. A., Faucon-Biguet, N., Berrard, S., Borbely, A. A., and Mallet, J. (1986). *J. Neurosc. Res.*, **16**, 167.
7. Chiang, C.-M., Chow, L. T., and Broker, T. R. (1993). In *Methods in molecular biology* (ed. B. A. White), Vol. 15, p. 189. Humana Press Inc., Totowa.
8. Sellner, L. N., Coelen, R. J., and Mackenzie, J. S. (1992). *Nucleic Acids Res.*, **20**, 1487.
9. Moore, R. E., Sheperd, J. W., and Hoskins, J. (1990). *Nucleic Acids Res.*, **18**, 1921.
10. Frohman, M. A., Dush, M. K., and Martin, G. R. (1988). *Proc. Natl Acad. Sci. USA*, **85**, 8998.
11. Dumas Milne Edwards, J. B., Delort, J., and Mallet, J. (1991). *Nucleic Acids Res.*, **19**, 5227.
12. Delort, J., Dumas, J. B., Darmon, M. C., and Mallet, J. (1989). *Nucleic Acids Res.*, **17**, 6439.
13. Uhlenbeck, O. C. and Gumport, R. I. (1982). In *The Enzymes*, Vol. XV, p. 31.
14. Chien, A., Edgar, D. B., and Trela, J. M. (1976). *J. Bacteriol.*, **127**, 1550.
15. Dumas Milne Edwards, J. B., Delort, J., and Mallet, J. (1993). In *Methods in molecular biology* (ed. B. A. White) Vol. 15, p. 365. Humana Press Inc., Totowa.
16. Wada, K. N., Wada, Y., Ishibashi, F., Gojobori, T., and Ikemura, T. (1992). *Nucleic Acids Res.*, **20**, 2111.
17. Smith, T. F., Waterman, M. S., and Sadler, J. R. (1983). *Nucleic Acids Res.*, **11**, 2205.
18. Martin, F. H. and Castro, M. (1985). *Nucleic Acids Res.*, **13**, 8927.
19. Rosenfeld, M. G. (1991). *Genes Dev.*, **5**, 897.
20. Sambrook, J., Fritsch, E. F., and Maniatis, T. (ed.) (1989). *Molecular cloning: a laboratory manual* (2nd edn). Cold Spring Harbor Press, Cold Spring Harbor, NY.

# 7

# Quantification of DNA and RNA by PCR

JIE KANG, JOCHEN E. KÜHN, PETER SCHÄFER, ANDREAS IMMELMANN, and KARSTEN HENCO

## 1. Introduction

The ability to quantify low copy numbers of a specific nucleic acid target in limited amounts of sample is important in both basic and applied medical research. The combination of PCR and TGGE (temperature gradient gel electrophoresis) provides an exact and reliable system for such quantification (1).

The PCR/TGGE system uses an internal standard which is identical, except for a single nucleotide, with the authentic target. Due to identical priming sites and almost identical sequences, the ratio between target and standard remains constant during amplification. Following extensive cycling, labelled standard DNA is added to the PCR amplification products. After a thermal denaturation and renaturation step, the labelled molecules have formed homoduplexes with the amplified standard and heteroduplexes with the amplified target. Subsequently, TGGE is applied to separate the homo- and heteroduplexes. The homoduplex migrates faster than the heteroduplex, as the latter partially denatures at a certain temperature due to the mismatch. The initial copy number of the target nucleic acids is easily calculated taking into account the ratio of intensities of hetero- and homoduplexes and the initial copy number of the standard. *Figure 1* shows the schematic strategy of this quantification system. For quantification of RNA molecules, an internal standard RNA is used to control the efficiency of the cDNA synthesis prior to the PCR amplification.

In addition to quantification, PCR/TGGE detects PCR artefacts and target mutants. Thus, PCR/TGGE is ideal for the detection and characterization of viral infections, allowing quantification of virus titres, transcriptional activities, viraemic, or silent states, and the control of therapy by means of detection of mutated resistant virus populations. 'False negative' results are directly visible. Examples of the application of PCR/TGGE in the field of biotechnology and medical diagnosis are discussed in this chapter.

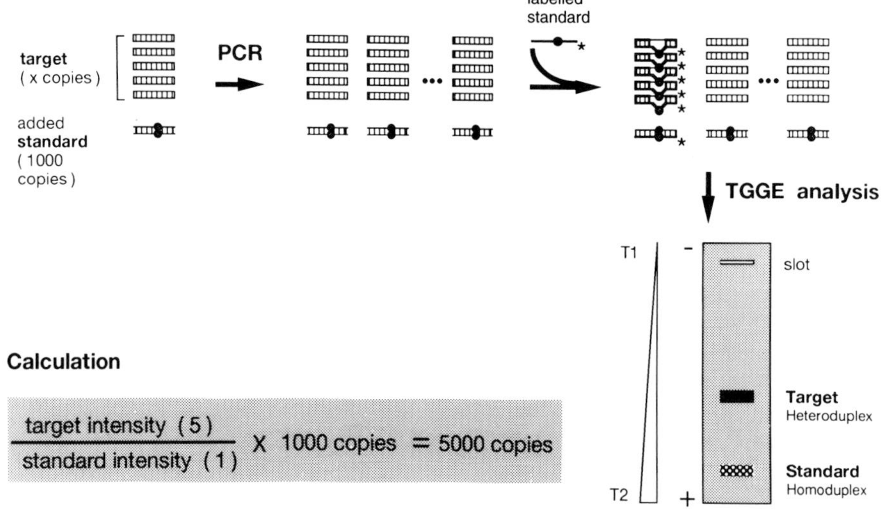


**Figure 1.** Schematic illustration of the strategy for PCR/TGGE quantification.

## 2. TGGE analysis

Our quantification system consists of two steps:

(a) co-amplification of the target and standard sequences
(b) TGGE analysis (which is performed according to the manufacturer's instructions; QIAGEN GmbH)

For TGGE analysis the standard DNA should be labelled on one strand only. This is carried out using a 'fill-in' reaction with [$\alpha$-$^{32}$P]dATP (2). The specific activity obtained should be 0.5–1 μCi/pmol standard. Non-radioactive labelling procedures such as by fluorescent dyes are also suitable. *Protocol 1* details the TGGE analysis procedure.

### **Protocol 1.** TGGE analysis

*Equipment and reagents*

- TGGE system (QIAGEN GmbH)
- X-ray film (Kodak X-Omat AR)
- 30% acrylamide stock (acrylamide:bisacrylamide, 30:0.5)
- Urea p.A. (Merck)
- 1 M MOPS, pH 8.0
- 0.5 M EDTA, pH 8.0
- 40% glycerol
- 4% Ammonium persulphate (freshly prepared)
- TEMED
- 2 × loading buffer: 8 M urea, 0.4 M MOPS pH 8.0, 10 mM EDTA, 0.02% bromophenol blue, 0.02% xylene cyanol
- 1 × electrophoresis buffer: 0.2 M MOPS, pH 8.0, 1 mM EDTA
- 10% ethanol, 0.5% acetic acid
- 2% glycerol

*Method*

1. Prepare a polyacrylamide gel by mixing:
   - urea 21.6 g
   - 30% acrylamide stock solution 12 ml
   - 1 M MOPS 0.9 ml
   - 0.5 M EDTA 90 μl
   - 40% glycerol 2.25 ml

   Bring to 44.6 ml with double distilled water (total volume of gel solution = 45 ml). Make sure that the urea is dissolved completely. Then add 340 μl 4% ammonium persulphate and 75 μl TEMED. Allow the gel to polymerize.
2. Prepare the samples by combining:
   - PCR product 4 μl
   - Labelled standard DNA (1 ng/μl) 1 μl
   - 2 × loading buffer 5 μl

   then denature at 98°C for 5 min and renature at 50°C for 15 min.
3. Run TGGE in 1 × electrophoresis buffer at 300 V for about 2 h with a temperature gradient between 20°C and 60°C. The bromophenol blue should have migrated 7–9 cm from the slots.
4. Fix the gel by incubation in 10% ethanol/0.5% acetic acid for 15 min.
5. Incubate the gel with 2% glycerol for 10 min and dry at 50°C for 3 h.
6. Expose the gel to X-ray film and evaluate the developed film densitometrically.

# 3. Quantification of DNA

## 3.1 Synthesis of standard DNA

Target fragments containing 100–300 bp are most suitable for TGGE analysis; a typical target for PCR/TGGE quantification is shown schematically in *Figure 2a*. Primers P1 and P2 (*Figure 2*) are used for quantification analysis. Primer Pq is exclusively used to construct the standard sequence. For easy cloning of the standard, all primers contain a 5′-overhang restriction site. In order to increase the difference in the electrophoretic mobility of homo- and heteroduplex, a GC-clamp of 20–30 nt (a stretch of GC base pairs that are thermodynamically very stable) should be added to the 5′-end of primer P2 (3).

The point mutation within the standard sequence can be conveniently introduced by primer Pq carrying a mismatched base. To exclude any potential differences in the amplification efficiency between standard and target, it is important to check that the base change in the standard sequence will not

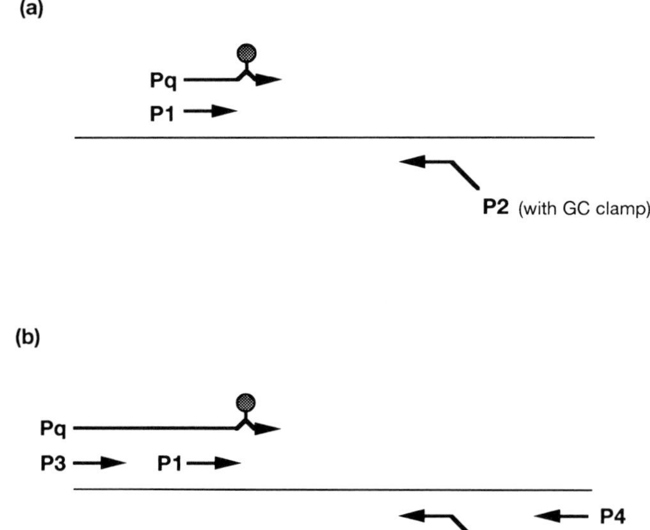


**Figure 2.** Schematic illustration of primers used for PCR/TGGE quantification. (a) A typical target of 100–300bp; (b) target for nested PCR.

create or destroy an inverted repeat. Repeats in the amplified sequence that form hairpin structures may influence amplification kinetics (unpublished data).

After PCR amplification using primers Pq and P2, the standard DNA fragment is cloned into a plasmid vector. Standard DNA is prepared as plasmid DNA, purified by HPLC, and then its concentration in μg/μl is determined spectrophotometrically (assuming that 1 $A_{260}$ corresponds to 50 μg/ml for double-stranded DNA). The DNA concentration in copies/μl is then calculated using the molecular weight of the standard DNA. Problems with PCR sensitivity arise when dealing with nucleic acid targets of poor quality or very low concentration. In such cases we suggest cloning a longer standard sequence, that may be amplified by using primers Pq and P4, for use in nested PCR (*Figure 2b*).

## 3.2 Determination of plasmid copy number

As an example, the PCR/TGGE method has been used to determine, quantitatively, the copy number of plasmid pUC19 per *E. coli* strain HB101 genome during a bench-scale fermentation process. Defined segments of the ampicillin resistance gene (*amp* fragment) of pUC19 and the pyruvate kinase I gene of HB101 (PK fragment) were selected as plasmid and genomic template fragments, respectively. Plasmid pBR322 containing a transition from A to

G within the *amp* fragment was used as the standard for plasmid quantification. The PK standard for genome quantification was constructed using a primer carrying an A instead of a T. The analysis was performed according to *Protocol 2*. Both calibrated standards (*amp* and PK) are present during the entire assay procedure. Therefore, all limitations inherent to the quantification will affect both the target and the standard to the same extent. In our approach, a series of standard dilutions are analysed with a fixed amount of target allowing the detection of potential false results due to contamination. The expected result is a gradual shift in band intensity from heteroduplex to homoduplex.

## **Protocol 2.** Determination of plasmid copy number

*Equipment and reagents*

- Thermal cycler (e.g. Perkin–Elmer)
- *amp* standard DNA ($5.4 \times 10^9$ copies/μl)
- PK standard DNA ($1.8 \times 10^7$ copies/μl)
- 10 × lysis buffer (5% NP40, 5% Tween 20)
- Restriction enzymes *Rsa*I, *Hae*III (10 U/μl)
- 10 × digestion buffer: 0.5 M Tris–HCl pH 8.0, 0.1 mM $MgCl_2$, 1 M NaCl
- Primer sets (10 pmol/μl each)
- 10 × PCR buffer: 0.1 M Tris–HCl pH 8.3 (at 72°C), 0.5 M KCl, 15 mM $MgCl_2$, 1% gelatin
- *Taq* polymerase (5 U/μl, Amersham)
- dNTPs (solution containing each dNTP at 2 mM)
- Mineral oil
- TGGE system (QIAGEN)

A. *Preparation of templates*

1. Prepare a culture of *E. coli* carrying the plasmid.
2. Prepare a standard solution by combining:
   - *amp* standard DNA 30 μl
   - PK standard DNA 30 μl
   - water 240 μl
3. Prepare four serial threefold dilutions of the standard solution from step 2.
4. Aliquot 10 μl of each standard dilution and 10 μl of the undiluted standard solution to five separate 0.5 ml microcentrifuge tubes.
5. Measure the $A_{550}$ of the *E. coli* culture (approximately 0.4–1.5).
6. Prepare a lysis mixture by combining:
   - cell culture 105 μl
   - 10 × lysis buffer ($A_{550}$ value × 84) μl
   - water to ($A_{550}$ value × 420) μl
7. Transfer 10 μl of the lysis mixture (from step 6) to each of the five PCR tubes from step 4.
8. Lyse the cells by heating to 95°C for 5 min.

**Protocol 2.** *Continued*

9. Allow to cool to room temperature then add to each tube:
   - 10 × digestion buffer 6 μl
   - *Rsa*I 1 μl
   - water 33 μl
10. Incubate for 1 h at 37°C.

B. *PCR*

1. Prepare separate PCR mixes for the plasmid and the genomic templates. In each case, set these up as follows, using appropriate primers to amplify the respective template. Combine in microcentrifuge tubes:
   - primer set [a] 5 μl
   - 10 × PCR buffer 50 μl
   - *Taq* polymerase 2.5 μl
   - dNTPs 5 μl
   - *Hae*III [b] 1 μl
   - water 136.5 μl
2. Incubate at 37°C for 30 min to degrade any contaminating template DNA then heat to 85°C for 5 min to inactivate *Hae*III.
3. To amplify the plasmid template, transfer 2 μl of the *Rsa*I-digested sample (part A, step 10) to a new 0.5 ml microcentrifuge tube and add:
   - water 58 μl
   - plasmid PCR Mix (part B, step 1) 40 μl

   To amplify genomic template, to the remaining *Rsa*I-digested sample (part A, step 10), add 40 μl of genomic PCR mix (part B, step 1).
4. Overlay with 100 μl mineral oil.
5. Perform PCR at 93°C, 5 min, then 35 cycles of (92°C, 1 min; 55°C, 1 min; 72°C, 1 min)
6. Perform TGGE analysis of the PCR products according to *Protocol 1* except use a temperature gradient of 40°C to 60°C.

[a] The plasmid and genomic PCR mixes differ in the primers required to amplify the respective template.
[b] *Hae*III cuts within the sequences to be amplified and therefore destroys any contaminating template DNA within the PCR reagents.

*Figure 3* shows the autoradiographic evaluation of samples analysed at 2, 4, 5, 6, and 7 h of fermentation. The calculation of copy number of pUC19 per HB101 genome during the fermentation was as described in ref. 4. The resulting data are illustrated in *Table 1*. To determine the reproducibility of the PCR/TGGE quantification system, we performed 25 independent analyses

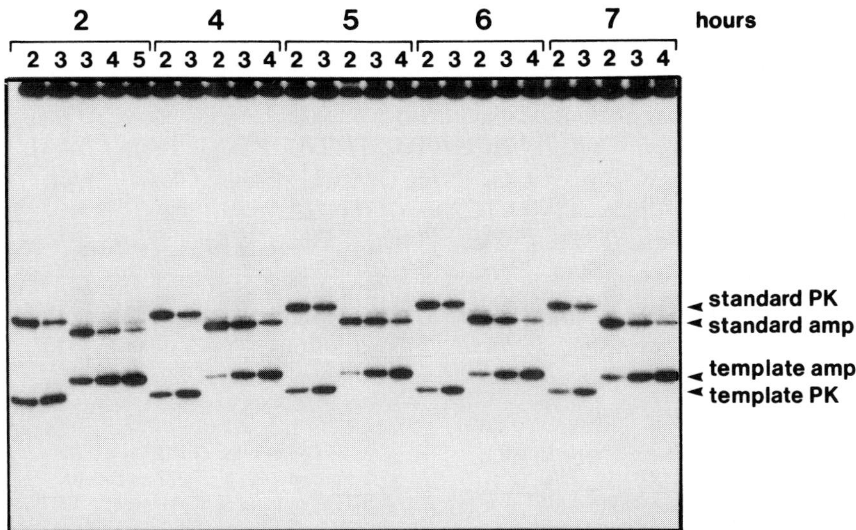


**Figure 3.** Autoradiographic evaluation of *Escherichia coli* cultures analysed after 2, 4, 5, 6, and 7 h fermentation. The numbers of the lanes refer to the standard solutions used. Lane 2: $1.8 \times 10^9$ copies/μl of *amp* standard, $6 \times 10^6$ copies/μl of PK standard; lane 3: $6 \times 10^8$ copies/μl of *amp* standard, $2 \times 10^6$ copies/μl of PK standard; lane 4: $2 \times 10^8$ copies/μl of *amp* standard, $6.7 \times 10^5$ copies/μl of PK standard; lane 5: $6.7 \times 10^7$ copies/μl of *amp* standard, $2.2 \times 10^5$ copies/μl of PK standard.

**Table 1.** Plasmid copy number per genome during fermentation

| Fermentation time (h) | $A_{550}$/ml | Plasmid copies/genome |
|---|---|---|
| 2 | 0.2 | 102 |
| 3 | 0.4 | 154 |
| 4 | 0.8 | 232 |
| 5 | 1.1 | 275 |
| 6 | 1.4 | 302 |
| 7 | 1.6 | 327 |
| 8 | 1.8 | 889 |

(data not shown). The quantitative evaluation results in a variability of $\leq$15%.

## 3.3 Determination of DNA copy number in clinical samples

A quantitative nested PCR approach (*Protocol 3*) has been established for the determination of HCMV (human cytomegalovirus) target sequences in clinical samples. The strategy is illustrated schematically in *Figure 2b*. The detection limit of the method was estimated to be less than five copies of the target sequence. The CMV standard was constructed using primer Pq carrying an A to G exchange. The primer sets used in the quantification analyses have the following nucleotide sequences: Primer set I: 5′-TCCAACACCCACAGTACCCGT; 5′-CGGAAACGATGGTGTAGTTCG. Primer set II: 5′-CCGGATCCCGCCGCCCGCCCCGCGCCCGCCGCGGCAGCACC TGGCT; 5′-GTAAACCACATCACCCGTGGA.

**Protocol 3.** Determination of HCMV DNA copy number in clinical samples

*Equipment and reagents*

- Thermal cycler (e.g. Perkin–Elmer)
- CMV standard DNA (10 copies/μl)
- Primer set I (1 pmol/μl each)
- Primer set II (10 pmol/μl each)
- 10 × PCR buffer: 0.1 M Tris–HCl pH 8.3 at 72°C, 0.5 M KCl, 15 mM $MgCl_2$, 1% gelatin
- Restriction enzyme *Cfo*I (2 U/μl)
- *Taq* polymerase (5 U/μl, Amersham)
- dNTPs (solution containing each dNTP at 2 mM)
- Mineral oil
- TGGE system (QIAGEN)

*Method*

1. Prepare DNA from blood or other clinical specimens.
2. Prepare a DNA mixture by combining:
   - CMV standard DNA 2 μl
   - blood DNA (or other specimen) 8 μl
3. Incubate at 100°C for 5 min to increase the amplification efficiency.
4. For each DNA mixture, prepare two PCR mixes (a primary mix and a nested mix) by combining:

| | *Primary mix* | *Nested mix* |
|---|---|---|
| • primer set I | 2 μl | |
| • primer set II | | 2 μl |
| • 10 × PCR buffer | 2 μl | 8 μl |
| • dNTPs | 2 μl | 8 μl |
| • *Taq* polymerase | 0.2 μl | 0.3 μl |
| • *Cfo*I[a] | 1 μl | 1 μl |
| • water | 2.8 μl | 60.7 μl |

5. Incubate at 37°C for 30 min then heat at 85°C for 5 min to inactivate *Cfo*I.
6. Transfer 10 μl of the primary mix to the DNA mixture (from step 2), overlay with 100 μl mineral oil and incubate the tubes in a thermal cycler at 93°C for 5 min, then for 20 cycles of (92°C, 1 min; 55°C, 1 min; 72°C, 1 min).

7. Add 80 μl of nested mix (from step 4) to the lower phase of each PCR tube, spin briefly in a microcentrifuge, and incubate in a thermal cycler again for 35 cycles of (92°C, 1 min; 55°C, 1 min; 72°C, 1 min).
8. Perform TGGE analysis with a temperature gradient of 20°C to 60°C according to *Protocol 1*.

[a] *Cfo*I cuts within the PCR template sequence and can be used as a decontamination enzyme.

When PCR is used for diagnostic applications, the possibility of false negative results due to non-specific inhibitors has to be carefully considered. Furthermore, a number of human pathogens (e.g. herpesviruses) establish life-long latency after primary infection. Thus, in HCMV infections the presence of low levels of viral target sequences in the leucocytes of latently infected individuals has to be discriminated from the state of viraemia during acute primary or recurrent infections. Quantitative determination of target molecules by the PCR/TGGE system may be helpful to differentiate between acute

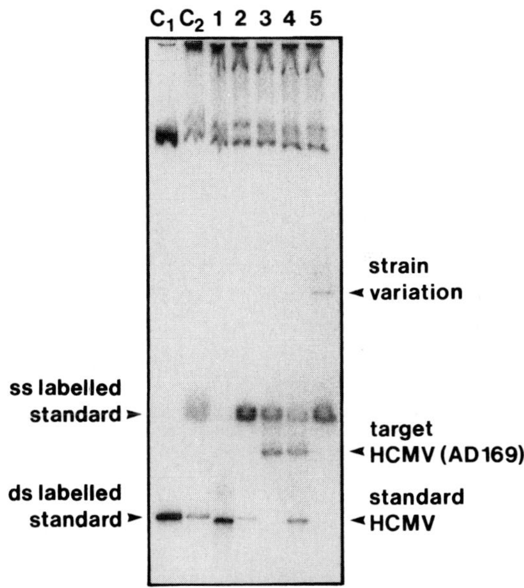


**Figure 4.** PCR/TGGE quantification in a HCMV diagnostic test using 20 copies of standard CMV ('cut-off'). **C**, control lanes with labelled standard, but without PCR products; **$C_1$**, native labelled standard; **$C_2$**, denatured/renatured labelled standard. **1**, no infection (target copies $\ll$ 'cut-off', no 'false positive'). **2**, 'false negative' (PCR reaction inhibited). **3**, viraemia in a patient experiencing an acute primary HCMV infection (high amounts of HCMV target DNA, target copies $\gg$ 'cut-off'). **4**, similar concentration of target and standard during clinically asymptomatic reactivated HCMV infection. **5**, viraemia with a HCMV strain differing within the amplified sequence from HCMV AD169 sequence.

and latent infections. The co-amplification of a defined number of standard sequences in each PCR permits a 'cut-off' approach in which the sensitivity of PCR is adjusted to a desired level. This should be useful to reduce unwanted positive PCR results, for example those due to the presence of low levels of the target sequences in latently infected healthy individuals. In addition, false negative results can be directly detected. As well as quantitative determination of target sequences, strain variations within the amplified region can be visualized. An example of results of the quantitative HCMV PCR obtained in clinical specimens are given in *Figure 4*.

# 4. Quantification of RNA

## 4.1 Synthesis of standard RNA

Standard RNA is used as a control for quantification of RNA templates, both for cDNA synthesis and PCR amplification. A general scheme for the synthesis of standard RNA is presented in *Figure 5* and procedures are described in *Protocol 4*. Primer sequences located on separate exons are chosen in order

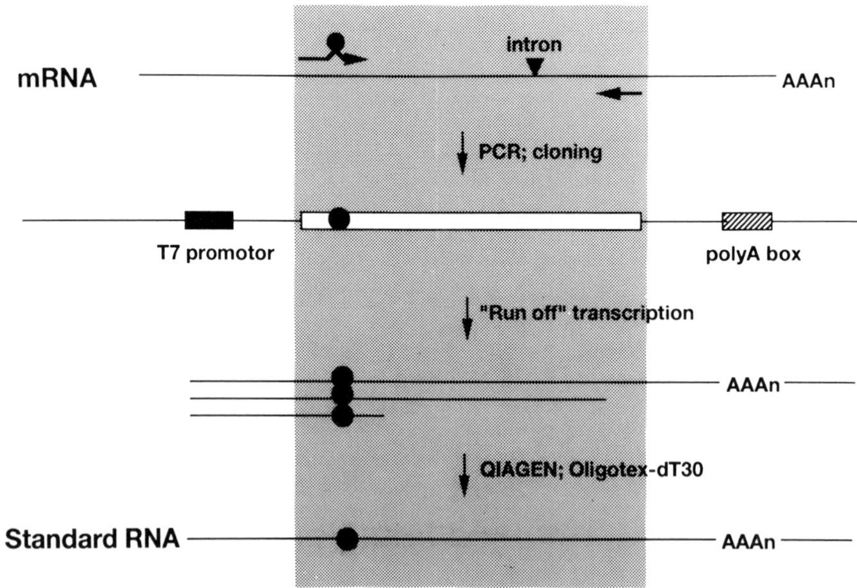


**Figure 5.** Schematic illustration of the strategy for synthesis of standard RNA. The PCR reaction is carried out using primer sequences located on separate exons within poly(A)$^+$ RNA containing the target RNA sequence. One of the primers contains a base exchange to produce the point mutation in the standard sequence. The PCR product is cloned into the vector pTA consisting of a T7 promoter and a poly(A) box. Following 'run-off' transcription the reaction mixture is purified through a QIAGEN column to remove the vector DNA. The full-length RNA transcripts are selected using Oligotex-dT.

to distinguish between PCR products derived from cDNA and products derived from contaminating genomic DNA. The cloning vector used, plasmid pTA, has been constructed by cloning a poly(A) fragment of 89 bp into the *Pst*I and *Hin*dIII sites of pTZ18R. Following 'run-off' transcription with T7 RNA polymerase, the reaction mixture is purified through a QIAGEN column to remove the vector DNA. Subsequently, the full-length RNA transcripts are selected using Oligotex-dT (*Protocol 4*).

## **Protocol 4.** Synthesis of standard RNA

### *Reagents*

- Restriction enzyme *Bgl*I (10 U/μl, BRL)
- 5 × 'run-off' buffer: 0.2 M Tris–HCl, pH 8.0, 0.125 M NaCl, 40 mM $MgCl_2$, 10 mM spermidine-$(HCl)_3$
- 100 mM DTT
- rNTPs (solution containing each rNTP at 8 mM)
- T7 RNA polymerase (50 U/μl, BRL)
- Phenol:chloroform:isoamyl alcohol (25:24:1)
- Chloroform
- QIAGEN-tip 20 (QIAGEN)
- Buffer QA: 400 mM NaCl, 50 mM MOPS pH 7.0, 15% ethanol
- Buffer QAT: QA containing 0.15% Triton
- Buffer QRU: 6 M urea, 900 mM NaCl, 50 mM MOPS pH 7.0, 15% ethanol
- Isopropanol
- 70% ethanol
- TE: 10 mM Tris–HCl pH 7.5, 1 mM EDTA
- Oligotex-dT suspension (10% (w/v), QIAGEN)
- 5 M NaCl
- Wash buffer: 10 mM Tris–HCl pH 7.5, 150 mM NaCl, 1 mM EDTA
- Elution buffer: 5 mM Tris–HCl pH 7.5

### *Method*

1. Linearize 5 μg of the vector pTA containing the standard fragment with 10 units *Bgl*I in 50 μl containing the appropriate restriction buffer at 37°C for 1 h.
2. For the 'run-off' reaction, combine in a microcentrifuge tube:
   - linearized pTA DNA (0.1 μg/μl) 10 μl
   - 5 × 'run-off' buffer 10 μl
   - 100 mM DTT 2.5 μl
   - 8 mM rNTPs 2.5 μl
   - T7 RNA polymerase 1 μl
   - water 24 μl
3. Incubate at 37°C for 1 h.
4. Add 50 μl water and 1 vol. phenol:chloroform:isoamyl alcohol (25:24:1). Vortex and centrifuge in a microcentrifuge (13000 *g*) to separate the phases.
5. Remove the aqueous supernatant and extract once with an equal volume of chloroform.
6. Adjust the volume of the aqueous phase to 1 ml by adding buffer QA.

**Protocol 4.** *Continued*

7. Equilibrate a QIAGEN-tip 20 with 1 ml buffer QAT.
8. Apply the supernatant from step 6 on to the QIAGEN-tip and allow it to enter the resin by gravity flow.
9. Wash the QIAGEN-tip with 5 × 1 ml buffer QA.
10. Elute the RNA transcripts with 1 ml buffer QRU into a clean tube.
11. Add 1 vol. isopropanol and spin in a microcentrifuge for 30 min.
12. Wash the RNA pellet with 70% ethanol, dry it briefly, and redissolve in 100 μl TE.
13. Heat at 65°C for 5 min, add 10 μl 5 M NaCl and 50 μl Oligotex-dT suspension.
14. Incubate the mixture at 37°C for 5–10 min.
15. Centrifuge the mixture at 13000 *g* for 5 min and remove the supernatant by pipetting.
16. Wash with 2 × 500 μl wash buffer.
17. Elute with 2 × 25 μl elution buffer preheated to 65°C.
18. Combine the eluates and measure $A_{260}$. Calculate the concentration in copies/μl of RNA standard assuming that 1 $A_{260}$ = 40 μg/ml and knowing the molecular weight of the RNA.

## 4.2 Quantification of mRNA

As an example, *Protocol 5* describes the quantification of interleukin-2 (IL2) mRNA, but this procedure can easily be applied to the quantification of other mRNAs. Two sets of primers are used in a nested amplification strategy (*Figure 2b*). For IL2 mRNA, the primers have the following nucleotide sequences: primer set I: 5′-CCGGATCCATTTGAATGGAATTAATAA-TTAC; 5′-CACCTGCAGTTTAGTTCCAGAACTATTACG. Primer set II: 5′ - CCAGTCGACGCCCGCCGCGCCCGCGCCCGCCGCCGCAAGAA-TCCCAAACTCACC: 5′-CCGAATTCATTTAGCACTTCCTCCAGA.

**Protocol 5.** Quantification of IL2 mRNA

*Equipment and reagents*

- IL2 standard RNA (2 × $10^3$ copies/μl)
- Total RNA (40 ng/μl). This is the RNA sample for which the IL2 mRNA concentration is to be determined.
- Primer set I (1 pmol/μl each)
- Primer set II (10 pmol/μl each)
- 5 × RvT buffer: 0.25 M Tris–HCl pH 8.2 (at 42°C), 0.5 M NaCl, 30 mM $MgCl_2$, 50 mM DTT)
- 10× PCR buffer: 0.1 M Tris–HCl pH 8.3 at 72°C, 0.5 M KCl, 15 mM $MgCl_2$, 1% gelatin.

- Restriction enzyme *Hae*III (2 U/μl)
- *Taq* polymerase (5 U/μl, Amersham)
- dNTPs (solution containing each dNTP at 2 mM)
- AMV reverse transcriptase (1 U/μl, Boehringer Mannheim)
- Mineral oil
- TGGE system (QIAGEN)

*Method*

1. For each sample of total RNA, prepare an RNA mixture containing:
   - IL2 standard RNA 5 μl
   - total RNA (40 ng/μl) 5 μl
2. For each RNA mixture prepare two PCR mixes as follows:

| | *Primary mix* | *Nested mix* |
|---|---|---|
| • Primer set I | 1 μl | |
| • Primer set II | | 5 μl |
| • 5 × RvT buffer | 4 μl | |
| • 10 × PCR buffer | | 8 μl |
| • dNTPs | 2 μl | 8 μl |
| • *Taq* polymerase | 0.2 μl | 0.3 μl |
| • *Hae* III[a] | 1 μl | 1 μl |
| • water | 0.8 μl | 57.7 μl |
| • RNA mixture | 10 μl | |

Incubate at 37°C for 30 min and then heat at 85°C for 5 min.

3. Add 1 μl AMV reverse transcriptase to the primary mix (from step 2) and overlay with 100 μl mineral oil.
4. Incubate the tubes in a thermal cycler at 42°C, 30 min; then 93°C, 5 min; then for 10 cycles of (92°C, 1 min; 55°C, 1 min; 72°C, 1 min).
5. Transfer the nested mix (from step 2) to the lower phase of the PCR tube, centrifuge briefly, and incubate in a thermal cycle for 35 cycles of (92°C, 1 min; 55°C, 1 min; 72°C, 1 min).
6. Perform TGGE analysis, with a temperature gradient of 30°C to 60°C, according to *Protocol 1*.

[a] *Hae*III cuts within the sequences to be amplified and therefore destroys any contaminating template DNA within the PCR reagents.

Using *Protocol 5*, we have quantified IL2 mRNA in human lymphocytes before and after induction with phytohaemagglutinin (PHA). *Figure 6* shows that IL2 mRNA after 26 h PHA induction was expressed to a level 18-fold higher than uninduced.

## 5. Summary

The PCR/TGGE system using the internal standardization strategy provides a reliable and precise method by which to quantify the DNA/RNA copy

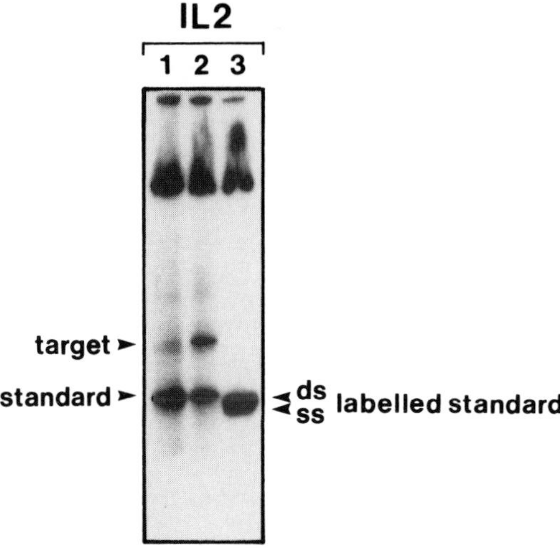


**Figure 6.** Quantification of IL2 mRNA in human lymphocytes. Lane 1: total RNA from uninduced cells; lane 2: total RNA from cells 26 h after PHA-induction; lane 3: negative control.

number in biological specimens. The target nucleic acid is co-amplified in the presence of a known copy number of a standard sequence and with identical efficiency, and subsequent TGGE analysis resolves amplified target and standard molecules. The initial copy number of the target is then easily calculated by measuring the ratio of intensities of target to standard bands by densitometry.

This system provides some significant advantages for diagnostic application. By using the internal standard, any failure in the PCR reaction is directly detected, thus avoiding 'false negative' results. In addition, a distinct copy number of the standard can be used for a 'cut-off' approach in which the sensitivity of PCR is adjusted to a desired level. Depending on the clinical situation and type of specimens tested, this should help to reduce unwanted positive PCR results, for example, those due to the presence of low levels of the target sequence in latently-infected healthy individuals. During TGGE, the homoduplex of the standard sequence always migrates faster than all heteroduplexes potentially formed between the target and the labelled standard. Mutations in the target sequence are therefore indicated by one or more shifted heteroduplex bands on TGGE. Viral strains (resistant mutants) selected for by events such as chemotherapeutic intervention may also be detected. PCR/TGGE can thus determine both the copy number and the presence of variants in a single procedure.

## Acknowledgements

We thank Jutta Harders and Thorsten Klahn for computer calculations of the thermodynamic DNA stabilities, and Christiane Mester, Susanne Welters, and Monika Heibey for excellent technical assistance.

## References

1. Henco, K. and Heibey, M. (1990). *Nucleic Acids Res.*, **19**, 6733.
2. Sambrook, J., Fritsch, E. F., and Maniatis, T. (1989). *Molecular cloning: A laboratory manual.* (2nd edn.). Cold Spring Harbor Laboratory Press, Cold Spring Harbor.
3. Myers, R. M., Sheffield, V. C., and Cox, D. R. (1989). *PCR technology—principles and applications for DNA amplification*, p. 71. Stockton Press, New York.
4. Kang, J., Immelmann, A., Welters, S., and Henco, K. (1991). *Biotech. Forum Europe*, **8**, 590.

# 8

# PCR MIMICs: competitive DNA fragments for use in quantitative PCR

PAUL D. SIEBERT and DAVID E. KELLOGG

## 1. Introduction

Competitive PCR has emerged as a valuable method for determining the relative levels of mRNAs; for example, the change in mRNA expression levels or to estimate the actual number of mRNA molecules. Perhaps the greatest advantage of competitive PCR is that useful data can be obtained during either the exponential phase or plateau phase of PCR amplification (1). In other methods, such as co-amplification of the target gene with an endogenous housekeeping gene, the PCR products must be obtained before the plateau phase of the reaction occurs. This is necessary for both the housekeeping gene and the target gene, which often exhibit different amplification kinetics.

In competitive PCR, a DNA fragment containing the same primer sequences as the target fragment is allowed to compete in the same tube with the target for primer binding and amplification. Experimentally, PCR reaction tubes containing the target samples are spiked with a dilution series of the competitor fragment. When the molar ratio of PCR products generated from target and competitor is equal to one, the amount of target is equal to the competitor. Since the amount of competitor is known, the amount of target can thus be determined.

The competitor fragment can be either a DNA sequence or a synthetic RNA that contains the same primer binding sequences as the target gene. In the latter case, the titration of the target and competitor is conducted during the reverse transcriptase step of cDNA synthesis. In order to differentiate between the PCR products generated from the target and the competitor, the competitor can be engineered to be slightly larger or smaller than the target (2). A unique restriction site can be added or removed from the target (2, 3) and the PCR products digested with a restriction enzyme before gel electrophoresis. Alternatively, procedures for physically distinguishing competitor and target sequences of identical size may be used, such as the PCR/TGGE procedure described in Chapter 7.

Generation of these types of competitor fragments involves techniques that are technically demanding and time consuming. Other types of competitive DNA fragments can be used that have the same primer binding sequences but a completely different intervening sequence. These types of competitor fragments can be easier to generate. For example, Uberla (4) used the target gene primers with low annealing stringency to amplify genomic DNA from another species. In this case, a product was chosen whose size differed from that of the target and was then purified from the agarose gel. Siebert and Larrick (5) ligated primer sequences to the ends of a restricted DNA fragment. This chapter describes an additional method for constructing competitive PCR fragments that is simple, rapid and reliable. The method involves amplification of a heterologous DNA fragment with a pair of composite primers. The composite primers contain the target primer sequences contiguous to a sequence that anneals to the heterologous DNA fragment. During amplification the target primer sequences are incorporated into the products. We refer to such heterologous competitor fragments as PCR MIMICs because the fragment 'mimics' the target gene for primer annealing and amplification. A clear advantage of PCR MIMICs is that they cannot form heteroduplexes with the target, a result which can complicate analysis. This chapter describes the experiments necessary for validating the method and provide protocols for performing competitive PCR.

## 2. Generation of PCR MIMICs

The generation of PCR MIMICs is achieved by two successive PCR amplifications of heterologous DNA, as described in *Protocol 1* and as shown in *Figure 1*. For the first PCR, two composite primers are used. One composite primer contains the upstream target primer sequence linked to 20 nucleotides that anneal to one strand of a heterologous DNA fragment. The other composite primer contains the downstream target primer sequence linked to 20 nucleotides that anneal to the opposite strand of the heterologous DNA fragment. The two composite primers and a small quantity of the heterologous DNA are added to a PCR reaction. During amplification of the DNA fragment, the target primer sequences are incorporated into the PCR product. A dilution of this PCR product is used to perform a second PCR using the shorter target primers. This ensures that the complete target primer sequences have been incorporated into the PCR product. At this point, the PCR product is purified by passage through a spin column, which removes PCR reaction components, primers, and other artefacts such as primer–dimers.

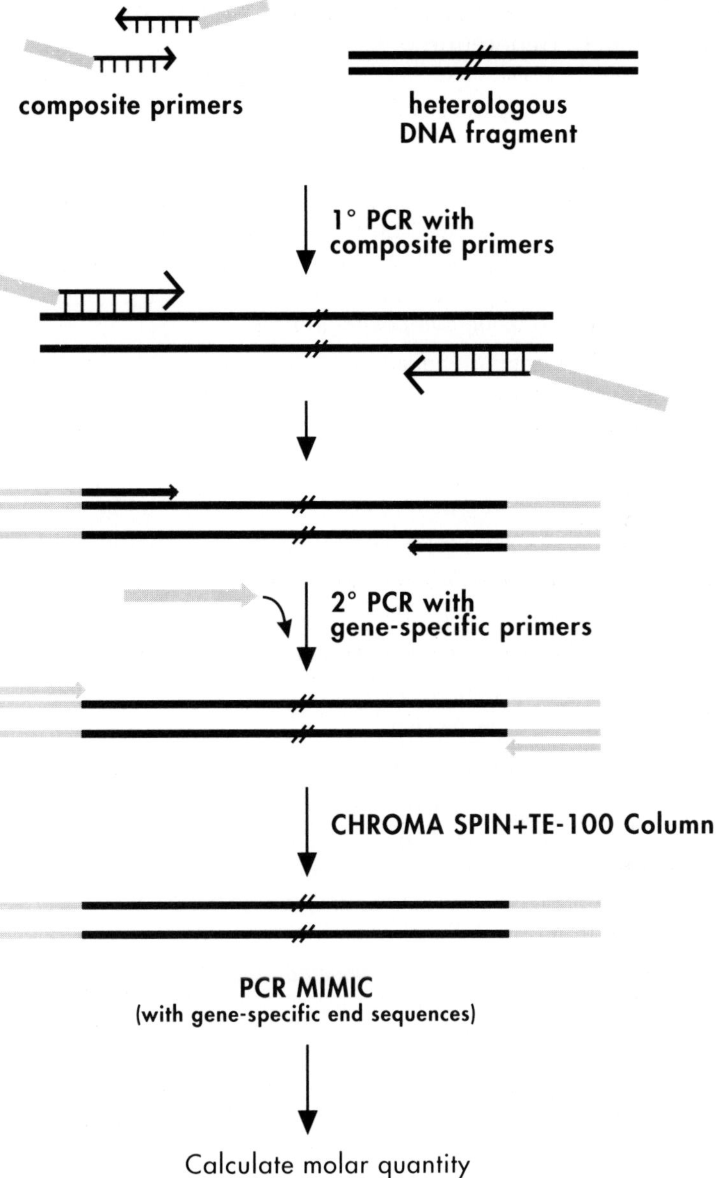


**Figure 1.** Flowchart describing the generation of competitive PCR MIMICs.

**Protocol 1.** Generation of PCR MIMICs

*Equipment and reagents*

- DNA size markers: *Hae*III digest of $\phi$X174 (e.g. Clontech)
- 10 × PCR buffer: 100 mM Tris–HCl pH 8.3, 500 mM KCl, 15 mM $MgCl_2$, 0.2% gelatin
- 50 × dNTP mix (10 mM of each dNTP)
- AmpliTaq® DNA Polymerase (Perkin-Elmer)
- Chroma Spin-100+TE columns (Clontech)
- DNA thermal cycler (e.g. Perkin-Elmer, model 4800)

A. *Primary PCR using composite primers*

1. To a PCR reaction tube, add:

| | |
|---|---|
| • distilled water | 37.6 μl |
| • 10 × PCR buffer | 5 μl |
| • 50 × dNTP mix | 1 μl |
| • neutral DNA fragment (2 ng) | 4 μl |
| • each composite primer (20 μM each) | 1 μl + 1 μl |
| • AmpliTaq® DNA Polymerase (5 U/μl) | 0.4 μl |
| • total volume is | 50 μl. |

2. Perform 16 cycles of PCR in an automated thermocycler:
   - denature (94°C for 45 sec)
   - anneal (60°C for 45 sec)
   - polymerize (72°C for 90 sec)
   - perform the final polymerization step for an additional 7 min
3. Run 5 μl of the reaction on a 1.8% EtBr–agarose gel (take care; EtBr is carcinogenic). If a strong band of the expected size is obtained, proceed; if not, perform another 4 cycles of PCR.

B. *Secondary PCR using gene-specific primers*

1. Remove 2 μl of the primary PCR reaction (from step 2 above) and dilute to 200 μl with water.
2. Add 2 μl of this dilution to a PCR reaction tube containing:

| | |
|---|---|
| • distilled water | 83.4 μl |
| • 10 × PCR buffer | 10 μl |
| • 50 × dNTP mix | 2 μl |
| • each gene specific primer (20 μM) | 2 μl + 2 μl |
| • AmpliTaq® DNA polymerase | 0.6 μl |
| • total volume is | 100 μl |

3. Perform 18 cycles of PCR using the same cycle parameters as for the primary PCR (part A, step 2).

4. Run 5 μl of the reaction on a 1.8% EtBr–agarose gel. If a strong band of the expected size is obtained, proceed; if not, perform another four cycles of PCR.

C. *Purification of the MIMIC*

1. To remove primers and reaction components, pass the reaction through two Chroma Spin−100+TE columns. Load 45 μl into each of two pre-spun columns and centrifuge following the manufacturer's protocol.
2. Check the quality of the MIMIC by running 5 μl on a 1.8% EtBr–agarose gel. Make sure that the salt concentration of the sample is adjusted to that of the agarose gel. If primers are still detected, repeat the chromatography step.

## 2.1 Determination of PCR MIMIC yield

The quantity of PCR MIMIC is determined by measuring the absorbance at 260 nm. Alternatively, an aliquot of the MIMIC can be run on a gel and the intensity of the bands compared with a dilution series of known quantities of DNA markers. These procedures are described in *Protocol 2*. The yield of PCR MIMIC is typically 0.5–1 μg, large enough for hundreds of competitive PCR experiments. A single MIMIC construction is therefore sufficient for a complete series of experiments. Any inaccuracies in the initial MIMIC yield determination will be consistent for all experiments and will not affect quantification of relative changes in mRNA levels.

**Protocol 2.** Determination of MIMIC yield

*Equipment and reagents*

- UV spectrophotometer
- MIMIC dilution solution: 50 μg/ml glycogen solution, molecular biology grade (Boehringer Mannheim) in 10 mM Tris–HCl pH 7.5, 0.1 mM EDTA

*Method*

The yield of MIMIC can be determined by either of two methods:

A. *Spectrophotometry*

1. Take one-half of the MIMIC and dilute in a minimum volume of distilled $H_2O$. The volume of the cuvette must not exceed 100 μl or the absorbance will be too low to measure accurately.
2. Measure $A_{260}$.

**Protocol 2.** *Continued*

3. Calculate yield as follows: ng/ml = $A_{260}$ (dilution factor) × (0.05) × (1000).
4. Convert the mass quantity to a molar quantity using the approximation that 1 ng of a 300 bp DNA fragment is equal to $5.4 \times 10^3$ amol (1 attomole (amol) = $10^{-18}$ moles and 1 mole = $6 \times 10^{23}$ molecules). This conversion factor is derived by using an average molecular weight of 310 g/mol for each nucleotide.
5. Dilute a portion of the MIMIC to 100 amol/μl in MIMIC dilution solution. Store the concentrated stock at −20°C where it will remain stable for at least 1 year.

B. *Gel electrophoresis*

1. Find the DNA marker band that is closest in intensity to that of the PCR MIMIC band. From the amount of marker DNA in that lane and the ratio of the size of the band with respect to the total size of the DNA marker, estimate the yield of MIMIC in nanograms. Typically the yield of MIMIC is 4–15 ng/μl.
2. Continue the calculation according to part A, steps 4 and 5.

## 2.2 Choice of heterologous DNA sequence

PCR MIMICs are constructed such that the size of the PCR product is slightly larger or smaller than the PCR product generated from the target gene. PCR MIMICs of different sizes can be made simply by choosing the appropriate distance on the heterologous DNA fragment to which the 20-nucleotide portion of the composite primers anneal.

The choice of heterologous DNA is important. For the heterologous DNA fragment we use a 600-bp *Eco*RI/*Bam*HI restriction fragment of the viral oncogene v-*erb*B. We chose this heterologous DNA fragment for two reasons:

- The complete nucleotide sequence is known.
- Computer analysis of the nucleotide sequence revealed no long stretches of Gs and Cs nor large internal homologies—thus making it a favourable choice as a PCR target. Any DNA fragment that exhibits such a neutral structure can be chosen.

Of course, if the native target itself is a suboptimal choice as an amplifiable region due to extreme C +/C content or secondary structure, this selection method could fail.

# 3. Validation of the PCR MIMIC strategy

## 3.1 Efficiency of amplification

Although PCR MIMICs share the same primer binding sites as the target template, the intervening DNA sequences differ, making it possible for the MIMIC and target to be amplified at different efficiencies. It is therefore necessary to show that their amplification efficiencies are similar. This is achieved by a kinetic analysis of the type shown in *Figure 2*. In the example shown, a MIMIC for glyceraldehyde-3-phosphate dehydrogenase (G3PDH) was generated. Approximately equal molar quantities of a G3PDH cDNA fragment and a G3PDH MIMIC were added to a single PCR reaction along with a small amount of [$\alpha$-$^{32}$P]dCTP. When PCR products could first be visualized on an agarose gel, aliquots were removed after each cycle for a total of seven cycles. The EtBr–agarose profile is shown in *Figure 2*. The bands corresponding to the target and MIMIC were then excised from the gel and the amount of radioactivity quantitated using a scintillation counter. The data were plotted as $\log_{\text{target c.p.m.}}$ and $\log_{\text{MIMIC c.p.m.}}$ versus cycle number (*Figure 2*). The linear portion of the curves have very similar slopes indicating that the G3PDH target and MIMIC share very similar amplification efficiencies.

## 3.2 Competitive PCR

An example of a competitive PCR experiment using a MIMIC is shown in *Figure 3*. Tenfold serial dilutions of the G3PDH MIMIC were co-amplified with a constant amount of a cloned G3PDH cDNA fragment (*Protocol 3*). Again the reactions contained a small amount of [$\alpha$-$^{32}$P]dCTP. A small portion of the products were then resolved on a 1.6% EtBr–agarose gel; the EtBr–agarose gel profile is shown in *Figure 3*. The data obtained by radioactive counting of the excised bands was then plotted as the log of the ratio of target to MIMIC c.p.m. versus log of the amount of MIMIC added to the PCR reaction (*Figure 3*).

The molar amount of target added to the reaction is equal to the molar amount of MIMIC when the ratio of their products becomes equal (log ratio 1:1 = 0). After taking into consideration the differences in size between the MIMIC and target, this point was reached at 0.075 attomoles—which is in very close agreement with the amount of target DNA added to the PCR reaction (0.1 amol). *Protocol 3* describes the titration of a constant amount of cDNA with 10-fold serial dilutions of the MIMIC. The results obtained from *Protocol 3* enable the setting up of a 'fine-tuned' two-fold MIMIC dilution series for the quantitative PCR experiment (see Section 3.3).

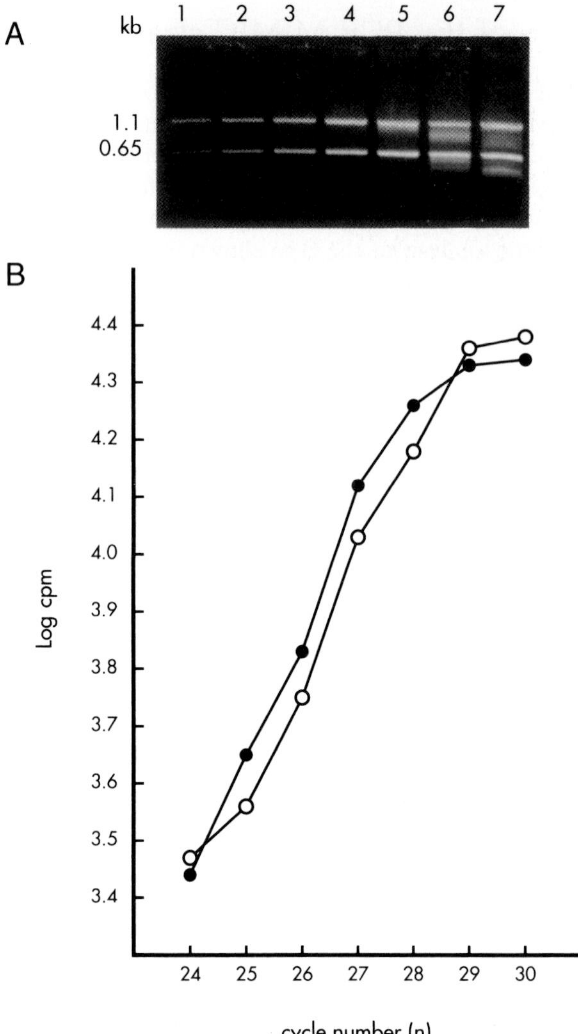


**Figure 2.** Kinetics of amplification of the G3PDH target cDNA and PCR MIMIC. An aliquot of 0.1 amol each of G3PDH cDNA and MIMIC were added to a PCR reaction together with [$\alpha$-$^{32}$P]dCTP. Starting at 24 cycles and after each of six additional cycles, 5% portions of the reaction were removed and resolved on a 1.8% EtBr–agarose gel. (A) EtBr–agarose gel profiles. (B) Following gel electrophoresis, the bands corresponding to the target (1.1 kb) and MIMIC (0.65 kb) were excised from the gels and the amount of radioactivity determined by scintillation counting. The data are plotted as a function of the log c.p.m. vs. cycle number. (●) G3PDH target; (○) G3PDH MIMIC.

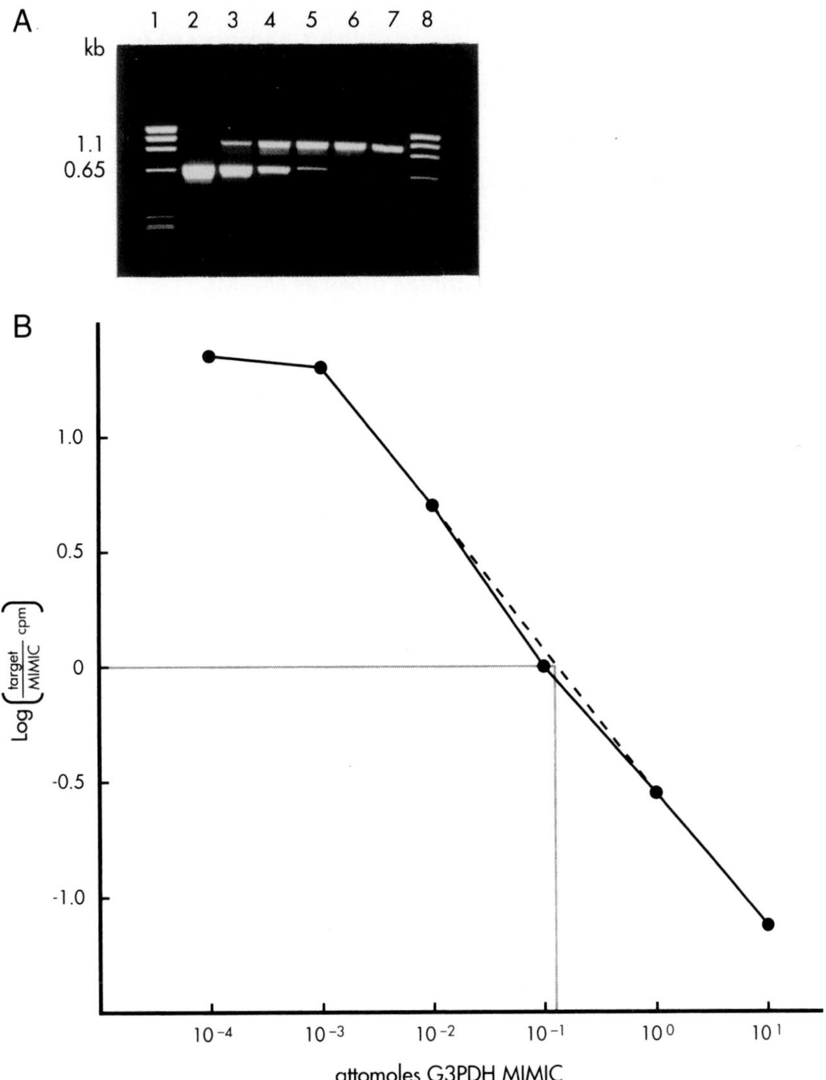


**Figure 3.** Titration of a G3PDH PCR MIMIC with a constant amount of a cloned G3PDH cDNA. An aliquot of 0.1 amol G3PDH cDNA was added to a PCR reaction containing 10-fold serial dilutions of a G3PDH MIMIC. [$\alpha$-$^{32}$P]dCTP was included in the reaction to quantify the PCR products. After 30 cycles, 10% portions of the products were resolved on a 1.8% EtBr–agarose gel. (A) Lanes 1–7: $10^1$, $10^0$, $10^{-1}$, $10^{-2}$, $10^{-3}$, and $10^{-4}$ amol G3PDH MIMIC, respectively. Lanes 1 and 8 contain a *Hae*III digest of $\phi$X174 as size markers. The positions of the 1.1 kb G3PDH target and 0.65 kb G3PDH MIMIC PCR products are indicated. (B) Plot of the data shown in (A). Following electrophoresis, the bands corresponding to the target and MIMIC were excised from the gel and the radioactivity determined by scintillation counting. The data is plotted as a function of the log (target c.p.m./MIMIC c.p.m.) against amol MIMIC added to the PCR reactions.

**Protocol 3.** Setting up a preliminary 10-fold PCR MIMIC dilution series

*Equipment and reagents*

See *Protocol 1* and *Protocol 2* (MIMIC dilution buffer).

*Method*

1. Label eight 0.5 ml microcentrifuge tubes as $M_1$–$M_8$.
2. Add 18 μl of MIMIC dilution solution to each tube.
3. Starting with the MIMIC stock solution ($M_0$) make the following 10-fold dilution series:[a]

| *Solution* | *Concentration (amol/μl)* | *Dilution* |
|---|---|---|
| $M_0$ | 100 | MIMIC stock solution |
| $M_1$ | 10 | add 2 μl $M_0$, mix[b], change pipette tip |
| $M_2$ | 1 | add 2 μl $M_1$, mix, change pipette tip |
| $M_3$ | $10^{-1}$ | add 2 μl $M_2$, mix, change pipette tip |
| $M_4$ | $10^{-2}$ | add 2 μl $M_3$, mix, change pipette tip |
| $M_5$ | $10^{-3}$ | add 2 μl $M_4$, mix, change pipette tip |
| $M_6$ | $10^{-4}$ | add 2 μl $M_5$, mix, change pipette tip |
| $M_7$ | $10^{-5}$ | add 2 μl $M_6$, mix, change pipette tip |
| $M_8$ | $10^{-6}$ | add 2 μl $M_7$ |

4. Set up PCR reactions in 50 μl reaction volumes according to *Protocol 1* containing a constant amount of cDNA and 2 μl of dilutions $M_2$–$M_7$.[c]
5. Perform PCR using the cycling parameters given in *Protocol 1.*

[a] The dilution series is stable at −20°C for at least 4 months.
[b] Mix by pipetting up and down several times.
[c] For very abundant gene transcripts, it may be necessary to use dilution $M_1$. For very rare gene transcripts it may be necessary to use dilution $M_8$.

## 3.3 Quantitative analysis of changes in IL-1β mRNA

To examine the ability of competitive PCR to accurately measure small changes in specific mRNA levels, we performed competitive PCR with cDNA derived from 0.5 μg and 2 μg of total RNA. This would imitate a fourfold induction. In this experiment we prepared a PCR MIMIC for human IL-1β. A kinetic study similar to the one shown in *Figure 2* was performed, which indicated that the IL-1β target and MIMIC amplified with similar efficiencies (data not shown).

To determine the appropriate amount of IL-1β MIMIC to use in the PCR amplification, a preliminary titration experiment was performed in which IL-

1β cDNA derived from 2 μg of total RNA was amplified in the presence of 10-fold serial dilutions of the IL-1β MIMIC. The results of this experiment enabled us to provide a fine-tuned, twofold dilution series as described in *Protocol 4* of the IL-1β MIMIC. Competitive PCR reactions were then performed with cDNA derived from both 0.5 μg and 2 μg of total RNA.

**Protocol 4.** Setting up the experimental 2-fold PCR MIMIC dilution series [a]

*Equipment and reagents*

See *Protocol 3.*

*Method*

1. Label eight 0.5 ml microcentrifuge tubes as $2M_1$–$2M_8$ (the number 2 preceding the M indicates that these tubes belong to the 2-fold dilution series to avoid the two series becoming confused).
2. To tube $2M_1$, add the appropriate amounts of MIMIC and dilution solution to give the desired ratio of MIMIC to target.
3. Make the 2-fold dilution series as follows. [b]
   - $2M_2$ = add 5 μl $2M_1$, mix [c]
   - $2M_3$ = add 5 μl $2M_2$, mix
   - $2M_4$ = add 5 μl $2M_3$, mix
   - $2M_5$ = add 5 μl $2M_4$, mix
   - $2M_6$ = add 5 μl $2M_5$, mix
   - $2M_7$ = add 5 μl $2M_6$, mix
4. Add 2 μl aliquots of the dilution series to the PCR reactions. Use the same PCR reaction components and cycle parameters as given in *Protocol 1.*

[a] The two-fold dilution series is used to measure relatively small changes in mRNA levels (less than 20-fold). The starting point of the dilution series is derived from the results of the 10-fold dilution series. Start with an amount of MIMIC that will give a ratio of MIMIC to target PCR products of about 10:1.

[b] If several experiments are to be performed, increase the volume of the two-fold dilution series. The dilutions are stable at −20°C for at least 4 months.

[c] Mix by pipetting up and down several times

The EtBr-staining pattern obtained is shown in *Figure 4*. The sizes of the IL-1β target and IL-1β MIMIC PCR products were 0.80 and 0.65 kb, respectively. Following gel electrophoresis, the bands corresponding to the target and MIMIC were excised from the gel and the amount of radioactivity quantitated by using a scintillation counter. To correct for 'trailing' of unincorporated [$\alpha$-$^{32}$P]dCTP in the gel, two blocks of agarose, the same size as and

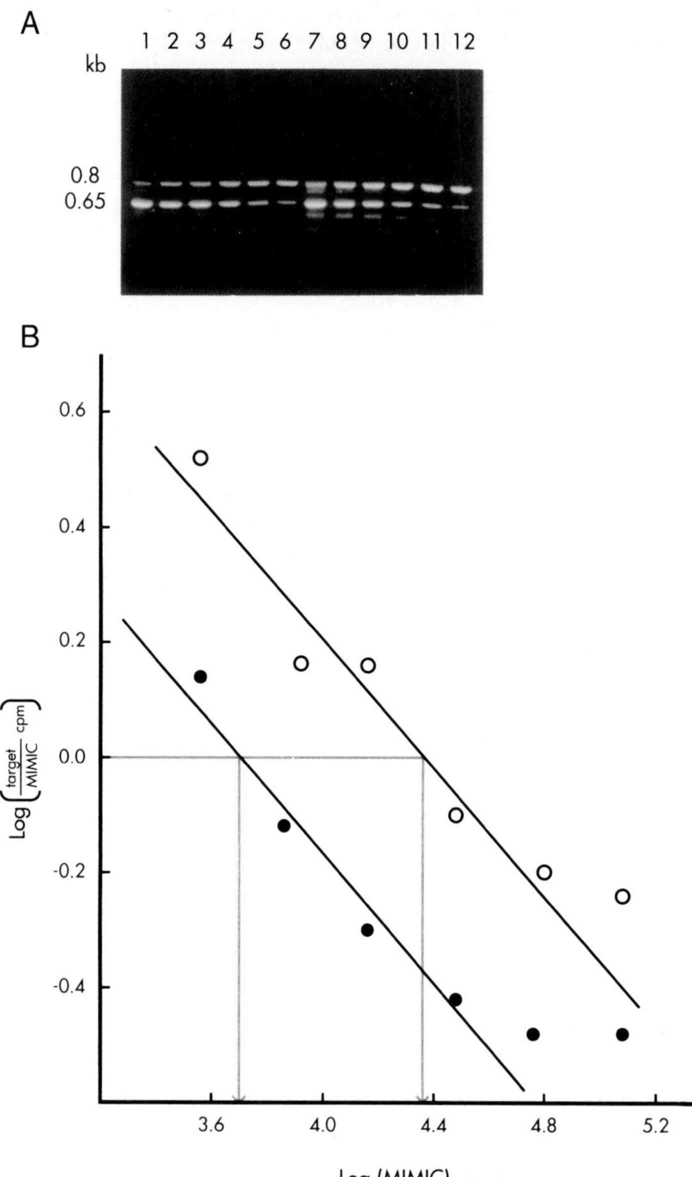


located slightly below two of the bands, were also excised and counted. The average amount of radioactivity in these control agarose blocks was subtracted from the values obtained from the PCR product bands. The log of the ratio of IL-1β target c.p.m. to IL-1β MIMIC c.p.m. was then plotted against the log of the molar amount of IL-1β MIMIC added to the reactions

**Figure 4.** Competitive PCR analysis of relative changes in IL-1β mRNA levels. 0.5 μg and 2 μg of human total RNA were used as templates for cDNA synthesis in four independent experiments. 10% portions of the cDNA were then amplified in the presence of twofold dilutions of the IL-1β MIMIC. [α-$^{32}$P]dCTP was included in the reaction to quantify the PCR products. 10% portions of the PCR products were then resolved on a 1.6% EtBr–agarose gel. (A) Lanes 1–6 and lanes 7–12 contain PCR products derived from 0.5 μg and 2 μg of RNA, respectively. Lanes 1 and 7 contain $20 \times 10^{-2}$ amol IL-1β MIMIC; lanes 1 and 8: $10 \times 10^{-2}$; lanes 3 and 9: $5 \times 10^{-2}$; lanes 4 and 10: $2.5 \times 10^{-2}$; lanes 5 and 11: $1.25 \times 10^{-2}$; lanes 6 and 12: $0.62 \times 10^{-2}$. (B) Quantitative analysis of the competitive PCR experiment shown in (A). The bands corresponding to the IL-1β target and MIMIC were excised from the agarose gel and the amount of radioactivity determined by scintillation counting. The log of the ratio of the corrected radioactivity of the target and MIMIC was then plotted against the log of (molecules) MIMIC added to the PCR reaction. The lines were drawn from a linear regression analysis of the data points excluding the value from the highest amount of MIMIC, which fell off the linear portion of the curve. The closed and open circles denote data derived from 0.5 μg and 2 μg of RNA, respectively.

(*Figure 4*). The amount of IL-1β cDNA was calculated by extrapolating from the intersection of the curve with $\log_1 = 0$, then down to the $x$-axis. The values obtained from the 0.5 μg and 2 μg RNA plots were $4.75 \times 10^3$ and $22.91 \times 10^3$ molecules, respectively, giving a change of approximately fivefold. The predicted value is fourfold.

As the actual number of target cDNA molecules added to the PCR reaction can be calculated, the number of mRNA molecules can in turn be calculated, assuming that the efficiency of cDNA synthesis is 100%. Of course, the actual efficiency must be less than this value. Thus, such a calculation would give the minimum number of mRNA molecules. While the number of mRNA molecules can only be estimated, it is possible to determine accurately the relative changes in mRNA levels.

# 4. Comments on cycle parameters and quantification of PCR yields

## 4.1 Number of amplification cycles

In competitive PCR, it is not necessary to assay PCR products exclusively during the exponential phase of the amplification. However, too few or too many cycles may make analysis of product yields more difficult. Too few cycles may lead to products which are hard to visualize. Too many cycles result in overloading of the agarose gel, thus obscuring the separation of the PCR MIMIC and target. The number of cycles should be determined using the 10-fold dilution series; a similar number of cycles should be used in the finer-tuned dilution series.

## 4.2 Quantification of PCR product yields

One of the simplest procedures for quantifying PCR product yields is to add a radioactively labelled dNTP to the PCR reaction. Another method is to label the 5′-OH group of one or both of the primers with $^{32}P$, which will be incorporated into the PCR products. After gel analysis, the band may be excised and its radiation signal measured in a scintillation counter. Alternatively, the gel may be dried and an autoradiogram generated, which can be scanned in a densitometer.

Southern blot hybridization with labelled, synthetic DNA probes designed to hybridize to an internal sequence of the target and MIMIC can also be used for quantification. The amount of radioactivity can then be determined either by densitometry of an autoradiogram or by excising and counting the signal from the hybridization membrane. Non-radioactive quantitation methods commonly include the use of biotinylated or digoxigenin-labelled primers in conjunction with the appropriate detection methods including densitometry. In addition to the above methods, several companies now offer gel video systems which can scan and quantitate EtBr-stained gel bands.

## 5. Extension of the MIMIC strategy

The method for generating competitive PCR MIMICs can be extended to enable generation of heterologous RNA MIMICs to explicitly control for the cDNA synthesis step. To generate an RNA MIMIC, an RNA polymerase promoter and poly(A)$^+$ tail can be incorporated into the PCR product. In this case, the composite primers would contain a promoter sequence on one primer and a poly(T) tail on the other. *In vitro* transcription of the PCR product will generate synthetic RNAs that contain the target primer sequences and a poly(A)$^+$ tail. RNA samples can then be titrated with the RNA MIMIC during the reverse transcriptase step. Transcriptional promoters have been successfully incorporated into PCR products via primer sequences (6), and recently competitive RNA fragments have been generated by this method (7).

## References

1. Nedelman, J., Heagerty, P., and Lawrence, C. (1992). *Computer appl. biol. sci.*, **8**, 65.
2. Gilliland, G., Perrin, S., Blanchard, K., and Bunn, H. F. (1990). *Proc. Natl Acad. Sci. USA*, **87**, 2725.
3. Becker-Andre, M. and Hahlbrook, K. (1989). *Nucleic Acids Res.*, **17**, 9437.
4. Uberla, K., Platzer, C., Diamantstein, T., and Blankenstein, T. (1991). *PCR Methods Applic.*, **1**, 136.
5. Siebert, P. D. and Larrick, J. W. (1993). *Biotechniques*, **14**, 244.
6. Horikoshi, T., Danenberg, K. D., Stadlbauer, T. H. W., Voldenandt, M., Shea, L. C. C., Aigner, K., Gustavsson, B., Leichman, L., Frosing, R., Ray, M., Gibson, N. W., Spears, L. P., and Danenberg, P. V. (1992). *Cancer Res.*, **52**, 108.
7. Heuvel, J. P. V., Tyson, F. L., and Bell, D. A. (1993). *Biotechniques*, **14**, 395.

# 9

# *In vitro* expression of proteins from PCR products

SCOTT A. LESLEY

## 1. Introduction

The primary advantage of *in vitro* translation of gene products over the *in vivo* expression of cloned genes is that *in vitro* translation allows the use of a defined template to direct *total* protein synthesis. There is a direct correlation between the sequence of the DNA added as a template for transcription and translation, and the protein produced without any background synthesis of endogenous proteins. This is a particular advantage when one would like to specifically radiolabel a protein of interest. The advent of PCR as a common tool for manipulating DNA sequences presents additional opportunities to accelerate such studies, by providing template for *in vitro* translations without the usual cloning and *in vivo* expression. By utilizing *in vitro* translation, the gene product can be amplified, translated, and assayed in a single day.

This chapter describes various *in vitro* translation systems available and their use in some common applications. It describes concerns unique to PCR-generated templates, such as primer requirements and PCR product purification. It is not the intention of this chapter to detail the amplification of products, but rather to provide the necessary information so that these amplification products can be translated and assayed.

There are several types of *in vitro* translation systems from both prokaryotic and eukaryotic sources. The *Escherichia coli* S30, rabbit reticulocyte, and wheat germ extracts are most commonly used and will be described here. Standard protocols for their preparation have been published (1–4) and commercial sources are also available. Any of the systems can be used with PCR products. One of the primary concerns when choosing a system is to determine whether there are any components which will interfere with the assay of the synthesized protein. For example, if the protein being studied is endogenous in the extract, it may overwhelm and interfere with the detection of the synthesized protein. Eukaryotic systems may require the use of a m7G(5′)ppp(5′) cap on the mRNA to achieve efficient expression (5). If post-translational processing is required, canine microsomal membranes can be added and work most efficiently in rabbit reticulocyte lysates. *E. coli* S30 extracts usually produce higher levels of protein. The system chosen is dependent upon the protein being examined and the assay to be performed.

Typical *in vitro* translation reactions will synthesize up to 0.05–1 μg of protein, and methods for increasing this level have been described (6, 7). Although the amount synthesized in a standard reaction is relatively small in comparison to *in vivo* expression levels, for many studies it is not necessary to produce large amounts of proteins. Sensitive radiolabel and immunological methods often exist for the detection and assay of synthesized proteins. These assays can be performed directly from the reaction mixture, avoiding any purification procedure.

*In vitro* translation reactions have been used for a variety of applications, the most common being the identification of gene products encoded by the template. Substitution of a radiolabelled amino acid in the extract will permit the synthesis of a labelled protein which is often examined to determine if it is of the correct molecular mass. Immunoassays can be used for identification and characterization of the translation product as well. *In vitro* translation systems can be used with PCR products to map epitopes of monoclonal antibodies (2, 8, 9), study protein–protein (2,10), and protein–DNA interactions (11) as well as for the rapid analysis of mutagenesis products (10–12).

## 2. Preparation of PCR products for translation

One of the most common mistakes when using *in vitro* translation systems is the omission of proper transcription and translation initiation and termination sequences. Sequences which should be added to the 5′-end of the upstream primer for use in the various *in vitro* translation systems are shown in *Table 1*. Efficient transcription and translation require that substantial upstream sequences be added to the coding region. The initial amplification need not necessarily use primers containing the full sequence required for transcription and translation; this additional sequence can be added by subsequent amplification of a portion of the product using primers which contain the transcription and translation signals.

### 2.1 Promoters

It is possible to use phage RNA polymerases such as T7, T3, or SP6 for transcription in both prokaryotic and eukaryotic translation systems. In addition to the high level of transcription which they provide, these polymerases are ideal because of their relatively short recognition sequences, which allow their direct incorporation in primers for amplification. In general, any of these phage RNA polymerase sequences can be used upstream of the translation initiation site. However, SP6 transcription may be low in the absence of

**Table 1.** Upstream transcription/translation sequences

All sequences are shown as 5′ → 3′ and correspond to 'sense' strand. Non-transcribed bases are shown in smaller font. ATG translation initiation codons are underlined.

| Extract | Promoter | Sequence |
|---|---|---|
| *E. coli* S30 | T7 [a] | GAATTCTAATACGACTCACTATAGGGTTAACTTTAAGAAG GAGATATACATATG |
| Rabbit reticulocyte | T3 | GGTACCGAATTAACCCTCACTAAAGGGATG |
| Rabbit reticulocyte | T3 [b] | CAGAGATGTTATTAACCCTCACTAAAGAATACAAGCTTGCTT GTTCTTTTTGCAGAAGCTCAGAATAAACGCTCAAC TTTGGCAGATCTACCATG |
| Wheat germ | T7 [c] | CCAAGCTTCTAATACGACTCACTATAGGGTTTTTATTTTTAA TTTTCTTTCAAATACTTCCACCATG |

[a] Sequence contains the 5′ untranslated region from phage T7 gene 10 (2).
[b] Sequence contains the translational enhancer from *Xenopus* globin gene from ref. 15 with T3 promoter sequence from ref. 13.
[c] Sequence contains the 5′ untranslated region from alfalfa mosaic virus (19).

significant sequence 5′ to the promoter (13). It may be preferable to use T7 in many cases since good *in vivo* expression systems are available if the PCR product is later cloned. Interestingly, synthesis of gene products in the absence of added phage polymerase has been observed from template containing a T7 promoter in an *E. coli* S30 system (2). This synthesis is probably due to either the formation of a fortuitous *E. coli* promoter or possibly end to end transcription of the PCR product by endogenous RNA polymerase. Such an effect may be template-dependent, so it is prudent to add the required phage polymerase to any reaction.

## 2.2 Translation initiation signals

Prokaryotic and eukaryotic translation systems have different translation initiation requirements. General schemes for translation initiation in prokaryotes and eukaryotes are shown in *Figure 1*. Prokaryotic translation initiation requires the presence of a Shine–Dalgarno sequence (AGGA) approximately nine bp upstream of the start (ATG) codon for the gene. Sequences surrounding this site, as well as the spacing, are important for gene expression. An example of an efficient ribosome binding site for *E. coli* is shown in *Table 1*. Eukaryotic translation initiation also requires the presence of 5′ untranslated sequences. The sequence GCC(A/G)CC immediately upstream from the start codon gives efficient translation (14). The presence of a translational enhancer (15) or 5′-capped message may also increase translation efficiency.

**A. Generalized scheme of prokaryotic translation initiation**

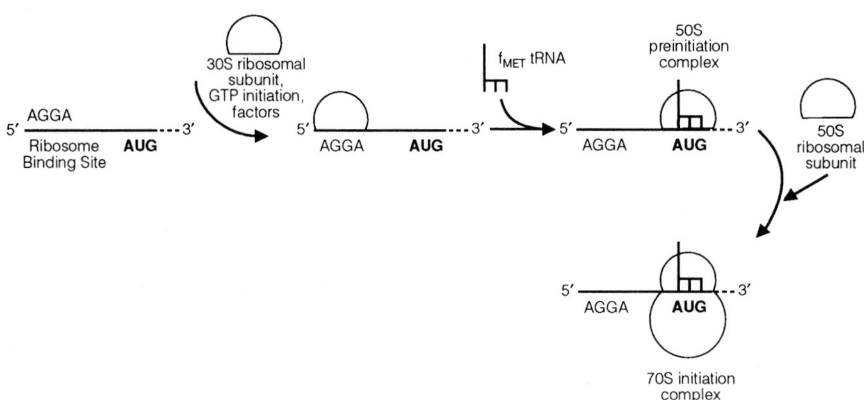


**B. Generalized scheme of eukaryotic translation initiation**

**Standard "Scanning" Initiations**

5′ $m^7$ $G_{ppp}$ NNGCCA/GCC**AUG**(G) . . . $AAA_{(N)}$3′

Kozak consensus sequence

Ternary Complex + 40S ribosomal subunit

+ CAP binding factors

Other factors?

**Internal Ribosome Entry**

5′ G . . . ACC**AUG**G . . . $AAA_{(N)}$ 3′

5′ **AUG**

48S initiation complex

60S ribosomal subunit

**AUG**

80S initiation complex

**Figure 1.** Generalized schemes of translation initiation. (A) Prokaryotic translation initiation involves the interaction of the 30S ribosomal subunit with the Shine–Dalgarno sequence in the 5′ untranslated portion of the mRNA. The $^{f}$Met initiator tRNA binds along with the 50S ribosomal subunit to form an active translation initiation complex. (B) Eukaryotic translation initiation involves the capped mRNA and initiation factors interacting with the 40S ribosomal subunit to form a 48S initiation complex. Addition of the 60S ribosomal subunits produces an active 80S initiation complex. The initiation AUG codon is shown in context with the consensus sequence for efficient translation. An alternate pathway for translation initiation of uncapped mRNA is also shown. An internal ribosome entry site, together with cellular factors binds to form the 48S initiation complex.

## 2.3 Translation termination signals

The 3′-end of the transcript should contain a stop codon (TAA, TAG, or TGA) to permit efficient termination of translation; these are usually incorporated in the downstream primer. For eukaryotic messages, the presence of a 3′-polyadenylation sequence can increase expression levels. Restriction digests of plasmids or PCR products are often useful for rapidly identifying regions of importance. Run-off transcripts of restriction fragments lacking a termination codon can be used, but the resulting translation products may incorporate numerous undesired amino acids at the carboxy terminus due to inefficient release of the message by the translation complex.

## 2.4 Purification of PCR products

PCR products used for translation should be of the highest possible purity. The translation extracts are extremely sensitive to contamination by salts and erroneous template. Final DNA precipitations should use sodium acetate rather than ammonium acetate, as the presence of excess ammonium ions in the DNA template can reduce translation efficiency. If significant amounts of contaminating amplification products are present, the desired DNA should be gel purified to remove them. A procedure for purifying PCR products for translation from a reaction lacking excessive non-specific amplification products is given in *Protocol 1*. It is important to remember that excessive plasmid template for the PCR reaction may contaminate the translation reaction; in this case, gel purification of the amplified product is necessary.

**Protocol 1.** Purification of PCR products for *in vitro* translation

*Reagents*

- 7.5 M ammonium acetate
- Isopropanol
- 0.3 M sodium acetate
- Anhydrous ethanol
- TE buffer: 10 mM Tris–HCl 7.8, 0.1 mM EDTA
- 70% (v/v) ethanol

*Method*

1. Add sufficient 7.5 M ammonium acetate to the completed PCR reactions to give a final concentration of 2 M.
2. Precipitate DNA by the addition of one volume of isopropanol and incubate at 25°C for 10 min.
3. Pellet the DNA by centrifugation in a microcentrifuge (13 000 r.p.m., 10 min) and discard the supernatants.
4. Dissolve the pellets in 100 µl of 0.3 M sodium acetate and reprecipitate by the addition of 2.5 vol. ethanol. Incubate at −20°C for 30 min.
5. Pellet the DNA by centrifugation in a microcentrifuge (13 000 r.p.m., 10 min). After centrifugation, discard the supernatant, and rinse the pellets with 70% ethanol.
6. Dry the pellets *in vacuo* for 10 min in a centrifugal evaporator and dissolve the dried pellets in 10–100 µl of TE buffer.

**Protocol 1.** *Continued*

7. Determine the DNA concentration of the purified PCR product by measurement of $A_{260}$ or estimate the concentration by comparison of its intensity of staining with ethidium bromide relative to known DNA markers following gel electrophoresis. An optimal concentration is 0.2–1.0 mg/ml.

# 3. *In vitro* translation of PCR products

## 3.1 Coupled transcription–translation in *E. coli* S30 extract

Purified PCR products can be used directly in an *E. coli* S30 extract to generate proteins. Transcription and translation are coupled in the system. In fact, protein synthesis is substantially greater when the DNA is transcribed in the reaction compared with addition of the cognate RNA to the reaction. This is probably due to the relative stability of a translated versus an untranslated mRNA in the extract. The S30 extract must be derived from a nuclease deficient *E. coli* strain (*rec*B,C *sbc* or *rec*D) to allow the use of linear DNA fragments without significant degradation. Protocols for extract preparation are given in refs 1 and 2, and commercial sources of extract can also be used (Promega).

A procedure for translation using the S30 system is given in *Protocol 2*. As translation systems are very sensitive to contaminants, it is imperative that only reagents of the highest purity be used. Magnesium ion contamination is especially important; a change of 2 mM in final concentration from the optimum can result in a dramatic decrease in activity. Contaminants from DNA preparations and enzyme additions are the most common cause of problems. Care should be taken to avoid introducing $Ca^{2+}$ into the system, as it may reactivate the nuclease used to prepare the extract and cause DNA template and mRNA degradation. Glycerol concentrations in phage polymerase preparations are often in the range of 50% and can have an inhibitory effect on translations. Additions of phage polymerases over 1 μl per reaction will inhibit activities; smaller amounts of polymerase are usually sufficient. When incorporating a radiolabelled amino acid, use a premix deficient in that amino acid. Although, there is still a contaminating level of endogenous amino acid in the S30, the specific activity of the radiolabel in the protein is substantially increased.

**Protocol 2.** *E. coli* S30 transcription/translation reaction

*Reagents*

- Premix: 87.5 mM Tris-acetate pH 8.0, 476 mM potassium glutamate, 75 mM ammonium acetate, 5 mM DTT, 20 mM magnesium acetate, 1.25 mM each of 20 amino acids, 5 mM ATP, 1.25 mM each of CTP, TTP, GTP, 50 mM phospho*enol*pyruvate (trisodium salt), 2.5 mg/ml *E. coli* tRNA, 87.5 mg/ml polyethylene glycol (8000 MW), 50 μg/ml folinic acid, 2.5 mM cAMP
- Radiolabelled amino acid (for example, translation grade [$^{35}$S]methionine at 10 mCi/ml)
- T7, SP6, or T3 RNA polymerase, approximately 20–40 U/μl
- S30 extract
- DNA template (see *Protocol 1*)

*Method*

1. Set up reactions in 0.5 ml tubes on ice, adding the components in the following order:
   - 20 μl premix[a]
   - 1–14 μl of purified PCR product in TE buffer containing approximately 1 μg of DNA (see *Protocol 1*)
   - 1 μl radiolabelled amino acid (optional)
   - 1 μl (40 units) phage RNA polymerase (dependent upon promoter used), add to the side of the tube
   - water to give a final volume of 35 μl
2. Add 15 μl of S30 and mix gently by flicking the side of the tube.
3. Centrifuge for 10 sec in a microcentrifuge to collect all the liquid at the bottom of the tube.
4. Incubate the reactions at 37 °C for 60 min. Completed reactions may be placed on ice prior to performing assays for synthesized product.

[a] Use premix containing all amino acids for non-isotopic synthesis or premix lacking appropriate amino acid for isotopic synthesis.

## 3.2 *In vitro* transcription of mRNA

The standard rabbit reticulocyte and wheat germ translation extracts require the addition of synthesized mRNA to serve as a translation template. Although coupled transcription/translation now exists for these systems (16, 17), a procedure for generating mRNA is described in *Protocol 3* for use in the standard reactions. The procedure described uses a Hepes-based buffer system rather than a Tris-based buffer system as this produces RNA that performs better in the translation systems. The PCR product used as a transcription template should contain a phage promoter at the 5′-end of the DNA fragment.

Transcription is more efficient if 6–9 bases are included upstream from the promoter sequence. If a run-off transcript is to be produced from a restriction digest of the PCR product, it is best to avoid using enzymes which leave a 3′-overhang. These overhangs can cause extraneous transcripts which can contain sequences complementary to the expected transcript (18). If such enzymes must be used, the 3′-overhang should be removed by treatment with Klenow fragment (18). PCR products should be prepared for transcription as described in *Protocol 1* using RNase-free components. The *in vitro* transcription procedure (*Protocol 3*) should yield approximately 200 μg of RNA for a 2000 bp transcript.

## **Protocol 3.** Preparation of RNA template by *in vitro* transcription

### *Reagents*

- 5 × transcription buffer: 400 mM Hepes–KOH pH 7.5, 10 mM-spermidine, 200 mM DTT, and 60 mM $MgCl_2$ (for T7 and T3 RNA polymerases) or 80 mM $MgCl_2$ (for SP6 RNA polymerase)
- rNTP solution: 25 mM each of rATP, rCTP, rGTP, rTTP
- RNasin® ribonuclease inhibitor (20–40 units/μl)
- Yeast inorganic pyrophosphatase (Sigma)
- T7, or SP6, or T3 RNA polymerase (approximately 80 U/μl)
- DNA template (see *Protocol 1*)

### *Method*

**NB**. Prepare reactions at room temperature, as the presence of spermidine can precipitate DNA at colder temperatures.

1. Set up reactions containing the following:
   - 20 μl of 5 × transcription buffer
   - 12 μl rNTPs[a]
   - 63 μl containing 3–10 μg of PCR product in nuclease-free water (see *Protocol 1*)
   - 5 μl containing 75 units RNasin® ribonuclease inhibitor, 0.5 units yeast inorganic pyrophosphatase, 175 units RNA polymerase
2. Gently mix the reaction by pipetting and incubate at 37°C for 2–6 h.
3. The RNA transcript can be used directly in rabbit reticulocyte translation reactions using 1–10 μl of a 1:10 dilution of the reaction.[b] Further purification of the RNA transcript by phenol extraction and ethanol precipitation can increase the yield protein synthesis.

[a] Some mRNAs show increased translation efficiencies when capped with m7G(5′)ppp(5′)G. To incorporate this analogue, reduce the final rGTP concentration to 0.6 mM and add m7G(5′)ppp(5′)G to a final concentration of 3 mM. The transcription yield in this case will probably be substantially lower.

[b] The PCR product can be removed after transcription by treatment with RQ1 RNase-free DNase, although this is not typically required for *in vitro* translation reactions.

### 3.3 Translation in rabbit reticulocyte lysate

A variety of rabbit reticulocyte lysates are available, including nuclease-treated, non-nuclease-treated, salts and DTT deficient, and coupled transcription/translation systems. The method given in *Protocol 4* is for nuclease-treated extracts. Care should be taken to avoid introducing $Ca^{2+}$ into the system as it may reactivate the nuclease used to prepare the extract and so cause mRNA degradation. The mRNA for translation should be prepared as described in *Protocol 3*. RNA transcripts should contain the proper translation initiation sequences as described in Section 2. Occasionally, reactions may need to be optimized by addition of KCl (40–70 mM final) to provide optimal expression of translation products. Reaction components should be stored at −70 °C and kept at 4 °C while preparing reactions.

**Protocol 4.** Translation in rabbit reticulocyte lysate

*Reagents*

- Rabbit reticulocyte lysate (nuclease-treated)
- RNasin® ribonuclease inhibitor, 20–40 U/μl
- Amino acid solution containing 1 mM each amino acid in water (lacking the amino acid to be incorporated as a radiolabel)
- Radiolabelled amino acid (for example, translation grade [$^{35}$S]methionine at 10 mCi/ml)
- mRNA template (see *Protocol 3*)

*Method*

1. Disrupt any secondary structure in the RNA by heating at 67 °C for 10 min and then immediately cooling on ice.
2. Set up reactions containing the following:
   - 35 μl rabbit reticulocyte lysate
   - 7 μl nuclease-free water
   - 1 μl (40 U) RNasin® ribonuclease inhibitor
   - 1 μl of a solution of 1 mM each amino acid
   - 4 μl radiolabelled amino acid
   - 2 μl (0.2–1.0 μg) RNA template
3. Incubate the translation reactions at 30 °C for 60 min. Completed reactions may be placed on ice prior to performing assays for synthesized product.

### 3.4 Translation in wheat germ lysate

Wheat germ extracts are available as a standard translation and a coupled transcription/translation system. Wheat germ extracts are often used, as described in *Protocol 5*, when reticulocyte extracts contain an endogenous activity which interferes with the assay of the product. Template requirements are similar to those for reticulocyte extracts.

**Protocol 5.** Translation in wheat germ extracts

*Reagents*

- Wheat germ extract
- RNasin® ribonuclease inhibitor (20–40 U/μl)
- Amino acid solution containing 1 mM each amino acid in water (lacking the amino acid to be incorporated as a radiolabel)
- Radiolabelled amino acid (for example, translation grade [$^{35}$S]methionine at 10 mCi/ml)
- 1 M potassium acetate
- mRNA template (see *Protocol 3*)

*Method*

1. Disrupt any secondary structure in the RNA by heating at 67°C for 10 min and immediately cooling on ice.
2. Set up reactions containing the following:
   - 25 μl wheat germ extract
   - 1 μl (40 U) RNasin® ribonuclease inhibitor
   - 4 μl of a solution of 1 mM each amino acid
   - 2.5 μl radiolabelled amino acid
   - 2 μl RNA (0.2–1.0 μg) template
   - 0–7 μl 1 M potassium acetate (template dependent)
   - nuclease-free water to give a final volume of 50 μl
3. Incubate the translation reactions at 25°C for 60 min. Completed reactions may be placed on ice prior to performing assays for synthesized product.

## 3.5 Assay of products

The most common means of monitoring protein synthesis is by the incorporation of radiolabelled amino acids. Typically, [$^{35}$S]methionine or [$^{14}$C]leucine is added to the reaction. Since only newly synthesized proteins incorporate the radiolabel, this allows specific detection of the template-encoded polypeptides. Both autoradiography of SDS–PAGE and immunoprecipitation can be used to detect and identify the translation products. Immunoblotting of SDS gels, another common assay of translation products, although less sensitive than isotope incorporation, usually permits detection of even poorly synthesized proteins, using high affinity antibodies. Furthermore, *in vitro* translation systems usually generate proteins which are correctly folded and retain their enzymatic activity. This enzymatic activity can be directly assayed as long as the extract proteins and buffers do not interfere. It is also possible to purify synthesized proteins from translation extracts. However, the relatively small amounts of protein synthesized and the inevitable losses during protein purification make this impracticable in most cases, unless a high yielding affinity purification system is available.

Protocols for SDS–PAGE analysis followed by autoradiography and for immunoprecipitation are given in *Protocols 6* and *7*. For immunoblot analysis, prepare samples as in *Protocol 6* then, after step 5, follow standard immunoblotting procedures (20).

### **Protocol 6.** Preparation of translation reaction for SDS–PAGE analysis

*Reagents*

- Acetone
- 4 × SDS sample buffer (50 mM Tris–HCl pH 6.8, 100 mM DTT, 2% SDS, 10% glycerol, 0.1% bromophenol blue)
- Gel fixing solution (20% methanol, 10% acetic acid)
- Isotopic enhancing fluor (e.g. Amersham)

*Methods*

1. To a 5 μl aliquot of the translation reaction, add 20 μl of acetone, and mix briefly.[a] Place on ice for 15 min.
2. Pellet the proteins by centrifugation in a microcentrifuge (13 000 r.p.m., 10 min).
3. Discard the supernatant and dry the pellet under vacuum in a centrifugal evaporator.
4. Dissolve the pellet in 20 μl of SDS sample buffer and heat at 100°C for 2 min.
5. Load samples on to an SDS polyacrylamide gel and electrophorese.[b]
6. After electrophoresis by SDS–PAGE, fix the acrylamide gel in a solution of 20% methanol, 10% acetic acid for 30 min.
7. After fixing, soak the gel in an isotopic enhancing fluor for 30 min. Rinse the gel briefly with water.
8. Dry the gel under vacuum and visualize the translation products by autoradiography.

[a] This precipitates the proteins and removes polyethylene glycol in the reaction which can interfere with SDS–PAGE analysis.

[b] Methods for SDS–PAGE analysis and post-electrophoresis treatment of gels are provided in another volume in this series (21).

**Protocol 7.** Immunoprecipitation of translation products

*Reagents*

- Acetone
- Binding buffer (20 mM Tris–HCl pH 7.9, 200 mM NaCl, 1 mM EDTA, 0.1 mM DTT)
- Polyclonal antisera
- Formalinized *Staphylococcus aureus* cells (Sigma)

*Method*

1. Remove 10 μl of translation reactions containing the radiolabelled protein. Add 40 μl of acetone to this portion to precipitate the protein.
2. Pellet the protein precipitate from step 1 by centrifugation in a microcentrifuge (13 000 r.p.m., 10 min). Discard the supernatant. The pellet contains the total proteins to be used as a reference for determining the efficiency of immunoprecipitation.
3. Remove a second 10 μl portion of the translation reactions. Add 40 μl of binding buffer.
4. Add 1–10 μl of rabbit polyclonal antisera and incubate at room temperature for 1 h.[a]
5. Pellets, the formalinized *S. aureus* cells from their storage buffer by centrifugation in a microcentrifuge (13 000 r.p.m., 2 min) and resuspend the pellets with an equal volume of binding buffer.
6. After antibody binding (step 4), add 10 μl of prepared *S. aureus* cells (step 5) to each 50 μl antibody binding reaction.[b] Incubate at 4°C for 30 min.[c]
7. Isolate the immunoprecipitates by centrifugation and wash the pellets twice with 400 μl of binding buffer. Analyse the immunoprecipitates by SDS–PAGE and autoradiography (see *Protocol 6*).

[a] Be sure to include controls (preimmune and no added sera, respectively) for reference.

[b] It may be necessary to titrate the amount of *S. aureus* added to provide the optimal precipitation conditions.

[c] During this incubation period, antibody complexes are bound to the protein A on the surface of the cells.

# 4. Applications of *in vitro* translation products

There are a variety of applications for *in vitro* translation of PCR products. The major advantage is that the DNA does not have to be cloned and expressed *in vitro* to examine the coding regions which are amplified. This can be of particular advantage when amplifying a gene from a cDNA library. The amplification product can be examined in a rapid assay to determine

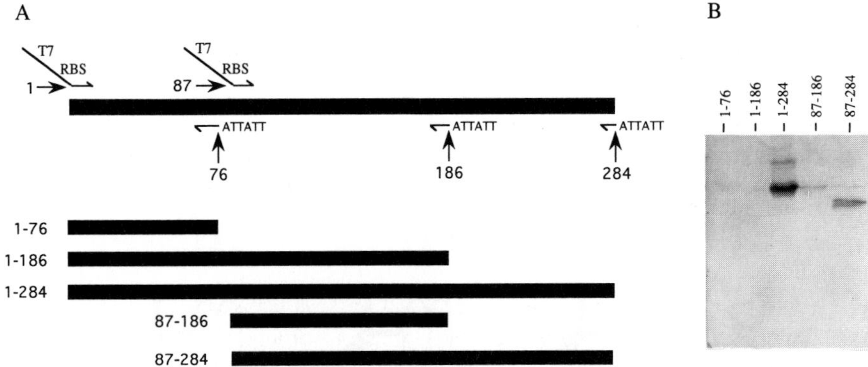


**Figure 2.** Epitope mapping of a monoclonal antibody to the *E. coli* transcription factor sigma-32 using *in vitro* translation products. (A) Location of primers and protein fragments produced. Arrows indicate the amino acid position for translation initiation in upstream (sense) primers. 5′ untranslated regions of upstream primers contain a phage T7 promoter and efficient ribosome binding site. Downstream primers contain sequence to encode two stop (TAA) codons in RNA transcripts from the PCR products. Arrows indicate the position of the last encoded amino acid. Full-length protein is 284 amino acids. Various primer combinations will produce products encoding amino acids 1–76, 1–186, 1–284, 87–186, and 87–284. (B) Immunoblot analysis of S30 reactions using PCR products. The monoclonal antibody reacts with proteins containing amino acids 1–284 and 87–284 but not with proteins containing amino acids 1–76, 1–186, or 87–186. This indicates that the epitope for the antibody lies near the C-terminus of the protein between amino acids 186 and 284. The S30 extract contains a small amount of the endogenous full-length protein. The presence of the truncated translation products can be confirmed by autoradiography of the translation reactions (data not shown).

whether a full-length protein has been produced. Since the PCR product has not been isolated by clonal selection, the translation template, and the protein produced, represents the PCR product population.

The ease with which DNA sequences are manipulated by PCR can be directly associated with protein manipulations via *in vitro* translations. Structure/function assays can be accelerated by the rapid assay of PCR-generated constructs. This may involve gross manipulations of the gene such as insertions, deletions, and chimera formation or the generation of point mutations.

A simple and useful example of the application of this technique is in the mapping of epitopes for monoclonal antibodies. A diagram illustrating the concept is shown in *Figure 2*. Both N-terminal and C-terminal deletions can rapidly and precisely be made by judicious choice of primers. Epitope mapping is normally performed by determining antibody reactivity with chemical cleavage products of the protein of interest. This limits the researcher to the sites and efficiency of the chemical cleavage reagents and requires that the protein be at least partially purified. Chemical and ezymatic protein cleavages

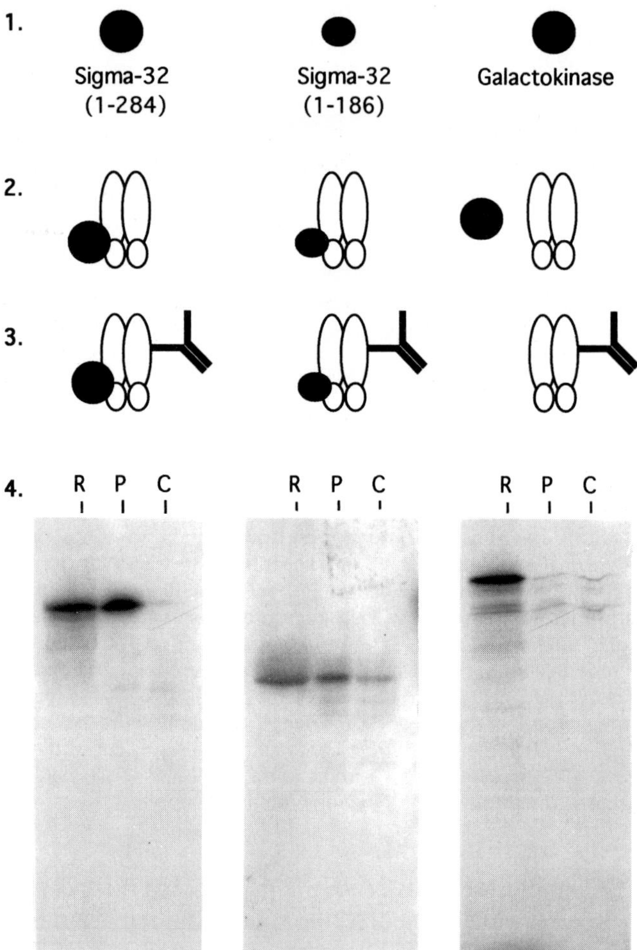


**Figure 3.** Immunoprecipitation of a multi-subunit complex. 1. Full-length (amino acids 1–284) or truncated (amino acids 1–186) transcription factor or a control (galactokinase) protein were synthesized in S30 containing [$^{35}$S]Met. 2. RNA polymerase was added to the reaction and the translation products allowed to bind. 3. Antibody to a polymerase subunit was used to immunoprecipitate the complex. 4. SDS–PAGE was used to examine the immunoprecipitates for the presence of the synthesized protein. (R) reaction containing labelled translation product, (P) immunoprecipitate using polymerase antibody, (C) control precipitation without polymerase antibody. Results indicate that both full-length and truncated forms of the sigma-32 protein bind polymerase.

are often incomplete, leaving ambiguity as to the identity of the reacting peptide. Producing protein fragments directly from the PCR product avoids this uncertainty.

Functional domain mapping can be carried out using the same procedure.

If functional activity can be measured directly in the extract, protein fragments can easily be made and assayed for activity. An example is shown in *Figure 3*. Enzyme subunit interactions can be probed using immunoprecipitation to isolate the complex. Since only the newly synthesized protein is radiolabelled, one can simply test whether the radioactivity co-precipitates with the complex. Similar types of studies can be performed with DNA-binding proteins by looking for the ability to specifically bind a DNA sequence.

Perhaps the greatest potential utility of these systems is in the rapid analysis of point mutations. More often than not, amino acid substitutions do not give the predicted or desired activities. *In vitro* translation assays provide a rapid means of screening mutants to identify those sites and substitutions which have important effects on protein activity.

## References

1. Chen, H. and Zubay, G. (1983). In *Methods in Enzymology* (ed. R. Wu, L. Grossman, and K. Moldave), Vol. 101, p. 674. Academic Press, London.
2. Lesley, S. A., Brow, M. A. D., and Burgess, R. R. (1991). *J. Biol. Chem.*, **266**, 2632.
3. Pelham, H. and Jackson, R. (1976). *Eur. J. Biochem.*, **67**, 247.
4. Anderson, C. *et al.* (1983). In *Methods in enzymology* (ed. R. Wu, L. Grossman, and K. Moldave), Vol. 101, p. 635. Academic Press, London.
5. Krieg, P. and Melton, D. (1984). *Nucleic Acids Res.*, **12**, 7057.
6. Spirin, A. S., Baranov, V. I., Ryabova, L. A., Ovodov, S. Y., and Alakhov, Y. B. (1988). *Science*, **242**, 1162.
7. Resto, E., Iida, A., Van Cleve, M. D., and Hecht, S. M. (1992). *Nucleic Acids Res.*, **20**, 5979.
8. Mackow, E. R., Yamanaka, M. Y., Dang, M. N., and Greenberg, H. B. (1990). *Proc. Natl Acad. Sci. USA*, **87**, 518.
9. Gulick, A. M., Lathrop Goihl, A., and Fahl, W. E. (1992). *J. Biol. Chem.*, **267**, 18946.
10. Tsirka, S. and Coffino, P. (1992). *J. Biol. Chem.*, **267**, 23057.
11. Wilson, T. E., Day, M. L., Pexton, T., Padgett, K. A., Johnston, M., and Milbrandt, J. (1992). *J. Biol. Chem.*, **267**, 3718.
12. Hammerle, T., Molla, A., and Wimmer, E. (1992). *J. Virol.*, **66**, 6028.
13. Logel, J., Dill, D., and Leonard, S. (1992). *Biotechniques*, **13**, 604.
14. Kozak, M. (1990). *Nucleic Acids Res.*, **18**, 2828.
15. Falcone, D. and Andrews, D. W. (1991). *Mol. Cell. Biol.*, **11**, 2656.
16. Beckler, G. S., Thompson, D., and Van Oosbree, T. V. (1992). In *Methods in molecular biology, in vitro transcription and translation protocols* (ed. M. J. Tymms). Humana, New Jersey.
17. Craig, D., Howell, M. T., Gibbs, C. L., Hunt, T., and Jackson, R. J. (1992). *Nucleic Acids Res.*, **20**, 4987.
18. Schenborn, E. T. and Mierendorf, R. C. (1985). *Nucleic Acids Res.*, **13**, 6223.

19. Kain, K. C., Orlandi, P. A., and Lanar, D. E. (1991). *Biotechniques*, **10**, 366.
20. Towbin, H., Staehelin, T., and Gordon, J. (1979). *Proc. Natl Acad. Sci. USA.*, **76**, 4350.
21. Hames, B. D. and Rickwood, D. (eds.) (1990). *Gel electrophoresis: a practical approach*. IRL Press, Oxford.

# 10

# PCR-based approaches to human genome mapping

J. F. J. MORRISON and A. F. MARKHAM

## 1. Introduction

Genetic maps are constructed by determining how frequently two markers are inherited together. These maps allow the ordering of DNA markers and give estimates as to the distances between these markers. Genetic maps differ from physical maps in that they can be used to site loci which are associated with disease. Physical maps, in contrast, specify nucleotide distances between markers along a chromosome. PCR is used to generate both types of map through the generation of sequence-tagged sites (STS) (1). Sequence-tagged sites are used to identify DNA segments in both genetic and physical maps, and usually comprise 200–500 bp of DNA sequence unique to that segment, sufficient to allow a PCR reaction to be performed. Thus these STSs can conveniently be identified by specific PCR primers which can be used to screen genomic DNA contained in yeast artificial chromosomes (YACs) or cosmids.

Generation of YAC or cosmid libraries is the first stage in physical mapping. After isolation of a clone, the heterologous DNA is sized on pulsed field gel electrophoresis (PFGE) and a restriction map produced, usually with a variety of rare cutter restriction enzymes. Sequence information derived from the YAC or cosmid ends can be produced conveniently using vectorette PCR (2). This generates a further set of PCR primers which can be used to rescreen the YAC library for more overlapping clones to generate a contig.

Once a YAC or cosmid contig is assembled, PCR methods based on repetitive sequences are frequently used to confirm quickly that putative overlapping clones actually contain the same sequences. Expressed sequences from these DNA segments can be isolated typically using such techniques as exon trapping, direct screening of cDNA libraries, or cDNA enrichment (if the likely source of the cDNA is known). In vertebrates, the 5′-end of genes often contain a high density of hypomethylated CpG residues or CpG islands (3). These sites are identified by the clustering of recognition sites for rare-cutting restriction enzymes and are confirmed by testing with methylation-sensitive and insensitive isoschizomers. Alternatively, coding sequences can be identified by interspecies cross-hybridization, as many sequences which are conserved between species, represent genes. This method is called exon connection or zoo blotting (4). The final objective is to generate sequence information by standard methodologies or by the use of automated sequencers.

The circle from physical mapping back to genetic mapping is squared by using YAC or cosmid contigs to generate highly polymorphic CA microsatellite repeat markers by subcloning fragments, screening with synthetic (CA) or (GT) oligonucleotides and sequencing regions flanking the CA repeats so that PCR primers can be designed (5). In this way the precision of the genetic linkage between a locus and an inherited disease gene is increased as a guide for additional physical mapping. Highly informative microsatellites, which are tightly linked to an inherited disease gene, are frequently invaluable clinically, as they allow accurate diagnosis in affected families without the need to characterize the specific mutation involved or without the absolute necessity to identify the gene responsible.

## 2. Chromosome localization

Once an STS is isolated, PCR can be used for chromosomal localization. A panel of rodent–human hybrid cell lines, each containing a whole or a portion of a human chromosome or chromosomes, is used as a template for the reaction using standard methodologies. Such chromosome mapping panels are available from the Coriell Institute NIGMS Human Genetic Mutant Cell Repository (Camden, New Jersey, USA) or from BIOS Laboratories. It is important to include total human DNA as a positive control, and the appropriate rodent DNA as a negative control, against cross-species gene homology.

Chromosome sublocalization of a new STS is now usually required for database registration (Genome Database, Johns Hopkins University, Baltimore, USA) and as a prerequisite for publication. Although this may occasionally be possible using hybrid cell lines containing only fragments of human chromosomes, the most convenient and general method involves fluorescent *in situ* hybridization (FISH) to metaphase chromosome spreads. Signal intensity considerations dictate that, currently, a positive signal is not reliably achieved with human inserts less than about 10 kb in size. In practice, cosmid or YAC clones are usually derived from a given STS, nick-translated and used as probes for FISH chromosome localization. Problems of cross-hybridization by the common repetitive elements, which will inevitably be found in such large probes, are overcome by pre-annealing with excess human (cot-1) DNA (commercially available, e.g. from Gibco/BRL). Usually, probes are labelled with biotin or digoxigenin and the non-isotopic label is detected with fluorescently labelled antibodies. All these reagents are readily available from a number of suppliers (e.g. Vector Laboratories Inc.).

# 3. Preliminary YAC analysis

## 3.1 Screening YAC libraries

YACs are vectors which can hold up to at least one megabase of cloned DNA. Most libraries have a minimum of 30 000 clones which have been arrayed in microtitre plates. With an average YAC insert size of say 300–400 kb, this means that the library represents three genome equivalents and gives a 95% statistical probability that a given sequence will be found on PCR screening for a single STS. Efficient methods for screening YAC libraries have been developed using PCR on DNA prepared from ordered arrays of pooled YAC clones (6–8). By judicious pooling strategies, PCR screening of whole libraries can rapidly be accomplished in as few as 50 individual PCR reactions. One human YAC library is commercially available from Clontech Laboratories. This library is configured in three formats (*Figure 1*). Format C consists of 96-well plates (344 plates in total) with individual clones in each well. Format B consists of four 96-well plates with each well containing a pool of the 96 clones from one format C plate. Format A consists of four tubes, each containing a pool of all the wells from one format B plate.

PCR screening starts with the four format A tubes. A positive result in one tube indicates the format B plate on which the clone is located. Screening of pools of rows and columns from the format B plate quickly identifies which pool contains the clone. Subsequently, the format C plate containing individual colonies is screened as for a format B plate. The positive control is human genomic DNA and the negative controls yeast DNA or water. Recently the human YAC libraries of the Centre d'Études du Polymorphisme Humain (CEPH) have become commercially available from Research Genetics Inc. in a pooled format for screening. This is valuable because these libraries contain a proportion of so-called 'mega-YACs' with very large human inserts which allow more rapid genome walking.

The YAC library described in ref. 8 may be particularly useful in that it contains very few non-contiguous or rearranged human sequences. It can be screened by contacting the MRC Human Genome Mapping Project Resource Centre (Clinical Research Centre, Harrow, UK). This library is screened by PCR initially in 40 pools each containing 864 individual YAC clones (i.e. the content of 9 × 96-well microtitre plates). Subsequent to obtaining a positive signal with one of the 40 primary master pools, nine separate PCR reactions on pools constituted from the individual microtitre plates which made up the primary pools will identify that microtitre plate holding the clone of interest. Rather than then performing 96 individual PCR reactions, testing '8 pooled rows and 12 pooled columns' from a microtitre plate allows an individual well to be identified by a further 20 PCR reactions (69 reactions in total).

Format A

4 x 1ml tubes
containing pooled material
from 96 wells
of a format B plate

A1 A2 A3 A4

4
PCR reactions

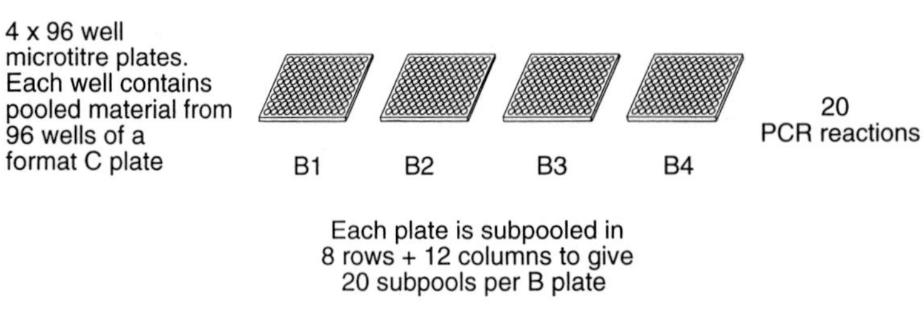


Format C

344 x 96 well
microtitre plates
containing
33024 individual
clones

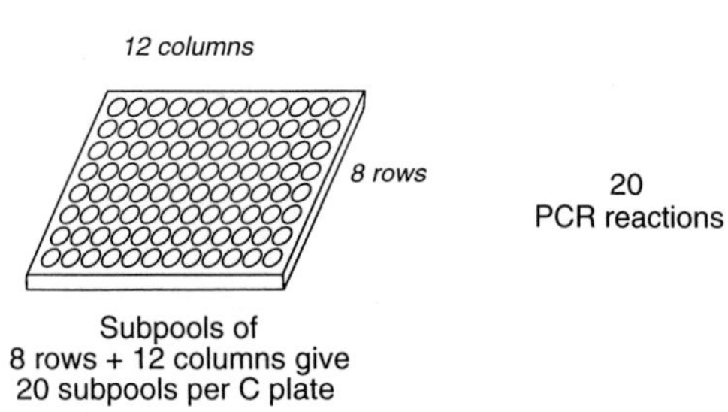


**Figure 1.** Schematic outline of a strategy used to screen a commercially-available YAC library. A single positive YAC colony is identified in 44 PCR reactions. See Section 3 of the text for a description of this strategy.

## 3.2 Isolation of yeast clones

Once an individual YAC clone of interest has been identified, preparation of high molecular weight DNA is performed in agarose plugs. This is essential to avoid shearing of high molecular weight DNA and has the advantage that restriction and subsequent ligation steps can be carried out without ethanol precipitation or column purification. The procedure is described in *Protocol*

*1.* General protocols for the preparation of YAC agarose plugs have also been published elsewhere (9, 10).

**Protocol 1.** DNA preparation from YAC clones[a]

*Reagents*

- SD medium: 0.7% yeast nitrogen base without amino acids, 2% glucose, 5.5 mg/100 ml adenine, and tyrosine, pH 7.0. Add 7 ml of 20% casamino acids (Difco filter sterilized) per 100 ml for double selection (−*ura*, −*trp*)
- SD agar: 2.2% Bactoagar in SD medium. Prepare SD medium lacking casamino acids, add the Bactoagar and autoclave. After cooling, add 7 ml of 20% casamino acids (filter sterilized) per 100 ml molten agar for double selection (−*ura*, −*trp*)
- YRB (yeast resuspension buffer): 1.2 M sorbitol, 20 mM EDTA, 10 mM Tris–HCl pH 7.5, 14 mM β-mercaptoethanol, 20 U/ml lyticase (Sigma)
- YLB (yeast lysis buffer): 0.1 M EDTA, 10 mM Tris–HCl pH 8.0, 1% lithium dodecyl sulphate
- TE buffer: 10 mM Tris–HCl pH 7.6, 1 mM EDTA
- LMT (low melting temperature) agarose: 1% LMT agarose in YRB

*Method*

1. Grow individual YAC colonies with selection in SD medium.
2. Streak out each YAC clone to be analysed on an SD agar plate and grow for 48 h at 30°C (11).
3. For each clone, inoculate a 10 ml SD liquid culture with one of the yeast colonies and shake (200 r.p.m.) overnight at 30°C.
4. Harvest the yeast cells by centrifugation (at room temperature at 1500 r.p.m. (2000 g) for 5 min).
5. Wash the cells in 50 mM EDTA pH 8.0 by centrifugation and resuspend to 3–5 × $10^8$ cells/ml in YRB.
6. Incubate at 37°C for 1 h, or until about 80% of cells are spheroplasts. Monitor by mixing a small aliquot of cells with 10% SDS on a microscope slide under a phase contrast microscope. Spheroplasts appear as dark ghosts compared to shiny, intact yeast cells.
7. Gently mix the cells with an equal volume of 1% LMT agarose in YRB at 37°C. Transfer to a plug mould set (BioRad).
8. When set, transfer the plugs to 2–3 volumes of YLB. Gently shake at room temperature for 1–2 h.
9. Incubate the plugs overnight with 10 vol. of fresh YLB at 40–50°C.
10. Rinse the plugs in TE and store at room temperature in YLB, or equilibrate the plugs in 0.1 M EDTA pH 9.0 and store at 4°C. This DNA is suitable for restriction mapping by PFGE (see Section 3.3, *Protocol 2*) or restriction digestion and ligation with vectorette sequences (see Section 4.2, *Protocol 3*).

[a] This protocol was designed for pYAC4 clones in *Sarccharomyces cerevisiae* AB1380 but will work for most YAC clones (11).

## 3.3 Pulse field gel electrophoresis (PFGE) of YACs

A number of protocols exist for PFGE. One which the authors have used successfully for the restriction mapping of YACs is given in *Protocol 2*.

### **Protocol 2.** Pulsed field gel electrophoresis

*Equipment and reagents*

- 1.5% Seakem agarose (Flowgen) in 0.5 × TAE buffer
- Agarase (New England Biolabs)
- Agarase/polyamine buffer (New England Biolabs)
- Dextran T40 (Sigma)
- Ammonium acetate (Sigma)
- TE buffer (see *Protocol 1*)
- 50 × TAE stock buffer. Prepare this by mixing 242 g Tris base, 57.1 ml glacial acetic acid, 100 ml 0.5 M EDTA pH 8.0 per litre. The final pH should be 8.3–8.5
- Millipore chlorine tablets
- DNA size markers (e.g. phage × DNA ladder agarose plugs; New England Biolabs)
- 1% LMP (low melting point) agarose in 0.5 × TAE
- YAC DNA plugs (prepare as *Protocol 1*)
- Ethidium bromide
- Waltzer PFGE apparatus (e.g. from Hoefer)
- Plug mould; a perspex mould which produces agarose plugs of 6 × 2 × 10 mm in size

*Method*

1. Boil the plug mould in 150 ml 0.25 M HCl, wash it twice with distilled water and finally boil in distilled water. Dry the mould and leave it on ice until needed.
2. Sterilize the PFGE apparatus by adding a Millipore chlorine tablet dissolved in 4 litres distilled water. Allow to circulate for 2 h. Wash out the apparatus twice with 4 litres of double-distilled water each time. Add 5.5 litres of 0.5 × TAE and cool to 18°C for 30 min.
3. Prepare a 1.5% agarose gel in 0.5 × TAE buffer and allow the gel to set. Add 20 ml of sterile TE to the surface of the gel and remove the comb.
4. Pre-equilibrate the YAC DNA plugs (prepared as in *Protocol 1*) in 0.5 × TAE. One variation is to add the dye, orange G, to the 0.5 × TAE buffer at 0.5 mg/ml to colour the plugs and hence facilitate their loading into the wells of the agarose gel (see step 5).
5. Place the equilibrated YAC plugs into the appropriate wells of the agarose gel. Ensure that the plugs are flush to the front of the wells. Blot the gel dry using tissues. Then seal the wells behind the plugs with hand-hot 1% LMP agarose in 0.5 × TAE.
6. The molecular size markers used on the gel should be chosen carefully. For YAC work, a phage λ DNA ladder (supplied commercially as an agarose plug, e.g. from New England Biolabs) can be used. Load the λ DNA plug into the well and seal it with 1% LMP agarose in 0.5 × TAE as in step 5. Use approx. 10 μg/ml λ DNA ladder.

7. Electrophorese the gel at 150 V; it should draw a current of approximately 280 mA. If the current is markedly different from this, prepare the running buffer again from the 50 × TAE stock. If it remains different the 50 × TAE should be discarded.
8. Electrophorese the gel for 30 h.
9. Stain the gel with 5.0 μg/ml ethidium bromide in water for 1 h. Destain in water for 2 h or overnight.
10. Visualize the DNA with UV and cut out the required band(s) from the gel. Wash the gel slices in polyamine/agarase buffer (New England Biolabs) four times for 30 min each time.
11. For each gel slice, add 1 U agarase (New England Biolabs) per 200 μl agarose and incubate at 42°C for 1–2 h. When the reaction is complete, the agarose will remain in solution at 4°C.
12. Add 5 μl Dextran T40 as a carrier. To precipitate the DNA, add ammonium acetate (to 2.5 M) and 2.5 vol. ethanol. Incubate at −70°C for 30 min. Spin for 15 min in a microcentrifuge and wash the pellet three times with 70% ethanol.
13. Resuspend the pellet in 15 μl TE buffer.

# 4. Vectorette libraries

## 4.1 Strategies of vectorette PCR

Traditional genome analysis requires the generation of DNA libraries in plasmid, phage, cosmid, or YAC vectors. These are then subcloned and specific DNA fragments isolated and arranged to form overlapping fragments. This process is time-consuming, and, for work on the human genome, was frequently confounded by the presence in human DNA of regions unclonable in phage or cosmids. Rearrangement of human DNA in YACs is also problematic. Vectorette PCR is a technique which enables many of these problems to be overcome.

Vectorette is a method for generating sequence information in a progressive manner and a known direction along any target DNA fragment or locus. Unlike previously described walking strategies, this approach requires no cloning or *in vivo* propagation of the target DNA. Hence problems with 'unclonable' sequences are avoided. A vectorette is an oligonucleotide cassette having a 'sticky end' for ligation and a constant mismatched portion in the middle. The mismatched portion of the bottom strand contains the same sequence as both the vectorette primer and the sequencing and nested primers.

Genomic DNA is restricted and site-specific vectorette units are ligated to the exposed ends. These are used as a substrate in PCR with primers

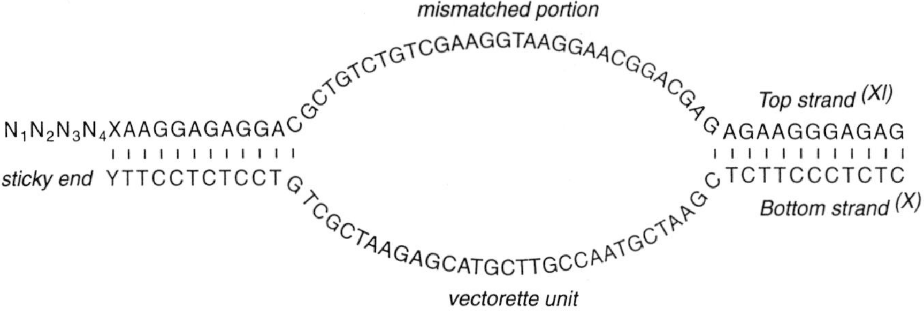


3′ - TCGCTAAGAGCATGCTTGCCAATGCTAAGC
*universal vectorette primer (VII)*

3′ - CCTCTCCTGTCGCTAAGAGCATGCTTGCCA
*nested vectorette primer (VIII)*

3′ - TTCCTCTCCTGTCGC
*vectorette sequencing primer (IX)*

**Figure 2.** This figure shows a vectorette unit with vectorette primer, nested vectorette primer and vectorette sequencing primer. Bases $N_1$, $N_2$, $N_3$, $N_4$, X and Y of the vectorette unit provide sticky ends. They are omitted from blunt end vectorettes.

corresponding to the known sequence and the vectorette. Two families of vectorette are available, each compatible with a number of restriction enzymes. Vectorettes 1 and 2 have the same basic design, but have different primer hybridization sites and so require the use of different vectorette, nested, and sequencing primers. Vectorette 2 has restriction enzyme sites incorporated into the bottom strand.

The first part of this chapter concentrates on a vectorette type 1 unit. To

**Figure 3.** Scheme to illustrate the construction and PCR amplification of a single vectorette library. (1) Digestion of target DNA with a restriction enzyme (R). A number of digests should be set up with different enzymes to maximize the chance of a site being within the GeneAmp™ PCR range. (2) Ligation of vectorette to the target DNA fragments, to form a vectorette library. (3a) In the first round of amplification, primer extension is from the initiating primer (IP) only. Amplification of the background is avoided since the vectorette primer (VP) will hybridize only the product of this first round. (3b) In the second and subsequent rounds of PCR, priming is from both the IP and the VP. (4) Sequencing of the purified amplification product using a sequencing primer hybridized to the vectorette downstream of the vectorette primer. For ease of sequencing, the amplification product may be rendered single-stranded with lambda exonuclease, if the vectorette primer is 5′ phosphorylated. For genome walking, steps 1–4 are repeated using the novel DNA sequence to synthesize a new initiating primer.

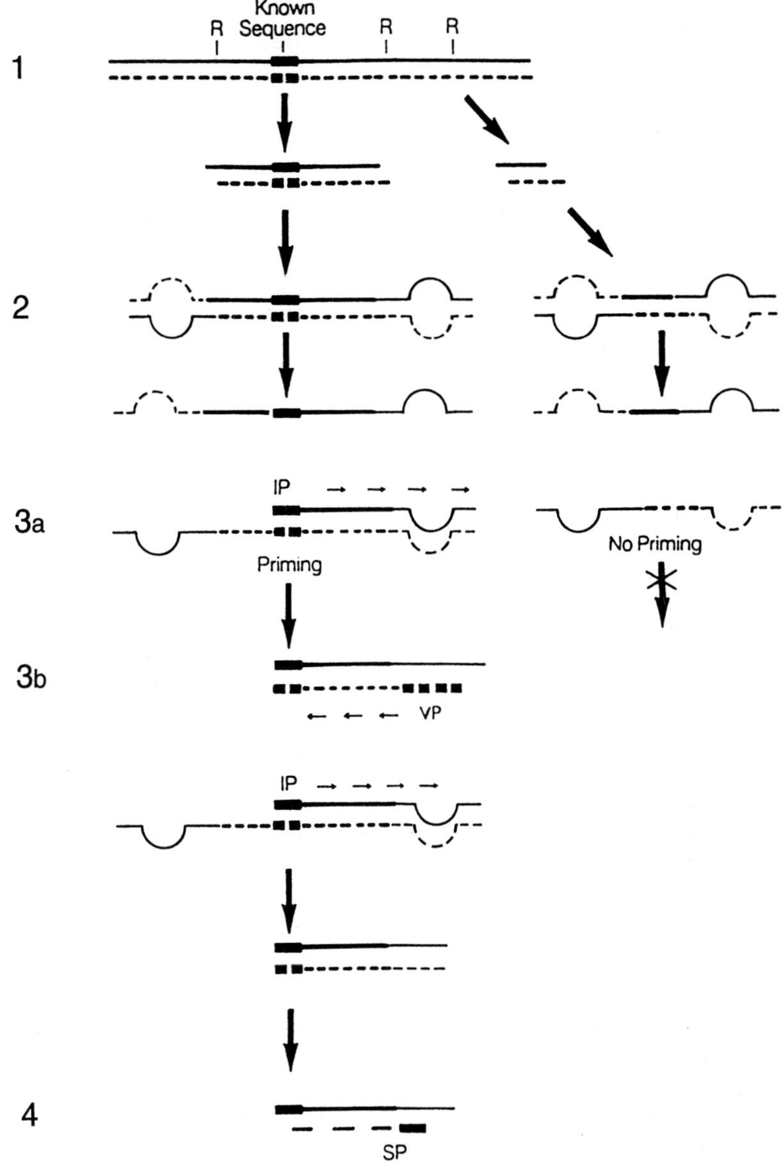


understand vectorette PCR, first consider a case where the target DNA is digested with *Eco*RI. The digested DNA is ligated to the appropriate complementary vectorette unit (the left-hand end of the vectorette unit shown in *Figure 2*). In the first round of PCR, a specific primer to whatever limited known sequence is available is used to initiate extension through an unknown region to the end of the flanking vectorette molecule (*Figure 3*). This produces a strand which is now complementary to the bottom strand of the vectorette unit shown in *Figure 2*. The vectorette itself is only partially double-stranded. A 'universal' vectorette primer, which has the same sequence as the bottom strand of the mismatched section of the vectorette, is also present in the reaction mixture. It has no complementary strand with which to anneal in the first round of PCR (see *Figure 2*). In the second round of PCR, this vectorette primer anneals to the vectorette-derived section of the newly synthesized strand and initiates extension by DNA polymerase as far as the 5′-end of the specific primer used in cycle 1. The fragment defined by the specific primer and the vectorette primer is further amplified by subsequent cycles of standard PCR (*Figure 3*). The ends of the fragment adjacent to the vectorette primer are sequenced using a vectorette sequencing primer. This provides new sequence information from which a new oligonucleotide primer can be synthesized to commence a further vectorette walk.

In practice, the target DNA can be digested with any one of a number of restriction enzymes, typically *Eco*RI, *Hin*dIII, *Bam*HI (compatible with *Bgl*II, *Bcl*I and *Sau*3A), *Cla*I (compatible with *Mae*II, *Hpa*II and *Taq*I), and blunt end cutters (*Alu*I, *Eco*RV, *Hae*III, *Pvu*II, *Rsa*I, *Sma*I), and ligated to the vectorette unit. A key feature of the sequence at the end of the vectorette is that, although it can be ligated to a particular sticky/blunt end, the original restriction site is not reformed in the target–vectorette construct (this is not the case for *Sau*3A/*Mbo*I and *Hpa*II/*Msp*I). This means that during vectorette library construction it is not strictly necessary to heat-inactivate the restriction enzyme before the next step of ligation to specific vectorette units containing the complementary sticky/blunt end (except for the enzymes mentioned as exceptions above). A consequence of this is that erroneously ligated target–target molecules which are non-contiguous in the human genome will be recut by the restriction enzyme. The target–vectorette constructs will not be cut. This increases the overall yield of target molecules which have a vectorette ligated at both ends and minimizes the generation of artefactual non-contiguous ligation products.

A further important feature of the vectorette technology is the possibility of constructing a panel of vectorette libraries made with different restriction enzymes and using them in parallel (*Figure 4*). In this way PCR products between a specific primer of known sequence and the nearest flanking sites for a series of particular restriction enzymes, are generated. In these circumstances, an internal control for the fidelity of the PCR amplification process is generated—the longest PCR product obtained should contain recognition sites at predicted points for the restriction enzymes used to generate shorter vectorette products (*Figure 4*). It is relatively straightforward to construct more than one vectorette library so as to perform this control step, and the time saved on not sequencing artefactual PCR products can be significant.

Reagents for the construction and analysis of vectorette libraries are available from Cambridge Research Biochemicals. *Figure 5* shows the sequence

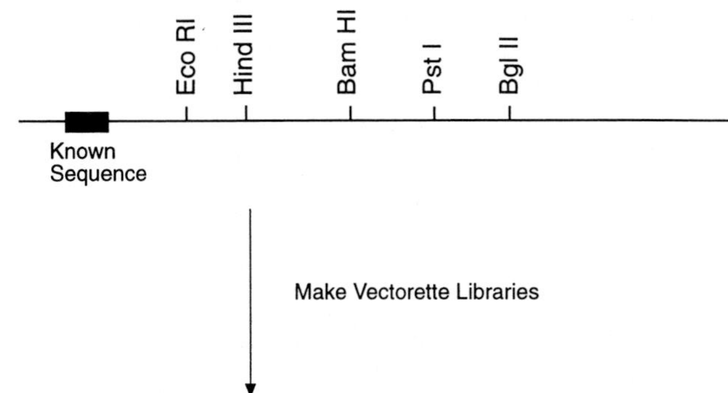


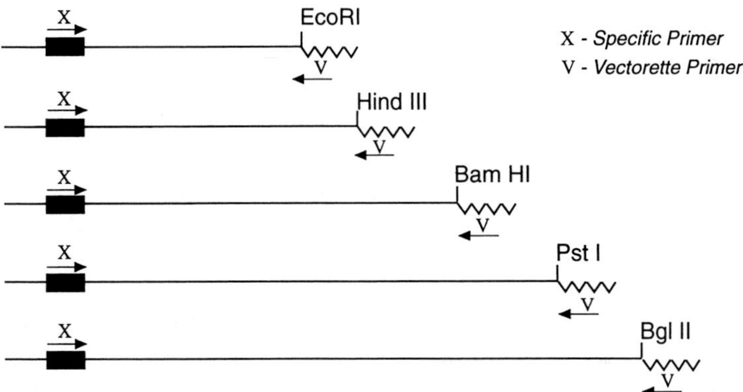


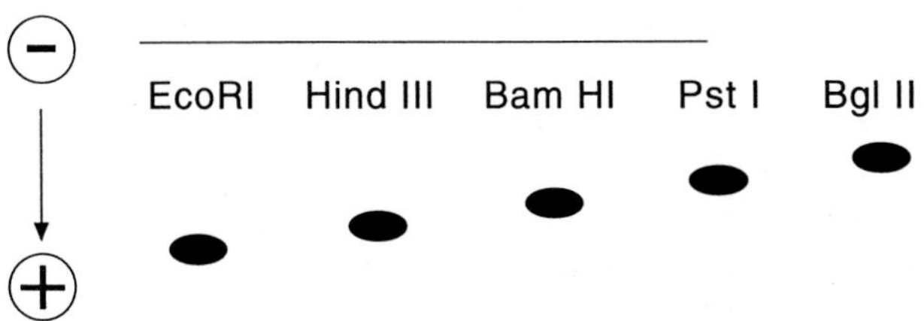


**Figure 4.** Scheme of construction and PCR amplification of a panel of vectorette libraries. The largest (*Bgl*II) product illustrated should be cut with *Eco*RI, *Hin*dIII, *Bam*HI, and *Pst*I at the predicted positions. (a) Digestion of target DNA in separate reactions with *Eco*RI, *Hin*dIII, *Bam*HI, *Pst*I and/or*Bgl*II; (b) PCR amplification of the libraries using the specific target DNA primer (X) and the vectorette primer (V); (c) agarose gel electrophoresis of the individual vectorette libraries.

**Table 1.** Vectorette I primers[a]

| | Sequence | Identity |
|---|---|---|
| Oligo I | CACCCGTTCTCGGAGCACTGTCCGACC | YAC left arm primary oligo |
| Oligo II | TCTTCAACAATTAAATACTCTCGGTAGCC | YAC left arm nested oligo |
| Oligo III | GTTGGTTTAAGGCGCAAG | YAC left arm sequencing primer |
| Oligo IV | ATATAGGCGCCAGCAACCGCACCTGTGG | YAC right arm primary oligo |
| Oligo V | ACCTGTGGCGCCGGTGATGCCGGCCAC | YAC right arm nested oligo |
| Oligo VI | GTCGAACGCCCGATCTCAAG | YAC right arm sequencing primer |
| Oligo VII | CGAATCGTAACCGTTCGTACGAGAATCGCT | Universal vectorette primer |
| Oligo VIII | ACCGTTCGTACGAGAATCGCTGTCCTCTCC | Universal nested vectorette primer |
| Oligo IX | CGCTGTCCTCTCCTT | Universal vectorette sequencing primer |
| Oligo X | CTCTCCCTTCTCGAATCGTAACCGTTCGTACGAGAATCGCTGTCCTCTCCTTY | Vectorette bottom strand |
| Oligo XI | $N_1N_2N_3N_4$XAAGGAGAGGACGCTGTCTGTCGAAGGTAAGGAACGGACGAGAGAA GGGAGAG | Vectorette top strand[b] |

[a] Vectorette 2 primers can be designed for the user's own specialized applications or purchased (Cambridge Research Biochemicals). A complementary range of vectorette 2 universal primer, nested primer and sequencing primer will be required.

[b] For blunt end applications residues $N_1$, $N_2$, $N_3$, $N_4$, X, are omitted as an overhanging sticky end is not required.

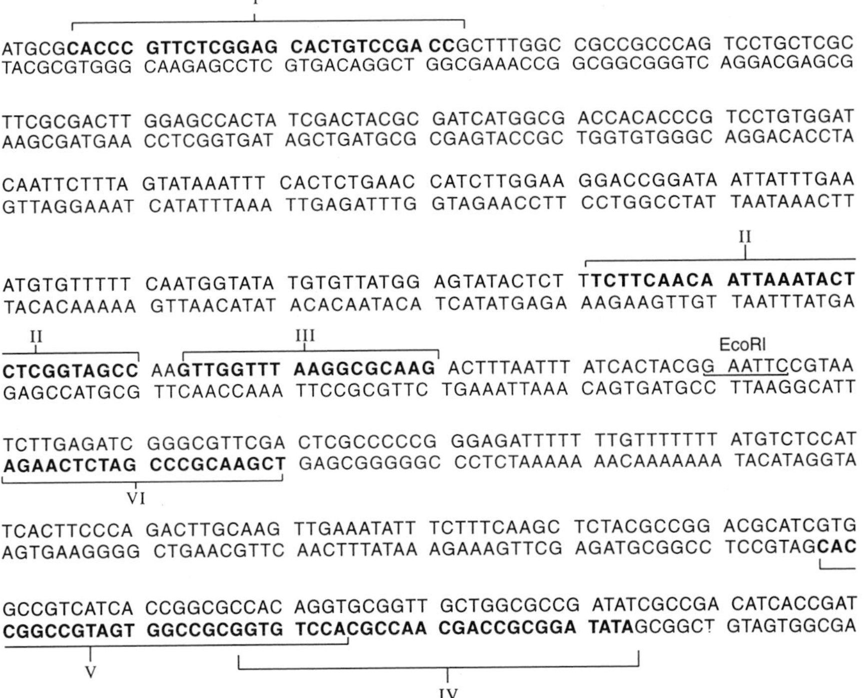


**Figure 5.** The sequence around the pYAC vector *Eco*RI cloning site. YAC sequences corresponding to left and right arm primary, nested and sequencing primers are in **bold** and their identity is shown in *Table 1*.

around the pYAC vector *Eco*RI cloning site and *Table 1* shows the primers required for use of vectorette type 1 units.

There are a wide number of applications for vectorette PCR. It may be used in genomic walking, sequencing of YAC and cosmid termini, mapping of introns in genomic DNA using cDNA sequence, sequencing of large clones without subcloning, and mapping of regions containing deletions, insertions, and translocations. For certain loci, the first round of vectorette PCR can produce various non-specific products. This problem can be overcome by performing a second round of PCR with nested primers with 'nesting' possible at either the specific primer end or the vectorette end or both (see Section 4.4).

There are certain advantages associated with YAC vectorette libraries over conventional technologies:

- There is no need to clone or subclone, which saves time and avoids problems with host–vector stability and unclonable sequences.
- Only small amounts of starting material are necessary as this method is PCR-based.

- The preparation of high purity DNA is unnecessary. For example, YAC vectorette libraries are routinely prepared without isolation of the YAC containing human DNA away from the total yeast genomic DNA.

## 4.2 Construction of vectorette libraries from YAC DNA

The procedures for construction of YAC vectorette libraries are given in *Protocol 3*.

### Protocol 3. Restriction digestion of YAC DNA and construction of vectorette libraries

*Reagents*

- 1 × ligation buffer: 50 mM Tris–HCl pH 7.6, 10 mM $MgCl_2$, 1 mM DTT
- 1 × restriction enzyme buffer. The composition of this buffer will vary depending on the specific restriction enzyme chosen
- Vectorette units. Anneal 5′ top strand phosphorylated and 5′ bottom strand non-phosphorylated vectorette units (Cambridge Research Biochemicals) in 10 mM Tris–HCl pH 7.5, 15 mM $MgCl_2$ by boiling and slow cooling. Store at 1 pM/μl at −20°C
- TE (see *Protocol 1*)
- 100 mM ATP
- T4 DNA ligase (5 U/μl)
- 10 × PCR buffer: 100 mM Tris–HCl pH 8.3 at 20°C, 50 mM KCl, 15 mM $MgCl_2$, 0.1% gelatin
- 50 mM dNTP mix (50 mM of all four dNTPs)
- Primers I, II, IV, V (*Table 1*)
- *Taq* DNA polymerase (Amplitaq, Perkin-Elmer Cetus, 2.5 U)
- 2% agarose gel in 1 × TBE
- 2% low melting temperature (LMT) agarose gel in 1 × TBE

*Method*

A. *Day 1*

1. Incubate the agarose plugs containing YAC DNA (3 μg/plug) overnight in 50 ml TE at 4°C on a slow roller. One third of a YAC plug is required per vectorette library (approx. 1 μg DNA).

B. *Day 2*

1. Incubate individual plugs for 1 h in 5 ml fresh TE at 4°C.
2. Incubate the plugs in 1 ml of the appropriate 1 × restriction enzyme buffer on ice for 1 h in a sterile 1.5 μl microcentrifuge tube.
3. Replace this buffer with 100 μl of fresh 1 × restriction enzyme buffer, add 20 units of restriction enzyme and incubate overnight at 37°C.[a]

C. *Day 3*

1. Replace the restriction enzyme buffer with 1 ml of 1 × ligation buffer and incubate for 1 h on ice.
2. Replace this buffer with 100 μl of fresh 1 × ligation buffer and add 3 μl vectorette units (3 pmol). Incubate at 65°C for 10 min to melt the plug. Cool to 37°C.

3. Add 1 μl 100 mM ATP and 10 U T4 DNA ligase. Incubate at 37°C for 2 h.
4. Add 300 μl water and vortex. Store in 50 μl aliquots at −20°C. These aliquots constitute vectorette libraries.
5. Carry out PCR analysis using the 'hot start' PCR technique in a total volume of 100 μl. To do this, in a microcentrifuge tube mix:
   - vectorette library (from step 4) — 5 μl
   - primers — 1 μM (100 pmol)
   - all four dNTPs — 50 μM each
   - 10 × PCR buffer — 10 μl
   - water — to 100 μl final volume

   Denature at 95°C for 5 min and then add 2.5 units *Taq* DNA polymerase. Then use 40 cycles of 95°C for 2 min, 60°C for 2 min, 72°C for 3 min, with a final extension step of 72°C for 10 min.
6. Sequence the YAC ends using the appropriate vectorette primer and a primer corresponding to the left-hand end (*Figure 5*, oligo I) or the right-hand end (*Figure 5*, oligo IV) of the YAC (2).

D. *Day 4*

1. Electrophorese an aliquot (10%) of the PCR products on a 2% agarose gel to check for homogeneity.
2. Select PCR products for analysis by monitoring the effect of *Eco*RI digestion.[b]
3. Digest 35 μl of the PCR reaction mixture with 25 U *Eco*RI in a volume of 40 μl at 37°C for more than 2 h.
4. Analyse on 2% LMP agarose gel and confirm expected size changes. The largest cut fragment can be recovered for use as a hybridization probe, or for sequencing using the vectorette sequencing primer IX (*Figure 2*). Samples of uncut PCR product confirmed as homogeneous in step 1 (or after a further cycle of nested PCR, see Section 4.4) may similarly be sequenced using YAC left arm (oligo III) or right arm (oligo VI) sequencing primers (*Figure 5*). PCR products deliberately not cut with *Eco*RI can of course also be sequenced with the vectorette sequencing primer IX as above.

[a] An alternative approach is to add 50 U of enzyme and incubate at 37°C for 2 h. This approach appears to be most efficient where blunt-end restriction enzymes such as *Alu*I, *Rsa*1, *Pvu*II and *Eco*RV are used. In these circumstances a set of digestions can be used with a single vectorette unit to generate several different vectorette libraries.

[b] YAC-vectorette PCR end products will be digested because they contain the original YAC *Eco*RI cloning site. The size of the digested fragment is dictated by the choice of PCR primers from the YAC vector left or right arms (2). For the primary and nested YAC left hand end primers shown in *Figure 5* (oligos I and II), the PCR products would be shortened by 285 or 69 bp respectively. For the primary and nested YAC right hand end primers (*Figure 5*, oligos IV and V) the PCR products would be shortened by 170 or 150 bp respectively on digestion with *Eco*RI.

## 4.3 Construction of vectorette libraries from other DNAs

Although YAC clones are an obvious source of genomic DNA for creating vectorette libraries, these can also be constructed directly from genomic DNA, or cosmid DNA, or bacterial DNA. The procedures are described in *Protocol 4*.

**Protocol 4.** Construction of vectorette libraries from other DNA sources

*Reagents*

- 1 × ligation buffer: 50 mM Tris–HCl pH 7.6, 10 mM $MgCl_2$, 1 mM DTT
- 1 × restriction enzyme buffer. The composition of this buffer will vary depending on the specific restriction enzyme chosen
- Vectorette units. Anneal 5′ top strand phosphorylated and 5′ bottom strand non-phosphorylated vectorette units (Cambridge Research Biochemicals) in 10 mM Tris–HCl pH 7.5, 15 mM $MgCl_2$ by boiling and slow cooling. Store at 1 pM/μl at −20°C
- TE (see *Protocol 1*)
- 100 mM ATP
- T4 DNA ligase (5 U/μl)
- 10 × PCR buffer: 100 mM Tris–HCl pH 8.3 at 20°C, 50 mM KCl, 15 mM $MgCl_2$, 0.1% gelatin
- 50 mM dNTP mix (50 mM of all four dNTPs)
- Primers I, II, IV, V (*Table 1*)
- *Taq* DNA polymerase (Amplitaq, Perkin Elmer-Cetus, 2.5 U)
- 2% agarose gel in 1 × TBE
- 2% low melting temperature (LMT) agarose gel in 1 × TBE

*Method*

1. In a sterile microcentrifuge tube, mix 1 μg human genomic DNA or cosmid DNA or 100 ng bacterial DNA with 5 μl 10 × restriction enzyme buffer and 20 U of the appropriate restriction enzyme. Make up to 50 μl with sterile distilled water. Incubate at 37°C for 1 h.
2. The following ligation reaction can be carried out in the same tube as the restriction digest without further manipulation.[a] Most restriction enzyme buffers are compatible with T4 DNA ligase. Add 3 μl (3 pmol) of the vectorette units corresponding to the restriction enzyme used in step 1, 1 U T4 DNA ligase, 1 μl 100 mM ATP, and 1 μl 100 mM DTT.
3. Incubate the microcentrifuge tube at 20°C for 60 min followed by 37°C for 30 min. Repeat three times in total. This temperature cycling re-digests any target DNA fragments which have ligated to each other rather than to vectorette units and therefore ensures optimal ligation of vectorettes to restriction fragment ends.
4. Add 200 μl sterile distilled water to each reaction mixture and store the vectorette library in 10 μl aliquots at −20°C.
5. Perform PCR as described in *Protocol 3*.

[a] If a four-base restriction enzyme is used whose recognition site is reformed on ligation (e.g. *Sau*3A *Msp*I), heat denature the restriction digest at 70°C for 10 min before ligation and incubate for 4.5 h at 20°C.

### 4.4 Nested PCR

With complex DNA it is sometimes necessary to perform a second round of PCR with an internal primer(s) to achieve complete specificity. PCR is carried out with 1 μM (100 pmol) of both specific and nested vectorette primers (see *Protocol 5*). In the specific case of YAC end sequencing, a nested YAC left arm (oligo II of *Figure 5*) or nested YAC right arm (oligo V of *Figure 5*) primer may be used in conjunction with the nested vectorette primer (oligo VIII of *Figure 5*) for additional specificity. In this case, use a 1/10 000 dilution of the first PCR reaction as a template.

**Protocol 5.** Nested PCR of YAC derived vectorette libraries

*Reagents*

- 10 × PCR buffer (see *Protocol 4*)
- Oligonucleotide primers I, II, V, VI, VII, IX (*Table 1*)
- dNTPs (1 mM each of all four dNTPs)
- 2% agarose gel in 1 × TBE
- *Eco*RI
- 10 × *Eco*RI digestion buffer.

*Method*

1. Dilute 2 μl of the primary vectorette PCR reaction to 400 μl with sterile distilled water.
2. Set up a nested PCR reaction containing 2 μl of the diluted primary PCR reaction mixture, 10 μl 10 × PCR buffer, 100 pmol of **either** primers II and VIII (YAC left arm primary reactions) or primers V and VIII (YAC right arm primary reactions), 10 μl dNTP mix, 75 μl sterile distilled water.[a]
3. Perform a further 35 cycles of hot start PCR as described in *Protocol 1*.
4. Analyse 10% aliquots of the PCR products by electrophoresis on a 2% agarose gel. Confirm the authenticity of the PCR products by digestion with *Eco*RI comparing the sizes of the uncut and *Eco*RI digested DNA.
5. Sequence the nested vectorette PCR products using the vectorette sequencing primer (oligo IX) and either the YAC left arm sequencing primer (oligo III) or the YAC right arm sequencing primer (oligo VI) by standard dideoxy sequencing techniques.[a]

[a] For details of the oligo primers, see *Figure 5*.

### 4.5 Using vectorette 2

Fragments greater than 1 kb are sometimes difficult to sequence completely. If this is essential and you are not experienced in routine subcloning, an alternative method to use when dealing with fragments of this size is shown

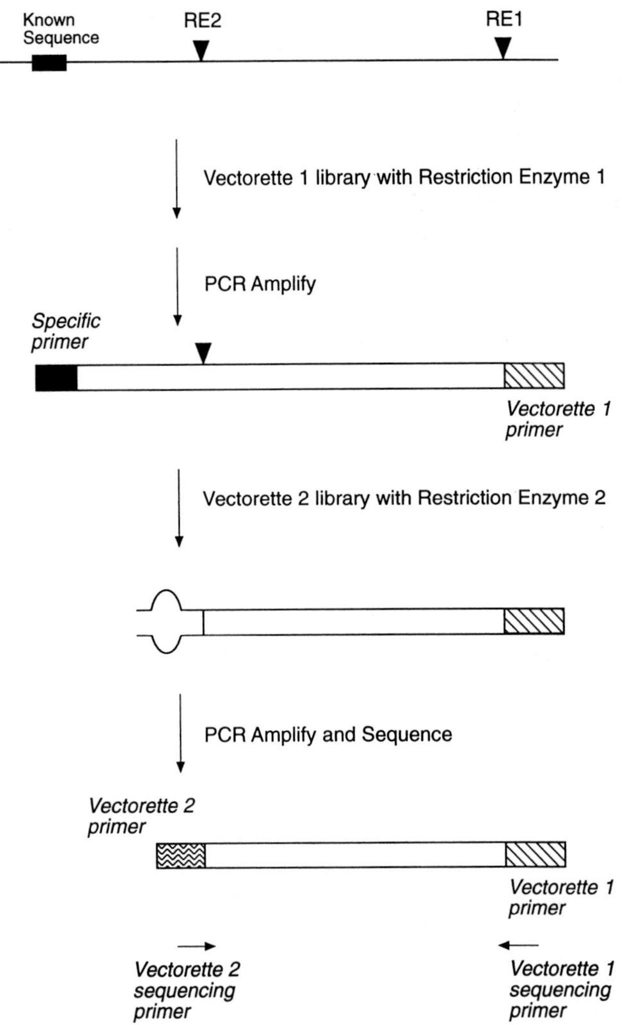


**Figure 6.** Schematic for the use of two different vectorette units for sequencing long fragments without subcloning. See Section 4.5 of the text for a description of this approach.

in *Figure 6*. The procedure is to amplify the vectorette DNA product by PCR then further digest with a second different restriction enzyme. Ligation with an appropriate alternative vectorette unit is then performed to yield a DNA product with a vectorette unit at each end. Both vectorette units are then used for further amplification and sequencing (*Figure 6*). The detailed procedures are described in *Protocol 6*.

It will be appreciated that the mismatched 'bubble' sequences in the vectorettes 1 and 2 are different. Thus the vectorette universal primers, nested primers, and sequencing primers are different between types 1 and 2. In the commercially available kits (Cambridge Research Biochemicals), the vectorette type 1 system is analogous to that described in *Figure 2*. The vectorette type 2 system contains multiple restriction sites within the bubble region for cloning of final PCR products. Of course all the vectorette libraries discussed above can be constructed with the type 2 vectorette with the proviso that only the appropriate universal primers, nested primers, and sequencing primers will then function.

**Protocol 6.** Use of two different vectorette units

*Reagents*

- 1 × ligation buffer: 50 mM Tris–HCl pH 7.6, 10 mM $MgCl_2$, 1 mM DTT
- 1 × restriction enzyme buffer. The composition of this buffer will vary depending on the specific restriction enzyme chosen
- Vectorette units: Anneal 5′ top strand phosphorylated and 5′ bottom strand non-phosphorylated vectorette units (Cambridge Research Biochemicals) in 10 mM Tris–HCl pH 7.5, 15 mM $MgCl_2$ by boiling and slow cooling. Store at 1 pM/μl at −20°C.
- TE (see *Protocol 1*)
- 100 mM ATP
- T4 DNA ligase (5 U/μl)
- 10 × PCR buffer: 100 mM Tris–HCl pH 8.3 at 20°C, 50 mM KCl, 15 mM $MgCl_2$, 0.1% gelatin
- 50 mM dNTP mix (50 mM of all four dNTPs)
- Primers I, II, IV, V (*Table 1*)
- *Taq* DNA polymerase (Amplitaq, Perkin-Elmer Cetus, 2.5 U)
- 2% agarose gel in 1 × TBE
- 2% low melting temperature (LMT) agarose gel in 1 × TBE

*Method*

1. Prepare a number of vectorette 1 libraries with different restriction enzymes as described in *Protocol 3*.
2. PCR amplify each library in 100 μl total volume with a specific primer and the vectorette 1 universal primer. Run 10 μl on an agarose gel to confirm the presence of a unique PCR product.
3. Select the vectorette library which has generated the largest PCR product. Precipitate the DNA generated in the PCR reaction by the addition of 2 vol. ethanol and let stand at −70°C for 30 min. Spin in a microcentrifuge for 10 min to pellet the DNA. Resuspend in 20 μl sterile distilled water.
4. Digest 5 μl of the DNA solution with the restriction enzyme producing the second largest vectorette library PCR product and ligate the appropriate vectorette 2 unit as described in *Protocol 3*.
5. Amplify the product using both the vectorette 1 and 2 primers.

## 4.6 Sequencing vectorette products

A full discussion of possible sequencing methods is beyond the scope of this chapter. However, we have found two methods to be both rapid and reliable

(12, 13). Both, in essence, involve the creation of single-stranded sequencing templates as well as allowing removal of interfering primers and dNTPs from preceding PCR reactions. Should difficulty be experienced in any cases, the option of subcloning vectorette products into M13 or pUC vectors for sequencing can be carried out instead. The first procedure involves streptavidin/biotin affinity purification and is described in *Protocol 7*. In the second approach (*Protocol 8*), PCR is performed with one primer 5′-phosphorylated with T4 polynucleotide kinase or by chemical synthesis. The strand in the PCR product which is 5′-phosphorylated can then be degraded by lambda exonuclease leaving the other strand intact. This provides a sequencing template for use with nested versions of the original phosphorylated primer.

## Protocol 7. Streptavidin/biotin affinity purification for vectorette sequencing

*Equipment and reagents*

- TEN: this is TE (see *Protocol 1*) containing 2 M NaCl
- 5′-biotinylated and unbiotinylated primers
- Dynabeads M-280 (Dynal)
- 0.15 M NaOH (freshly made)
- Magnetic particle concentrator (MPC) (e.g. Dynal)

*Method*

1. Perform PCR as described in *Protocol 2* with one primer (100 pmol, 1 μM) biotinylated at the 5′-end. Usually, to increase specificity, this will be the specific primer rather than the vectorette primer. In the context of YAC end sequencing of vectorette products, the left or right arm primary or nested primers (oligos I, II, IV or V; see *Figure 5*) may conveniently be biotinylated.
2. Wash 20 μl Dynabeads M-280 streptavidin with TEN. Resuspend these in 40 μl TEN. Incubate the washed Dynabeads with an equal volume of biotinylated primer PCR reaction product for 15 min at room temperature with gentle shaking.
3. Magnetize the beads using an appropriate magnetic particle concentrator (MPC) for 1 min.
4. Remove the supernatant and wash the beads three times with TEN using the MPC.
5. Displace the unbiotinylated DNA strand by denaturation by incubating with 1 ml freshly made 0.15 M NaOH for 5 min at room temperature.
6. Wash the beads twice with 1 ml water each time using the MPC. Resuspend the beads in water for use in the sequencing reaction.

7. Sequence the DNA using either Sequenase (USB) or thermal cycling methods (e.g. Promega). Where the specific primer is biotinylated (e.g. YAC right or left arm primers) then sequencing is performed with the vectorette sequencing primer (oligo IX; see *Figure 5*).

**Protocol 8.** Lambda exonuclease method for sequencing vectorette products

*Equipment and reagents*

- Lambda exonuclease 10 × buffer: 67 mM glycine–NaOH pH 9.4, 2.5 mM $MgCl_2$
- TE equilibrated phenol (see *Protocol 1* for composition of TE buffer)
- Vacuum evaporator (Speedvac, Uniscience Ltd)
- Lambda exonuclease (4 U/μl)

*Method*

1. Add an equal volume of TE equilibrated phenol to the PCR reaction mixture. Vortex and spin in a microcentrifuge for 5 min at 13 000 r.p.m.
2. Remove the aqueous layer, add 2 vol. ethanol and place at −70°C for at least 15 min.
3. Pellet the DNA in a microcentrifuge for 10 min and remove the supernatant.
4. Dry the pellet in a vacuum evaporator (Speedvac) and redissolve it in 21.5 μl water.
5. Add 2.5 μl 10 × exonuclease buffer and 1 μl lambda exonuclease (4 U/μl). Mix and incubate at 37°C for 30 min.
6. Add 25 μl water and perform two phenol extractions as in step 1.
7. Ethanol precipitate the DNA and dry as in steps 2–4.
8. Resuspend the pellet in 20 μl sterile distilled water and sequence. In the context of sequencing YAC vectorette end products oligos I or II (left arm) or IV or V (right arm) would be 5′-phosphorylated. Oligo III or VI, respectively, would be used as sequencing primers.[a]

[a] See *Table 1* and *Figure 5*.

# 5. Exon identification

## 5.1 Previous approaches

Several approaches for screening genomic DNA for exons have emerged.

(a) The first method is the relatively straightforward technique of exon connection based on zoo blots. As outlined in the introduction to this chapter

(Section 1), zoo blots exploit the conservation of expressed sequences between species. Exon connection is clearly described in the primary literature and will not be discussed further (14). For researchers who wish to try this and have limited access to a zoo, collections of DNA from different species can be purchased from several suppliers (e.g. BIOS Laboratories).

(b) Detailed protocols have been provided elsewhere for the detection of transcribed sequences within YACs on the basis of HTF island detection and cloning (15).

(c) The technique of using labelled purified YACs to screen cDNA libraries has also been fully documented (15, 16). This approach has proved robust and reliable. Obviously it is dependent upon availability of a range of appropriate cDNA libraries from different tissues or unknown tissue-specific transcribed sequences will be missed.

(d) To complement these basic methods, a range of further techniques have been developed for specific applications (reviewed in ref. 17). Protocols have been devised to detect the DNA coding for expressed genes in YACs or cosmids (18–20) and somatic cell hybrids (21, 22). The YAC and cosmid methods rely on immobilization of genomic DNA cloned in these vectors to a solid support such as a nylon filter. Alternatively the cloned genomic DNA has been biotinylated by nick translation or photo-biotinylation and then bound to streptavidin beads. Repetitive sequences are suppressed by hybridization with sonicated total human DNA or cot-1 DNA (BIOS). cDNAs are then hybridized to the immobilized genomic DNA. Non-specific binding is reduced by stringent washing. The retained specific cDNAs are eluted off and reamplified using nested vector primers. These cDNAs are either submitted to a further round(s) of selection or are cloned and analysed. The resulting enrichment can be of the order of several thousand-fold (20). Whether these methods prove to be more reliable than simple screening of cDNA libraries with whole YACs remains to be established in the long term. This method is described in detail in *Protocol 9*.

(e) Expressed human sequences have been detected in rodent somatic cell hybrid DNA using either *Alu* primers or 5′-intron splice sequence primers to initiate cDNA synthesis from the heterogenous nuclear RNA as template. The splice primers select immature, unspliced mRNA as a template which still retains species-specific repeat sequences in introns. A derived cDNA library was screened for human repeat sequences with total human DNA. Because of sequence divergence at the 5′-splice site, a mix of four hexamers was used (23) in proportion to the frequency of the splice site in sequenced genes (CTTACC, 50%; CTCACC, 30%; CCTACC, 10%; CCCACC, 5%). The *Alu* protocol (24) exploits the presence of human *Alu* repetitive elements in introns and 3′-untranslated regions of human transcripts. Heterogenous nuclear RNA was extracted and cDNA synthesis was primed separately by two oligonucleotides derived from conserved regions of human *Alu* repeats (5′TGCAGTGAGCCGAGAT; 5′TCT(C/T)(G/A)GCTCACTGCAA). The cDNA was converted to double-stranded DNA and cloned into a plasmid vector. Radiolabelled inserts were hybridized to somatic cell hybrid mapping panels to confirm their human origin. At least 80% of the clones thus obtained were of human origin. Both these approaches rely on the transcription of the heterologous human genes in the particular rodent cell background. It is probable that this will frequently fail to occur, particularly for genes whose expression is highly tissue-specific.

## **Protocol 9.** cDNA enrichment using cloned genomic DNA

*Reagents*

- Hybridization buffer: 750 mM NaCl, 50 mM sodium phosphate pH 7.2, 5 mM EDTA, 5 × Denhardt's, 0.05% SDS, 50% formamide
- YAC DNA (see *Protocol 2*) or cosmid DNA
- Biotin-16-dUTP (Boehringer-Mannheim)
- Reagents for nick translation (dATP, dGTP, dCTP, dTTP, BSA, DTT; see step 2)
- DNase I
- DNA polymerase I (20 U)
- Streptavidin-coated magnetic beads (Dynabeads, M-280, Dynal)
- Bead buffer (2 M NaCl, 10 mM Tris–HCl pH 7.5, 1 mM EDTA)
- 50 μg Denatured human cot-1 DNA (BIOS)
- 1% agarose gel in 1 × TBE buffer
- Phenol/chloroform (1:1)
- 2 × SSC
- 0.2 × SSC

A. *DNA preparation*

1. Prepare YAC DNA (see *Protocol 2*) or cosmid DNA.
2. Nick translate the DNA to incorporate biotin-16-dUTP. Do this in 50 μl total volume containing 1.2 μg cosmid DNA,[a] 50 mM Tris–HCl pH 7.6, 5 mM $MgCl_2$ 50 mg/ml BSA, 50 μM each dATP, dGTP, dCTP, 38 μM dTTP, 2 μM biotin-16-dUTP, 10 mM DTT, 5 ng DNase I, 20 units of DNA polymerase I. Incubate at 15°C for 3 h initially (the time for incubation may need to be optimized).
3. Denature 300 ng of the biotinylated DNA and bind to 0.5 mg streptavidin-coated magnetic beads (Dynal) in a volume of 50 μl bead buffer at 25°C for 1 h.
4. Pre-anneal the bound DNA in 50 μl hybridization buffer with 50 μg of denatured human cot-1 DNA at 42°C for 2 h.

B. *Amplification of cDNA inserts*

5. Start with about $10^8$ independent cDNA clones. Amplify the inserts in 100 μl reactions using PCR with primers specific to flanking vector sequences. A 'hot start' PCR technique is recommended. The cycling parameters may need to be optimized, but a good starting point is 35 cycles of 95°C for 1 min, 55°C for 1 min and 72°C for 2 min.

**Protocol 9.** *Continued*

6. Check an aliquot (5 μl) of the PCR product on a 1% agarose gel. Then phenol/chloroform extract and ethanol precipitate the amplified cDNA. Resuspend in water at a concentration of 1 μg/ml.
7. Reamplify 5–10 ng of the cDNA as in step 5 but with nested vector specific primers for a further 30–35 cycles. Purify the DNA as in step 6 and resuspend in water (~1 μg/ml).

C. *Hybridization*

8. To 50 μl biotinylated DNA immobilized on streptavidin-coated magnetic beads, add 50 μl of fresh hybridization buffer and 10 μl of the reamplified cDNA from step 7, previously denatured. Incubate at 42°C for 16 h.

D. *Separation and further amplification of specific cDNA*

9. To remove non-specific cDNA, wash the beads for 15 min with 500 μl each of 2 × SSC at 25°C, 2 × SSC at 68°C, and five times with 0.2 × SSC at 68°C.
10. Elute the specific cDNA from the beads by incubating the beads with 50 μl water at 70°C for 10 min.
11. Re-amplify the recovered specific cDNA and clone into a vector if required. Re-amplification of specific hybridized cDNAs may be performed using nested vector specific primers to further enhance specificity using the conditions indicated above under B. *Amplification of cDNA inserts.*
12. Repeat the whole process of hybridization 1–2 times to enrich the cDNA further. Use 30 μg of eluted and re-amplified material together with a new aliquot of biotinylated cosmid DNA (300 ng). If possible, co-amplification using specific primers for a cDNA known to be present in the cosmid or YAC should be included as an internal control to test the success of the enrichment strategy. The overall enrichment factor is usually 2–3 orders of magnitude per amplification step.

[a] When working with purified YAC DNA it is unlikely for practical reasons that more than 0.1–1 μg PFGE purified DNA will be readily available.

## 5.2 Exon trapping

Exon trapping relies on the cloning of genomic DNA into vectors containing splicing signals. The 5′-consensus splice donor sequence is AG/GTRAGT (where R = A or G) and the 3′-consensus acceptor sequence is NCAG/G (where N is any nucleotide). The presence of these elements in the genomic

### Subcloned Genomic DNA

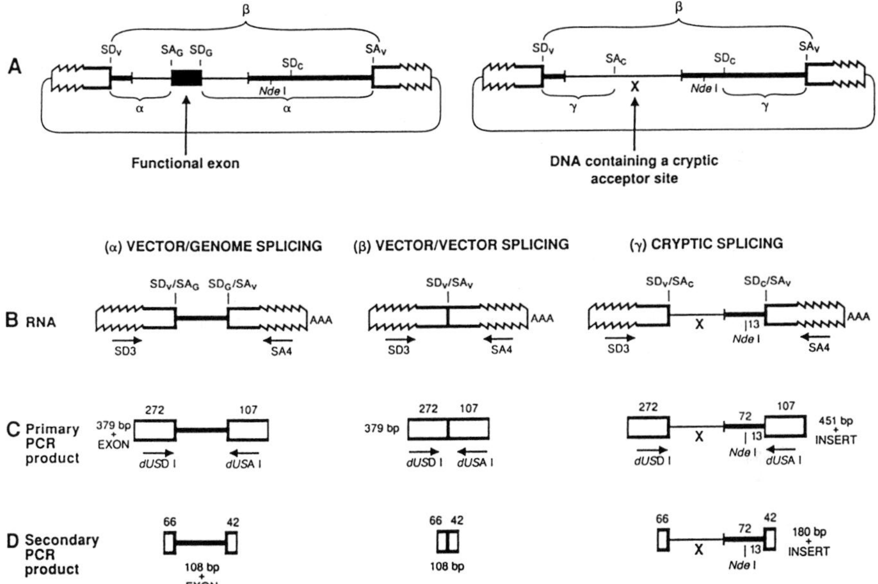


**Figure 7.** Schematic outline of the exon trapping procedure of Buckler *et al.*(26) using the mammalian expression vector pSPL1 in COS-7 cells. The consequences of cloning a functional exon or genomic DNA without exons (but with a cryptic splicing acceptor site) are compared. The various possible splicing events are illustrated. Commercially available PCR primers for primary and secondary amplification are shown; their sequences are shown in *Protocol 10*. SDV, vector splice donor site; SDG, genomic DNA splice donor site; SDC, genomic DNA cryptic splice donor site; SAG, genomic DNA splice acceptor site; SAC, genomic DNA cryptic splice acceptor site; α, vector/genome splicing; β, vector/vector splicing; γ, cryptic splicing. Reproduced with permission from Gibco/BRL.

sequence leads to appropriate splicing. One of the first exon trapping methods (25) used the vector to provide the splice donor site, and the genomic insert was tested for the presence of a splice acceptor site. This method produced a relatively high frequency of artefacts because of cryptic splice acceptor sites, where the splice site joins with an RNA sequence resembling a splice site. The method of Buckler *et al.* (26) and Hamaguchi (27) decreases this problem by requiring both donor and acceptor splice sites from the inserted genomic DNA sample to be functional in COS-7 cells. The exons thus isolated are amplified by reverse transcription and PCR. The exons can then be cloned to produce exon libraries (*Figure 7*). To avoid false positives, the inserts must still be screened for interspecies conservation, and hybridization to a discrete mRNA species on Northern blots. Procedures are described in *Protocol 10*.

## **Protocol 10.** Exon trapping

### *Reagents*

- 5 × First strand buffer: 250 mM Tris–HCl pH 8.3, 375 mM KCl, 15 mM $MgCl_2$
- 10 × PCR buffer: 500 mM KCl, 100 mM Tris–HCl pH 8.4, 15 mM $MgCl_2$ 0.01% gelatin
- Oligo SA4: (5′ CAC CTG AGG AGT GAA TTG GTC G); 20 μM stock solution in water
- Oligo SD3: (5′ GTG AAC TGC ACT GTG ACA AGC TGC); 20 μM stock solution in water
- Oligo dUSD1 (5′ CUA CUA CUA CUA GCG ACG AAG ACC TCC TCA AGG C); 20 μM stock solution in water
- Oligo dUSA1 (5′ CUA CUA CUA CUA GTC GGG TCC CCT CGG GAT TGG); 20 μM stock solution in water
- Vector pSPL1 (Gibco/BRL; see *Figure 8*). Linearize this vector with *Bam*HI and treat with alkaline phosphatase before use
- Vector pAMP10 (Gibco/BRL; see *Figure 9*)
- T4 DNA ligase (5 U/μl)
- Cloned genomic DNA (digested with *Bam*HI, ethanol precipitated and resuspended in water at; 1 μg/μl)
- 5×T4 ligase buffer: 150 mM Tris–HCl pH 7.8, 50 mM $MgCl_2$, 50 mM DTT, 2.5 mM ATP
- *E. coli* competent cells
- Agar plates (Unipath) containing 100 μg/ml ampicillin
- COS-7 cells (available from American Type Culture Collection; ATCC)
- Distilled water treated with diethylpyrocarbonate (DEPC)
- 0.1 m DTT
- 10 mM dNTP mix (mixture of all four dNTPs)
- 200 U/μl, reverse transcriptase (RNase H free; Superscript II, Gibco/BRL)
- 2 units/μl RNase H (Gibco/BRL)
- 2.5 U/μl *Taq* DNA polymerase (Amplitaq)
- 2% low melting temperature (LMT) agarose gel in 1 × TBE buffer
- Phenol:chloroform:isoamyl alcohol (25:24:1)
- 7.5 M ammonium acetate
- Glycogen
- 1 U/μl Uracil DNA glycosylase (Boehringer)
- 10× Uracil DNA glycosylase buffer (600 mM Tris-HCl pH8.0, 10 mM EDTA, 10 mM DTT, 1 mg/ml BSA)

A. *Exon expression in COS-7 cells and synthesis of cDNA*

1. In a microcentrifuge tube, mix:
   - T4 DNA ligase — 1 μl
   - 5 × T4 ligase buffer — 2 μl
   - genomic DNA (*Bam*HI digested) — 1 μg
   - Vector pSPL1 DNA (linearized and phosphatased) — 0.25 μg
   - water to — 10 μl total volume

   Incubate at 15°C for 12 h for ligation to occur.
2. Transform into *Escherichia coli* competent cells. Select on agar plates containing ampicillin (100 μg/ml).
3. Isolate recombinant plasmid from each colony and evaluate it by restriction enzyme digestion (typically with *Bam*HI) and gel electrophoresis for the presence and size of genomic inserts.
4. Transfect 1 μg plasmid DNA into COS-7 cells (ATCC) using standard protocols (28). It is possible to pool several clones in a single transfection using 0.5 μg plasmid DNA from each.[a] Culture for 2 days at 37°C in 5% $CO_2$.

5. Isolate and ethanol precipitate total RNA from transfected cells using standard protocols. Redissolve in DEPC-treated distilled water. In some cases products greater than 450 bp may be derived from transfected DNA or pre-mRNA molecules. If a large number of products in this size range occur then cytoplasmic RNA should be used.
6. Mix 1 μl of 20 μM oligo SA4 and 1–3 μg RNA. Adjust to 12 μl with DEPC-treated water.
7. Denature at 70°C for 5 min then place on ice for 1 min.
8. Add 4 μl of 5 × first strand buffer, 2 μl 0.1 M DTT, 1 μl 10 mM dNTP mix. Incubate at 42°C for 2 min.
9. Add 200 U reverse transcriptase and incubate for 30 min at 42°C and 5 min at 55°C.
10. Add 2 U of RNase H and incubate at 55°C for 10 min. Store this reaction mixture at −20°C.

B. *PCR amplification of cDNA*

1. For the initial PCR amplification across vector and genomic splice sites, mix:
   - cDNA mixture 10 μl
   - 10 × PCR buffer 10 μl
   - 10 mM dNTP 2 μl
   - 20 μM oligo 5AA 2 μl
   - 20 μM oligo 5D3 2 μl
   - sterile distilled water 70 μl

   Denature at 94°C for 5 min. Add 2 U *Taq* DNA polymerase and perform 40 cycles of 'hot start' PCR with denaturation at 94°C for 1 min, annealing at 60°C for 30 sec, and extension at 72°C for 60 sec. Perform a final extension at 72°C for 10 min.
2. Analyse 10 μl of the reaction mixture on a 2% LMT agarose gel.
3. Cut out the bands containing any putative exons (>379 bp, see *Figure 7* C). Purify these DNA fragments using Geneclean (Bio101), or an alternative method.
4. Digest with *Nde*I to discriminate real from cryptic splicing artefacts (see *Figure 7*), and ethanol precipitate.
5. Redissolve the DNA in 80 μl of water. Add 10 μl 10 × PCR buffer, 2 μl 10 mM dNTP mix, 2 μl of 20 mM oligo dUSD1,[b] 2 μl of 20 mM oligo dUSA1,[b] and water to 100 μl. Carry out PCR using the same cycle parameters as in the primary PCR reaction.
6. As an alternative, secondary PCR products may be subcloned into standard 5′-dT-ended vectors using primers identical of oligo dUSD1 and oligo dUSA1 with dT substituted for dU.

**Protocol 10.** *Continued*

7. Run 10 μl of the product on a 2% LMP agarose gel. No PCR product visible implies the presence of an *Nde*I site in the primary PCR product, usually implying cryptic splicing, or more rarely a genomic *Nde*I site.

C. *Cloning of secondary PCR products*

1. Remove the DNA fragment containing the putative exon from the LMP agarose by melting for 10 min at 65°C. Dilute the sample with three volumes of TE buffer.
2. Add 1 vol. phenol:chloroform:isoamyl alcohol (25:24:1). Mix and then centrifuge at room temperature for 5 min at 13 000 r.p.m. Remove the aqueous phase.
3. Add to the aqueous phase 0.5 vol. 7.5 M ammonium acetate, 2 vol. of ethanol and 20 μg glycogen as a carrier. Mix, place at −70°C for 15 min. Centrifuge at 13 000 r.p.m. for 15 min at 4°C. Wash the pellet with 70% ethanol, recentrifuge and dry the DNA.
4. Add on ice,
   - 10–100 ng purified secondary PCR product in 5 μl
   - 1 μl pAMP10 cloning vector
   - 1 μl 10 × Uracil DNA glycosylase buffer
   - 2 μl water
   - 1 μl uracil DNA glycolysase (1 U/μl)

   Mix and incubate for 30 min at 37°C.
5. Use 1–5 μl to transform 100 μl of *E. coli* competent cells, and plate on agar containing 100 μ/ml ampicillin.

D. *Exon confirmation*

1. Amplified products > 108 bp in size are candidate exons (see *Figure 7D*). Therefore, further test these by hybridization to the initial cloned genomic DNA to confirm their origin. Hybridization to specific mRNA species and tissue distribution can be determined by Northern analysis. Full length sequence information for the mRNA can be determined by screening the appropriate tissue cDNA library using the exon or probe.

[a] Specific protocols for lipofection or electroporation are available directly from Gibco BRL for use in conjunction with the pSPL vector.

[b] These primers contain dU residues which are internal to those in the primary PCR reaction. Treatment with uracil DNA glycosylase removes the uracil residues producing PCR products with 3′-single-stranded termini that are complementary to the single stranded termini of vector pAMP10 (*Figure 9*).

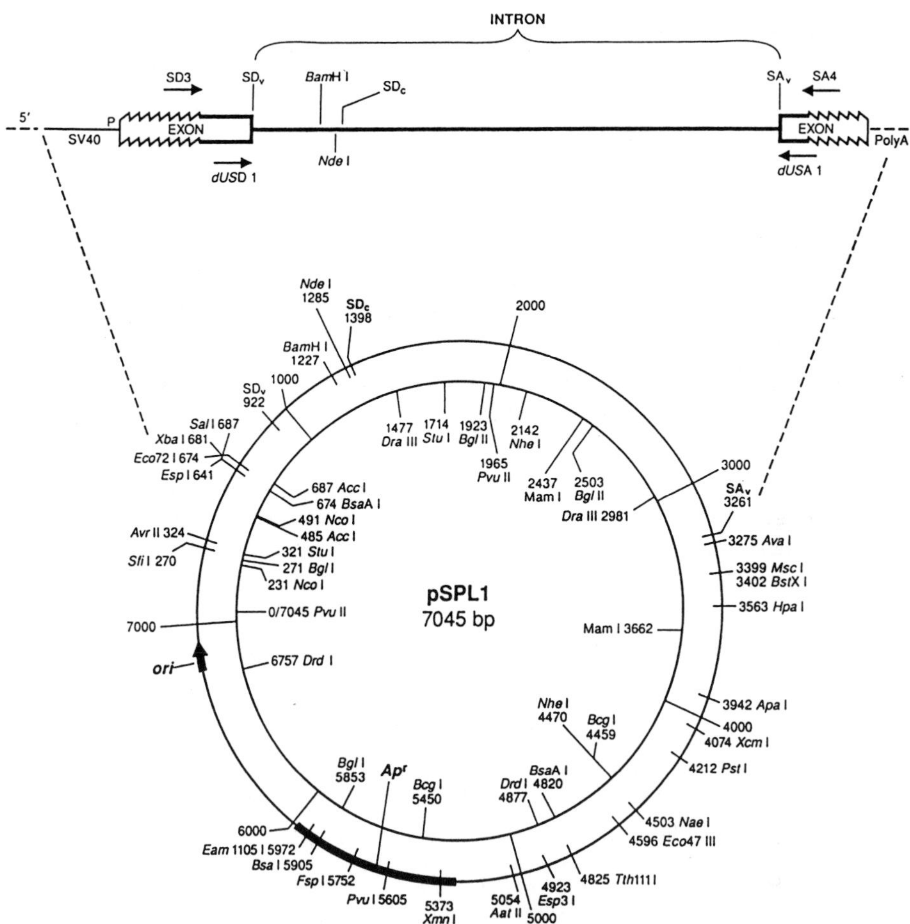


**Figure 8.** Map of pSPL1. Vector splice donor and acceptor sites are derived from the gene for the HIV-1 tat protein. ORI, SV40 origin of replication; SV40P, simian virus early promoter; SD3, oligonucleotide (5′ GTG AAC TCG ACT GTG ACA AGC TGC); SA4, oligonucleotide (5′ CAC CTG AGG AGT GAA TTG GTC G); dUSD1, oligonucleotide (5′ CUA CUA CUA CUA GCG ACG AAG ACC TCC TCA AGG C); dUSA1, oligonucleotide (5′ CUA CUA CUA CUA GTC GGG TCC CCT CGG GAT TGG); $SD_v$, vector splice donor site; $SD_c$, vector cryptic splice donor site; $SA_v$, vector splice acceptor site. (Reproduced with permission of Gibco/BRL).

An exon trapping system is now also commercially available from Gibco/BRL and pre-prepared Northern blots are commercially available from Clontech. The commercial system utilizes a mammalian expression vector containing the simian virus origin of replication (SV40) and the human immunodeficiency virus 1 *tat* intron and flanking exon sequences controlled by the early region promoter upstream of rabbit β-globin gene (*Figures 7* and *8*). Genomic DNA

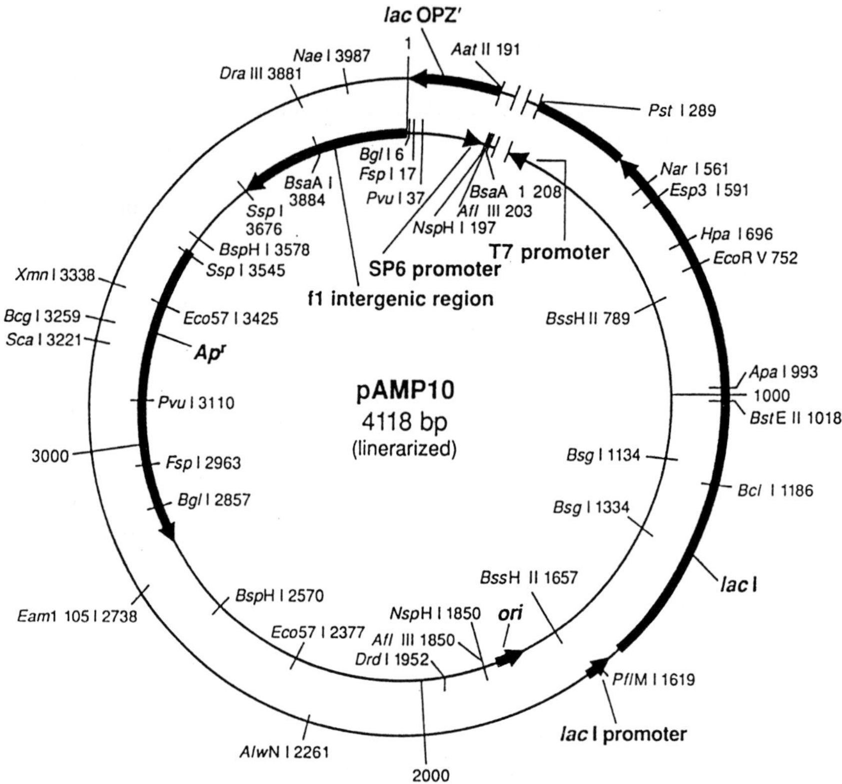


**Figure 9.** Map of pAMP10. On the circular map, enzymes that cleave pAMP10 once are shown on the outer circle, and restriction endonucleases that cleave twice are shown on the inner circle. In the lower section of the figure are shown the DNA sequences around the pAMP10 cloning site. Reproduced with permission of Gibco/BRL.

is cloned into the *Bam*HI site (the vector will accept 1–4 kb of DNA). The reporter gene is transcribed by the SV40 early promoter with a polyadenylation signal derived from SV40. Upon transfection of the plasmid construct into COS-7 cells, RNA transcripts are generated and the *tat* intron sequences are spliced to produce a polyadenylated cytoplasmic RNA. This only occurs when the insert is in a sense orientation. The pSPLI is modified to contain an *Nde*I site 3′ to the *Bam*HI site and a cryptic splice donor site. If correct splicing occurs the *Nde*I site will be excised from the transcript. If, however, the insert contains a cryptic splice acceptor, then the *Nde*I site will also be transcribed between the cryptic splice acceptor site and the vector cryptic splice donor site (see *Figure 7B*). After reverse transcription with oligo SA4 as an initiating primer and PCR using oligos SA4 and SD3, the possible product sizes are shown in *Figure 7C*. The product is then digested with *Nde*I.

Only cDNA containing a cryptic splicing will be digested. Nested PCR is performed on the digested DNA with oligos dUSD1 and dUSA1 which contain 5′-uracil residues. Absence of a PCR product implies an *Nde*I site in the primary PCR product, or more rarely a genomic *Nde*I site. After PCR, the product is digested with uracil DNA glyosylase to remove the uracil residues. This creates an alkali-sensitive, apyrimidinic (AP-DNA) site in the DNA. The AP-DNA sites are hydrolysed by alkali to destroy the section of the DNA strand which previously contained uracil. The product can now be cloned into vector pAMP10 which contains a single-stranded terminal cloning region complementary to the now single-stranded ends of the PCR product (*Figure 9*).

## Acknowledgements

We thank J. H. Riley, D. J. Ogilvie, P. J. Hedge, P. Elvin, A. Graham, J. E. N. Morten, C. R. Newton, M. McLean, and J. C. Smith for advice and support. We thank P. Hallett, Gibco/BRL for supply of exon trapping materials and protocols for evaluation prior to their commercial availability and for permission to reproduce *Figures 7–9*.

## References

1. Olson, M., Hood, L., Cantor, C., and Botstein, D. (1989). *Science*, **245**, 1434.
2. Riley, J., Ogilvie, D., Finniear, R., Jennes, D., Powell, S., Anand, R., Smith, J. C., and Markham, A. F. (1990). *Nucleic Acids Res.*, **18**, 2887.
3. Bird, A. P. (1987). *Trends Genet.*, **3**, 342.
4. Monaco, A. P., Neve, R. L., Colletti-Feener, C., Bertelson, C. J., Kurnit, D. M., and Kunkel, L. M. (1986). *Nature*, **316**, 336.
5. Yuille, M. A. R., Goudie, D. R., Affara, N. A., and Ferguson-Smith, M. A. (1991). *Nucleic Acids Res.*, **19**, 1950.
6. Albertson, H. M. (1990). *Proc. Natl Acad. Sci. USA*, **87**, 4256.
7. Kwiatkowski, Jr., T. J., Zoghbi, H. Y., Ledbetter, S. A., Ellison, K. A., and Chinault, A. C. (1990). *Nucleic Acids Res.*, **18**, 7191.
8. Green, E. D. and Olson, M. V. (1990). *Proc. Natl Acad. Sci. USA*, **87**, 1213.
9. Anand, R., Riley, J. H., Butler, R., Smith, J. C., and Markham, A. F. (1990). *Nucleic Acids Res.*, **18**, 1951.
10. Bentley, D. R. (1992). In *Techniques for the analysis of complex genomes* (ed. R. Anand), p. 113. Academic Press, San Diego. *See also* Riley, J. R., Ogilvie, D., and Anand, R. *ibid*. p. 59.
11. Lohr, D. (1988). In *Yeast: a practical approach* (ed. I. Campbell and J. H. Duffis), p. 125. IRL Press, Oxford.
12. Copley, C. G., Boot, C., Bundell, K., and McPheat, W. L. (1991). *Biotechnology*, **9**, 74.
13. Hultman, T., Stahl, S., Hornes, E., and Uhlén, M. (1989). *Nucleic Acids Res.*, **17**, 4937.

14. Kinzler, K. W., Nilbert, M. C., Vogelstein, B., Bryan, T. M., Levy, D. B., Smith, K. J., Preisinger, A. C., Hamilton, S. R., Hedge, P. J., Markham, A. F., Carlson, M., Joslyn, G., Groden, J., White, R. L., Miki, Y., Miyoshi, Y., Nishisho, I., and Nakamura, Y. (1991). *Science*, **251**, 1366.
15. Elvin, P., Butler, R., and Hedge, P. J. (1992). In *Techniques for the analysis of complex genomes* (ed. R. Anand), p. 156. Academic Press, San Diego.
16. Elvin, P., Slynn, G., Black, D., Graham, A., Butler, R., Riley, J. H., Anand, R., and Markham, A. F. (1990). *Nucleic Acids Res.*, **18**, 3913.
17. Hochgeschwender, U. (1992). *Trends Genet.*, **8**, 41.
18. Parimoo, S., Patanjali, S. R., Shukla, H., Chaplin, D. D., and Weissman, S. M. (1991). *Proc. Natl Acad. Sci. USA*, **88**, 9623.
19. Lovett, M., Kere, J., and Hinton, L. M. (1991). *Proc. Natl Acad. Sci. USA*, **88**, 9628.
20. Korn, B., Sedlacek, Z., Manca, A., Kioschis, P., Konecki, D., Lehrach, H., and Poustka, A. (1992). *Hum. Molec. Genet.*, **4**, 235.
21. Liu, P., Legerski, R., and Siciliano, M. J. (1989). *Science*, **246**, 813.
22. Corbo, L., Maley, J. A., Nelson, D. L., and Caskey, C. T. (1990). *Science*, **249**, 652.
23. Padgett, R. A. (1986). *Ann. Rev. Biochem.*, **55**, 1119.
24. Gubler, U. (1987). In *Methods in enzymology* (ed. S. L. Berger and A. R. Kimmel), Vol. 152, p. 325. Academic Press, New York.
25. Duyk, G. M., Kim, S., Myers, R. M., and Cox, D. R. (1990). *Proc. Natl Acad. Sci. USA*, **87**, 8995.
26. Buckler, A. J., Chang, D. D., Graw, S. L., Brook, J. D., Haber, D. A., Sharp, P. A., and Houseman, D. E. (1991). *Proc. Natl Acad. Sci. USA*, **88**, 4005.
27. Hamaguchi, M., Sakamoto, H., Tsuruta H., Sasaki, H., Muto, T., Sugimura, T., and Terada, M. (1992). *Proc. Natl. Acad. Sci. USA*, **89**, 9779.
28. Freshney, R. I. (ed.) (1987). *Culture of animal cells: a manual of basic techniques*, 2nd edn. A. R. Liss, New York.

# 11

# Fingerprinting of DNA and RNA using arbitrarily primed PCR

JOHN WELSH, MANUEL PERUCHO, MIGUEL PEINADO, DAVID RALPH, and MICHAEL McCLELLAND

## 1. Introduction

### 1.1 Applications of arbitrarily primed PCR

Arbitrarily primed PCR is a modification of PCR that provides an information-rich fingerprint of genomic DNA. This method is based on the selective amplification of genomic sequences that, by chance, are flanked by adequate matches to an arbitrarily chosen primer. If two template genomic DNA sequences are different, their arbitrarily primed PCR products display different banding patterns. Such differences can be exploited in ways largely analogous to the uses of restriction fragment length polymorphisms, including genetic mapping, taxonomy, phylogenetics, and the detection of mutations.

The most frequent use of arbitrarily primed PCR has been the detection of dominant polymorphic markers in genetic mapping experiments (1). Using DNA from a family of recombinant inbred mice, we and others have placed several hundred anonymous markers derived from arbitrarily primed PCR on the mouse genetic map (2–5). Arbitrarily primed PCR (including the method known as RAPD, ref. 1) has found wide use in the molecular genetics of plants. Sobral and colleagues have placed 200 markers on the sugar cane genetic map (6), while similar mapping projects are proceeding in several laboratories on many other plants, for instance, pine (7), and *Arabadopsis* (8). Genetic markers that are linked to a measurable phenotype can be detected with these methods by bulked segregant analysis (9). In another system, Kubota *et al.* (10) used arbitrarily primed PCR to detect X-ray-induced damage in fish embryos.

Polymorphisms detected by arbitrarily primed PCR can also be used as taxonomic markers in population studies of a wide variety of organisms (11, 12). In this context, the method is directly applicable to epidemiology and historical ecology. The application of arbitrarily primed PCR to the taxonomy and phylogeny of microorganisms, such as *Staphylococcus* and

*Borrelia burgdorferi*, is essentially the same and has been described elsewhere (12, 13).

Arbitrarily primed PCR can be used to detect somatic genomic changes that occur in anonymous sequences in certain tumours. Such somatic genetic alterations include allelic losses or gains (14) and deletion or insertion mutations of one or a few nucleotides (15) which occur during tumour development and progression. Finally, arbitrarily primed PCR can also be applied to RNA to detect differentially expressed genes (12, 16).

## 1.2 The arbitrarily primed PCR reaction

The standard PCR reaction involves the amplification of a sequence by primed DNA synthesis using two primers whose complementary sequences flank the desired sequence, followed by denaturation of the product, and further synthesis using the product from the first step as the template for the second step. When repeated, this process results in the exponential amplification of the sequence flanked by the two primer sequences. Normally, the primers that flank the desired sequence are annealed to the template DNA at relatively high stringency, so that only the desired sequence is amplified. High stringency during the primer annealing step ensures that the primers do not interact with the template DNA at positions where they do not match. Arbitrarily primed PCR, on the other hand, is based on the ability of PCR to generate a reproducible array of products when the annealing step is performed at low stringency. Under low stringency conditions, a single primer can interact with the template DNA at positions where the match is imperfect.

In arbitrarily primed PCR, a single oligonucleotide of arbitrary sequence is used to initiate DNA synthesis from sites along the template with which it matches only imperfectly. A few of these sites exist, stochastically, on opposite strands of the template, within several hundred nucleotides of each other. After two low stringency cycles, the sequence of the primer flanks a handful of anonymous sequences, which can be amplified by PCR. After a few low stringency cycles, the annealing temperature is raised and the reaction is allowed to continue under standard, high stringency PCR conditions. The products of this reaction are then resolved by gel electrophoresis. Although the parameters governing the actual priming and amplification events are quite complex and not fully understood, for a determined set of experimental conditions the pattern of DNA bands that are amplified is reproducible.

The large number of bands amplified with a single arbitrary primer generates complex fingerprints that reflect differences between different DNA templates. Because the priming events during the initial low stringency cycles depend on the nucleotide sequence of the primer, and this sequence has been chosen arbitrarily, the amplified sequences represent an arbitrarily chosen small sample of the template DNA. Fingerprinting using other primers allows the comparison of, firstly, the presence or absence and secondly, the molecular weights of an essentially limitless array of anonymous sequences. Application of this method to fingerprinting in various prokaryotic and eukaryotic systems has been demonstrated (1, 17–19).

The first part of this chapter describes the methods we have used for the DNA fingerprinting of large genomes (human, mouse, maize, etc.) to detect genetic differences between two or more templates. As an example of DNA fingerprinting, we describe here the methods we have used for the detection, isolation, and characterization of a polymorphic sequence that is frequently lost in colorectal cancer. All of these procedures can be generalized to other situations with minimal and obvious modifications. The second part of this chapter describes new methods based on arbitrarily primed PCR for the detection of differentially expressed genes. This method uses arbitrary priming in conjunction with reverse transcription and PCR to detect differences in the abundance of arbitrarily sampled messenger RNAs isolated from cells that have been subjected to different growth conditions or tissues that have developed from different developmental pathways.

## 2. Genomic fingerprinting of mammalian DNA

### 2.1 Application to cancer research

DNA fingerprinting by arbitrarily primed PCR provides a complementary molecular approach to the cytogenetics of solid tumours. The application of arbitrarily primed PCR allows an unbiased examination of genetic alterations of the cancer cell genome and has led to a better understanding of the genomic instability of cancer cells (15, 20). Tumour-specific somatic genetic alterations can be easily detected by comparing arbitrarily primed PCR fingerprints of DNA from normal and tumour tissue isolated from the same individual (14). Fingerprint bands reflecting these somatic mutations have been cloned and characterized.

Of special interest regarding the application of arbitrarily primed PCR to cancer research is that the amplified bands usually originate from unique sequences (rather than from repetitive elements). Furthermore the amplification is semi-quantitative, in that the intensity of an amplified band is proportional to the concentration of its corresponding template sequence. Therefore information on the overall allelic composition of the genome can be obtained by careful control of arbitrarily primed PCR. Thus, the degree of aneuploidy of a tumour cell genome is reflected in differences in the intensities of arbitrarily primed PCR bands, compared to those from the normal diploid genome from the same individual. These experiments require careful adjustment of the template DNA concentration, because the relative intensities of bands within a single arbitrarily primed PCR fingerprint are determined, to some extent, by the initial template concentration. It is recommended that each genomic DNA is fingerprinted at several concentrations, to reveal any products which show a significant template concentration dependence. The kinetics of arbitrarily primed PCR are discussed in more detail in Section 3.2.

## 2.2 Arbitrarily primed PCR

### 2.2.1 Genomic fingerprinting

Arbitrarily primed PCR produces a profile of arbitrarily amplified bands from the genome. The assay is performed under competitive conditions (50 or more different sequences are co-amplified with the same primer), providing the requirements for quantitative amplification. Consequently, it is possible to detect losses or gains in the number of copies of a sequence by changes in the intensity of a band in the arbitrarily primed PCR pattern. This property, in addition to the demonstrated ability of the arbitrarily primed PCR to detect polymorphisms, is very useful in many studies, for example investigations of the genetic events that occur in the tumour cell during its transformation from normal into cancer cell.

To carry out genomic fingerprinting using arbitrarily primed PCR (*Protocol 1*), the genomic DNA from frozen normal and tumour tissue of the same individual is diluted to 20 ng/ml and to 10 ng/ml in TE buffer. The measurement of DNA concentration is subject to several possible artefacts and is often not very accurate, so adjustments may be needed if the intensities of most of the arbitrarily primed PCR products are not uniform between DNA samples. This is one of the reasons why it is always wise to fingerprint in duplicate, using template DNA at concentrations differing by two- or four-fold. In *Protocol 1* 25 ng and 50 ng of template DNA are used in duplicate reactions for this reason.

**Protocol 1.** Arbitrarily primed PCR

*Equipment and reagents*

- Thermal cycler (e.g. Perkin-Elmer, model 480)
- PCR tubes
- Sequencing gel electrophoresis apparatus (40 cm long, 30 cm wide, 0.4 mm thick)
- Gel dryer
- Autoradiogram markers (e.g. Glogos II, Stratagene)
- X-ray film and exposure cassettes
- TBE buffer: 0.089 M Tris-borate, 0.025 M disodium EDTA pH 8.3
- 6% polyacrylamide, 8 M urea gel 1 × TBE buffer
- 10 × PCR buffer: 100 mM Tris–HCl pH 8.0, 500 mM KCl, 25 mM $MgCl_2$, 0.01% gelatin, 100 mM $MgCl_2$
- TE (1 mM Tris–HCl pH 7.5, 1 mM EDTA)
- 10 × dNTP mix (1.25 mM each dNTP)
- Genomic DNA from frozen normal and tumour tissue of the same individual (diluted to 20 ng/ml and to 10 ng/ml in TE)[a]
- Arbitrary primer (KpnX primer sequence: 5′-CTTGCGCGCATACGCACAAC-3′, concentration: 100 pmol/μl)
- [$\alpha$-$^{35}$S]dATP > 1000 Ci/mmol (5 μCi/reaction tube) or [$\alpha$-$^{32}$P]dCTP > 3000 Ci/mmol (1–2 μCi/reaction tube)[b]
- *Taq* polymerase (5 U/μl; Perkin-Elmer Cetus)
- Denaturing loading buffer: 95% formamide, 0.1% bromophenol blue, 0.1% xylene cyanol, 10 mM EDTA

*Method*

1. A reaction mixture for 20 tubes is prepared to compare ten genomes. Add components sufficient for 21 reactions in the following order:

| | *per reaction* | *total (×21)* |
|---|---|---|
| • distilled water | 16.05 μl | 337 μl |
| • 10 × PCR buffer | 2.5 μl | 52.5 μl |
| • 100 mM $MgCl_2$ [c] | 0.5 μl | 10.5 μl |
| • 10 × dNTP mix | 2.5 μl | 52.5 μl |
| • arbitrary primer [d] | 0.5 μl | 10.5 μl |
| • 5μCi α-[$^{35}$S]dATP | 0.5 μl | 10.5 μl |
| • 1 U *Taq* polymerase | 0.2 μl | 4.2 μl |
| • Total volume | 22.5 μl | 400 μl |

2. Distribute 22.5 μl to each reaction tube. Add 50 ng genomic DNA (2.5 μl of 20 ng/μl) to one set of tubes and 25 ng (2.5 μl of 10 ng/μl) to the other set of tubes. Add one drop of mineral oil to each tube.
3. Perform the reaction in a thermal cycler for:
   - five low stringency cycles (95°C, 60 sec; 50°C, 1 min; 72°C, 2 min) then
   - 30 high stringency cycles (95°C, 30 sec; 65°C, 30 min; 72°C, 2 min)
4. Dilute 10 μl of the complete reaction mix with 10 μl of formamide–dye buffer and incubate at 90–95°C for 3 min.
5. Immediately chill the solution on ice and load 2 μl on to an 8 M urea/ 6% polyacrylamide sequencing gel [e] and electrophorese for 6 h at 55 W.
6. Dry the gel under vacuum at 80°C and directly expose to an X-ray film at room temperature without an intensifier screen. [f] Stick three or more luminescent labels (autoradiogram markers) to the dried gel in order to localize the bands in the gel in case the isolation of one of the bands is desired.

[a] See Section 2.2.1 for comments on genomic DNA concentration.
[b] Bands are sharper when $^{35}$S is used.
[c] Considering the PCR buffer content, the final $MgCl_2$ concentration is 5 mM.
[d] Primers can also be used in pairwise combinations to give unique fingerprints.
[e] The samples with 50 ng and 25 ng DNA from each individual should be loaded side by side to act as controls for reproducibility.
[f] $^{32}$P labelled gels will require a longer exposure time if an intensifier screen is not used, but the bands are considerably sharper. $^{35}$S gels usually require an exposure of 2 or more days, while $^{32}$P gels require less than 24 h.

*Figure 1* shows the arbitrarily primed PCR analysis of 13 matched colorectal normal-tumour tissue DNA pairs using the KpnX primer (50) under the conditions described in *Protocol 1*. DNA fragments of sizes ranging from 100 to about 2000 nucleotides are reproducibly amplified. In this case, differences

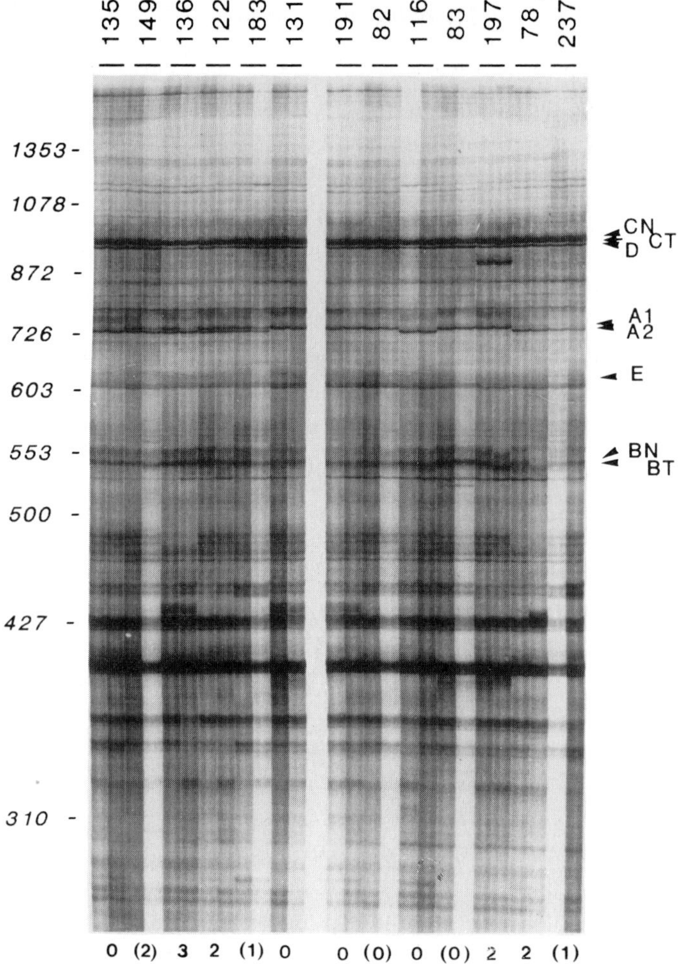


**Figure 1.** Arbitrarily primed PCR on human tumours. Primer KpnX and [$^{35}$S]dATP were used according to *Protocol 1*. A total of 13 pairs of the normal (left panel) and the tumour tissue DNA (right panel) from the same individual were analysed. The PCR fingerprints were resolved by denaturing polyacrylamide gel electrophoresis. Numbers above each lane are patient sample numbers. A molecular size marker (in bases) is shown at the left. Letters on the right indicate denomination of bands, as discussed in the text. Although only one concentration of genomic DNA (50 ng) is shown for each sample, it is strongly recommended that fingerprinting be performed using at least two different concentrations of each DNA sample.

in the banding patterns are apparent. Some of these differences represent polymorphisms in the human population, because they are present in both normal and tumour tissues from only some individuals.

Comparison of tumour versus normal tissue reveals four types of changes:

(a) new bands in the tumour tissue DNA, but not in the matching normal (or any other normal tissue DNAs);

(b) changes in molecular weight of amplified fragments, reflected by changes in the mobility of bands;

(c) increases in the intensity of a band;

(d) decreases in the intensity of a band.

Examples of most of these changes are shown in *Figure 1*.

The following sections describe the isolation and characterization of bands of interest identified by the arbitrarily primed PCR analysis. *Protocols 2–6* described in these sections list all of the procedures conducted to isolate and characterize bands A1 and A2 of *Figure 1*. These two bands correspond to two different alleles (length polymorphism) from the same locus.

### 2.2.2 Isolation and cloning of genomic DNA sequences amplified by arbitrarily primed PCR

Once a fingerprint has been generated, it is often necessary to purify and clone bands that display polymorphisms for further characterization or for use as biomarkers. In *Protocols 2* and *3*, we present methods that allow polymorphic bands to be purified, reamplified, and cloned.

**Protocol 2.** Gel purification and reamplification of the purified band

*Material and equipment*

- Scalpel or razor blade
- PCR and electrophoresis material as in *Protocol 1*

*Method*

1. Align the autoradiogram markers on the gel with their exposed images. Use a needle to mark in the dried gel the exact position of the band, then excise with a scalpel or razor blade.[a]
2. Place the excised portion of the gel (approx. 0.5–1 × 2–3 mm) in 50–100 μl of water and incubate at 60°C for 10–20 min to elute the DNA.
3. Reamplify 1 μl of the eluted DNA with the same primer and under the same conditions as in *Protocol 1*, except do not add extra $MgCl_2$ (final concentration of $MgCl_2$ is 1.5 mM instead of 5.0 mM). Perform 30 high stringency cycles (95°C, 30 sec; 65°C, 30 sec; 72°C, 2 min).
4. Analyse the PCR product by gel electrophoresis running the sample next to an arbitrarily-primed PCR product to verify its size and purity.

[a] Re-exposure of the gel will confirm the accuracy in the excision of the band.

If other bands are co-amplified, the desired band can again be cut from the gel and reamplified by PCR (*Protocol 2*). If the major product of the PCR is the appropriate band, this product can then be cloned (*Protocol 3*).

Some arbitrarily chosen oligonucleotide primers will contain recognition sites for restriction endonucleases that digest DNA to produce DNA fragments with staggered ends compatible with the cloning sites of commercially available plasmid or phagemid vectors. In this case, one can digest both the vector and re-amplified band DNA with the appropriate restriction enzyme(s), ligate, and transform using standard protocols. Frequently, however, arbitrarily chosen oligonucleotides do not contain recognition sites for restriction enzymes. *Protocol 3* is a procedure for cloning any reamplified PCR product into a plasmid or phagemid vector.

**Protocol 3.** Cloning of reamplified bands

*Equipment and reagents*

- Plasmid or phagemid vector DNA (10 ng/ml) (e.g. Bluescript II or pCR-script; Stratagene)
- *E. coli* cells competent for transformation. We use Epicurian Coli XL1-Blue cells (Stratagene)
- *Eco*RV or *Srf*I (Stratagene)
- 10 × Reaction buffer (1 M potassium, 250 Tris-acetate pH 7.6, 100 mM magnesium acetate 5 mM β-mercaptoethanol, 100 μg/ml BSA)
- T4 DNA ligase (GIBCO-BRL)
- 10 mM ATP
- LB agar microbiological plates supplimented with 50–70 μg/ml ampicillin
- LB medium
- 1.5 ml microcentrifuge tubes
- 0.5 ml microcentrifuge tubes
- 37°C water incubator
- 42°C water bath
- 65°C water bath
- 100 mM IPTG
- 2.0% X-gal in dimethylformamide
- Other material and reagents described in *Protocol 1*.

A. *Ligation, transformation, and colony selection*

1. In a 0.5 ml microcentrifuge tube, mix in the following order:
   - PCR product [a] — 2.0 μl
   - vector DNA at 10 ng/ml — 1.5 μl
   - 10 X reaction buffer — 1.0 μl
   - 10 mM ATP — 0.5 μl
   - restriction endonuclease (about 10 U) — 1.5 μl
   - T4 DNA ligase (about 4 U) — 1.0 μl
   - Water to a final volume of — 11 μl

   Incubate at room temperature for 2 h. Then incubate at 65°C for 10 min. Chill on ice.
2. In a prechilled 1.5 ml microcentrifuge tube, transform 60–100 μl of highly competent *Escherichia coli* cells by adding 2–4 μl of the ligation reaction and leaving on ice for 30 min.
3. Heat shock the transformed cells for 45 sec at 42°C. Then return the cells to ice.
4. Add 1.0 ml of LB medium to the 1.5 ml microcentrifuge tube, mix, and incubate at 37°C for 1 h.

5. During this 1 h incubation at 37°C, spread 40 μl of 100 mM IPTG solution and 70 μl of 2.0% X-gal solution on to each of several LB plates supplemented with 50–70 μg/ml of ampicillin.
6. After the 1 h incubation, spread 50–100 μl of the transformed cells on to each prepared LB plate. Incubate for at least 20 h at 37°C.
7. Pick white colonies into culture tubes containing 2–5 ml LB + ampicillin. Grow overnight at 37°C with aeration.
8. Add glycerol to 15% to an aliquot of each culture and freeze at −70°C for long-term storage.

B. *Reamplification of the insert by PCR*[b]

1. Mix 10 μl of the bacterial culture from step 6 above with 90 μl 1 × PCR buffer.
2. Boil for 30 sec then pellet the cells in a microcentrifuge (13 000 *g*).
3. Perform high stringency PCR according to *Protocol 2,* step 3 using the original primer.

[a] In general, extraction of the PCR product is not essential, but if cloning does not work the first time, purification by ethanol precipitation or extraction with any of the commercial kits available for this purpose (Bio101 Geneclean, Millipore cartridges, DS Primer Remover, etc.) should solve the problem.

[b] Alternatively, if one wishes to sequence single-stranded DNA mobilized from these cultures with helper phage, display the cell free supernatants from helper phage infected cultures electrophoretically. Analyse by sequencing only those phagemids which contain inserts. We routinely analyse 60 white colonies per day using this method.

### 2.2.3 Southern hybridization of arbitrarily primed PCR gels

In many cases, hybridization analysis may be necessary to confirm that the cloned band corresponds to the band visualized in the arbitrarily primed PCR fingerprint. A dried arbitrarily primed PCR gel can be transferred to a blotting membrane (*Protocol 4*). Then it is possible to demonstrate if the probe hybridizes to the expected band by Southern hybridization using the cloned band as probe.

**Protocol 4.** Transfer of arbitrarily primed PCR DNA from gel to a blotting membrane[a]

*Equipment and reagents*

- Nitrocellulose or nylon blotting membrane (e.g. Bio-Rad Zeta-Probe and Amersham Hybond)
- 3MM Whatman paper
- 3 M NaCl, 0.5 M Tris–HCl pH 7.4
- 2 × SSPE: 0.036 M NaCl, 0.02 M sodium phosphate pH 7.7, 2 mM EDTA
- UV crosslinker (e.g. Stratalinker, Stratagene)

**Protocol 4.** *Continued*

*Method*

1. Cut out the part of the dried gel that is going to be transferred to the nitrocellulose or nylon membrane. Cut at least 10 sheets of Whatman 3MM paper and the blotting membrane to the same size as the gel segment.
2. Dip four sheets of paper and the dried gel in 3 M NaCl, 0.5 M Tris–HCl pH 7.4 and lay them on a glass plate with the gel on top.
3. Soak the blotting membrane with water and lay this on top of the gel. Place one wet sheet of 3MM paper and five or more sheets of dry 3MM paper on top of the membrane.
4. Cover with another glass plate and a weight (~0.5 kg). Leave for 2 h to overnight.
5. Wash the blotting membrane with water and 2 × SSPE. Irradiate with UV to cross-link the DNA. To do this, either use a UV cross-linker (such as the Stratalinker, Stratagene) or place the membrance face down on clear plastic wrap on a transilluminator for 15 min.
6. Dry the membrane (a few minutes at 40–70°C) and expose to X-ray film.

[a] Using the procedure given in this protocol, 10–50% of the radioactive material is transferred to the blotting membrane. The transfer efficiency is not the same for all the bands but depends on size: bands larger than about 1500 bp will transfer with lower efficiency.

The probe for the hybridization step is prepared directly from the purified PCR product (from *Protocol 2*) or from a plasmid obtained by a miniprep. If the purified PCR product is to be used, It must first be extracted with any of the conventional methods used for purification of PCR products. Ethanol precipitation does not always work well, especially with small PCR products, whereas filtration cartridges or the various precipitation methods do work well (see Section 2.2.2). About 20–100 ng of purified PCR product needs to be labelled. Unincorporated nucleotides are removed by elution of the probe using resin columns such as NucTrap probe purification columns (Stratagene). Hybridization is performed using standard methods (21).

This type of experiment is especially informative when using arbitrarily primed PCR gels with $^{35}$S-labelled DNA fragments as the efficiency of the transfer to blotting membranes can then be easily monitored by following the radioactivity during transfer. Subsequent hybridization to a $^{32}$P-labelled probe yields distinguishable signals over the background $^{35}$S-radioactive bands. Specific hybridization by the probe to the band of the correct size is evidence of successful cloning. The presence of the background $^{35}$S-bands facilitates the identification of new hybridizing bands. Cross-hybridization with other bands amplified with the same or different arbitrary primers can also yield interesting information.

## 2.3. DNA sequencing by PCR-based cycle sequencing

A cycle sequencing method that uses a double-stranded PCR product as template and does not require sample purification has been developed and is described in *Protocol 5* (20). The substrate is generated by PCR using two primers that anneal to the vector sequence. Alternatively the cloned product in *Protocol 3* can be rescued as a phagemid for ssDNA sequencing.

A typical example of results of a cycle sequencing reaction is shown in *Figure 2*. Other protocols have been developed (22).

### **Protocol 5.** Cycle sequencing

*Equipment and reagents*

- *M13* sequencing primers (*−20* and *reverse* primers)[a] or any other primers that allow the amplification of the plasmid insert by PCR
- Material for PCR as described in *Protocol 1*
- One of the primers used for PCR amplification (*Protocol 6*) (100 pmol/μl)
- 5 μCi [α-$^{32}$P]dCTP
- 2 × ddNTP solutions prepared in 1 × PCR buffer (*Protocol 1*).
    - 2 × ddGTP 80 μM
    - 2 × ddATP 500 μM
    - 2 × ddTTP 500 μM
    - 2 × ddCTP 400 μM
- Other equipment and reagents required are described in *Protocol 1*

*Method*

1. Perform PCR amplification with 1 μl of the supernatant of the boiled overnight culture (from *Protocol 2*, step 3) using the following cycling profile: 30 sec at 95°C, 30 sec at 55°C and 1.5 min at 72°C for 30 cycles for the *M13 −20* and *reverse* primers.[b]
2. Prepare a mix (in this case for eight sequencing reactions) containing:

| | *per reaction* | *total (× 9)* |
|---|---|---|
| • water | 42.5 μl | 382.5 μl |
| • 10 × PCR buffer | 5.0 μl | 45.0 μl |
| • 10 × dNTP[c] | 0.5 μl | 4.5 μl |
| • sequencing primer | 1.0 μl | 9.0 μl |
| • 5 μCi [α-$^{32}$P]dCTP[d] | 0.5 μl | 4.5 μl |
| • 2 U *Taq* polymerase | 0.4 μl | 3.6 μl |
| • total volume | 50.0 μl | 400.0 μl |

3. Distribute 50 μl to each of eight microcentrifuge tubes.

**Protocol 5.** *Continued*

4. Add 1–2 μl of the PCR product from step 1. Mix and distribute 10 μl into each of four PCR tubes (one of each of the four ddNTPs reactions).
5. Add 10 μl of the appropriate ddNTP solution[e] to each sequencing reaction to give a final volume of 20 μl.
6. Perform 30–35 cycles of PCR with a thermal profile identical to that used in step 1.[f]
7. Dilute the final PCR product two- to four-fold with formamide denaturing buffer, heat at 90–95°C for 3 min and cool on ice.
8. Analyse 2 μl aliquots on a denaturing 6% polyacrylamide/8 M urea sequencing gel at 60 W with loadings that run for 2 h and for 5 h.
9. Dry the gel under vacuum at 85°C and expose to a X-ray film at room temperature without an intensifier screen.

[a] We have used the *M13 −20* and *reverse* primers, but if the cloning site is very close to the position where these primers anneal to the vector, other primers further from the cloning site should be used.

[b] For cycle sequencing, no purification of the band amplified during step 1 is needed. In a few cases, plasmid that has lost the insert can result in a second PCR band smaller than the expected one. If the artefactual band is very short (<50 bp) it will not interfere significantly with the cycle sequencing, but larger bands will produce a second ladder that can make the reading of the insert sequence difficult. In this case, gel purification of the PCR band is recommended.

[c] Note that the 10 × dNTP (1.25 mM each) mix is the same as the one used for arbitrarily primed PCR (*Protocol 1*) but for sequencing the final concentration of dNTP is 20-fold less.

[d] The sharpness of the $^{32}$P-labelled bands is notably improved if exposure is done at room temperature and without an intensifier screen.

[e] Optimal results may require small adjustments of ddNTPs concentrations. The final concentration of dNTPs is about 7–8 μM due to the 6.25 μM of freshly added dNTPs and the carry over of dNTPs from the PCR. Thus, the dNTP:ddNTP ratios are approximately 1:5 for G, 1:30 for A and T, and 1:25 for C.

[f] Under these conditions, essentially only one of the strands is extended and incorporates [α-$^{32}$P]dCTP. The concentration of the primers carried over from the previous PCR is very low relative to the newly added primer (1:50–1:100) and does not interfere with the sequencing if the reaction is carried out for a sufficient number of cycles.

## 2.4 Chromosomal localization of cloned sequences

The appropriate choice of standard somatic cell hybrid template DNA permits rapid chromosomal localization of polymorphic bands by arbitrarily primed PCR. As an example of this method, arbitrarily primed PCR fingerprinting of somatic cell hybrid DNA using a primer that identifies an interesting polymorphism related to cancer allows identification of the human chromosome from which the polymorphism originated. Thus for bands A1/A2 (see Section 2.2.1) the experimental approach was as follows.

(a) Specific PCR amplification of the isolated band.

Once the sequence of the cloned band is known, specific PCR primers

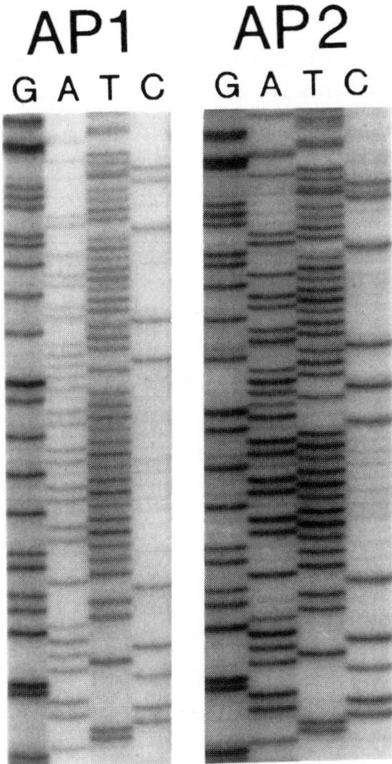


**Figure 2.** Cycle sequencing. Ladder obtained from PCR amplification directly from transformed bacteria. The sequences are from clones derived from bands A1 and A2 in *Figure 1*.

are designed to amplify these sequences directly from genomic DNA. For arbitrarily primed PCR A1/A2 bands (see Section 2.2.1), the primers used were 5′ GGGGCTAGTCACCCACATTA 3′ and 5′ TGGGGAATGTGACGGTCAAT 3′. PCR is performed according to *Protocol 2*, step 3, using high stringency conditions, in this case 35 cycles of 95 °C, 30 sec; 60 °C, 30 sec; 72 °C, 90 sec. Single copy sequences will produce a unique band. The identity of this band with the previously isolated band must be confirmed by sequencing and/or RFLP analysis.

(b) PCR amplification from rodent/human hybrids

DNA panels of rodent/human somatic cell hybrids are available from PCR amplifiable panel I from BIOS and the NIGMS mapping panel II from Coriell Institute for Medical Research, Camden, New Jersey. PCR is performed with 50–100 ng of genomic DNA from the different hybrids using specific PCR primers and the PCR products are analysed by non-denaturing 6% polyacrylamide gel electrophoresis. Because each cell line contains only one or a few human chromosomes, successful amplification of the specific product will indicate that the sequence of interest is located in one of the chromosomes present in this/these cell line(s). In the present example, the specific product can be observed only in human DNA (positive control) and hybrid 8 which contains human chromosomes 8, 17, and 18. However, as chromosomes 8 and 18 are also present in other hybrids that did not show amplification, it was concluded that this sequence is located in chromosome 17 (14). A more precise localization requires linkage experiments.

## 3. RNA fingerprinting by arbitrarily primed PCR

There are many biological situations where differential gene expression results in distinguishable phenotypes, for example, tissue and cell types, responses to hormones, growth factors, stress and the heterologous expression of certain genes. Several methods exist for detecting differential gene expression and cloning differentially expressed genes that do not rely on a biological assay of phenotype. Most of these methods fall into two general categories: subtractive hybridization and differential screening. Each of these approaches has its strengths and weaknesses.

- *Subtractive hybridization* is technically very challenging, and has been used successfully by only a few laboratories. The difficulty of subtractive hybridization methods probably derives from the nature of hybridization kinetics. Abundant genes hybridize faster and to greater level of completion than low abundance genes, making them more amenable to subtractive methods. Unfortunately, many interesting regulatory genes are of the low abundance type, where hybridization is difficult to drive to completion. A further problem with subtractive methods is that exhaustive hybridization between one RNA population (or cDNA equivalent) and a driver from another population is generally necessary if there is to be any hope of avoiding sequences that are not differentially expressed. The problem is that significant differences in expression that do not fall into the 'all-or-nothing' category will be missed.
- *Differential screening* suffers from similar drawbacks. In a typical differential screening experiment, radioactive probe is made from cDNA from two cell types, for example, and used to screen a cDNA library prepared from one of the two. Occasionally, clones from the library hybridize to one or the other but not to both probes. Unfortunately, low abundance messages do not yield sufficient probe mass to allow favourable hybridization kinetics.

Methods that rely on a biological assay of gene function have been very useful, particularly in the area of cancer research. However, appropriate bioassays for most genes do not exist.

We have developed a novel approach to the detection and cloning of differentially expressed genes. This method is based on a modification of arbitrarily-primed PCR (23). Here, we describe the methods and rationale behind the adaptation of arbitrarily primed PCR to the fingerprinting of

RNA, thereby facilitating the detection of RNA transcripts which are differentially expressed. A similar method has been developed by Liang and Pardee (16).

## 3.1 RNA arbitrarily primed PCR (RAP–PCR)

In RAP–PCR, first strand synthesis is initiated from an arbitrarily chosen primer at those sites in the RNA that best match the primer. It appears to be the case that the 3′ seven or eight nucleotides in the primer are the most important. However, genomic fingerprinting experiments indicate that the more 5′ sequences also have an effect on which sequences amplify. DNA synthesis from these priming sites is performed by reverse transcription. Second strand synthesis is achieved by adding *Taq* polymerase and the appropriate buffer to the reaction mixture. Once again, priming occurs at the sites where the primer finds the best matches. Poorer matches at one end can be compensated for by very good matches at the other. The consequence of these two enzymatic steps is the construction of a collection of molecules that are flanked at their 3′- and 5′-ends by the exact sequence (and complement) of the arbitary primer. These then serve as templates for high stringency PCR amplification, which can be performed in the presence of radioactive label such that the products can be displayed on a sequencing-type polyacrylamide gel, as in arbitrarily primed PCR fingerprinting of genomic DNA (24, 25). Because priming is arbitrary in the first step, the lack of poly(A)$^+$ tails in most bacterial RNAs is also not a problem for this method (26). The RAP–PCR procedure is described in *Protocol 6*.

**Protocol 6.** RNA arbitrarily primed PCR (RAP–PCR)

*Equipment and reagents*

- 96-well format thermocycle
- 0.2 ml reaction tubes
- An arbitrarily chosen primer. We suggest trying the *M13* sequencing primer or reverse sequencing primer.
- Multichannel micropipette
- 2×DNase I treatment mixture (20 mM Tris–HCl, pH 8.0, 20 mM $MgCl_2$, 40 U/ml DNase I
- First strand reaction mixture: 10 mM Tris–HCl pH 8.3, 5.0 mM KCl, 4 mM $MgCl_2$, 20 mM DTT, 0.2 mM each dNTP, 1 μM primer, 0.5 U murine leukaemia virus reverse transcriptase (MuLVRT)
- Second strand reaction mixture: 10 mM Tris–HCl pH 8.3, 25 mM KCl, 2 mM $MgCl_2$, 1 μM primer (as in first-strand reaction mix), 0.2 mM each dNTP, 0.2 μCi/μl [α-$^{32}$P]-dCTP, 0.1 U *Taq* polymerase (AmpliTaq, Cetus)
- Kodak AR-5 X-Omat X-ray film
- DNase I (Boehringer Mannheim Biochemicals)
- Denaturing loading buffer and other components (see *Protocol 1*)

*Method*

1. Prepare total cellular RNA by guanidinium thiocyanate–caesium chloride centrifugation or guanidinium thiocyanate–acid phenol–chloride extraction (27). Dissolve the final RNA pellet in 100 μl of water.

**Protocol 6.** *Continued*

2. Add 100 μl of 2 × DNase I treatment mixture and incubate at 37 °C for 30 min. Phenol extract and ethanol precipitate.
3. Prepare the treated RNA at two concentrations of about 20 ng/μl and 4 ng/μl by dilution in water.
4. Add 10 μl first strand reaction mixture to 10 μl RNA at each concentration. Allow the reaction to proceed at 37 °C for a time period 30 min to 2 h.
5. Add 20 μl of the second strand reaction mixture to each first strand synthesis reaction. Cycle through one low stringency step (94 °C, 5 min; 40 °C, 5 min; 72 °C 5 min) followed by 40 high stringency steps (94 °C, 1 min; 40 °C, 1 min; 72 °C, 2 min).
6. Add 2 μl of each reaction to 10 μl of denaturing loading buffer, and electrophorese on a 4% or 6% polyacrylamide sequencing-type gel containing 40% to 50% urea in 0.5 × Tris–borate–EDTA buffer.
7. Wrap the gel in plastic film or dry the gel and autoradiograph.

It should be clear that, given two or more RNA populations, differences in the fingerprints will result when corresponding templates are represented in different amounts. However, while these fingerprints contain anywhere from 40–150 bands, typical RNA populations for eukaryotic cells have complexities in the tens of thousands of molecules. Therefore, searching for a *particular* differentially expressed gene would be futile. Rather, the method is more appropriate for problems where many differentially expressed genes are anticipated. This is not so great a limitation as it might initially appear. First of all, many developmental and pathological phenomena are accompanied by many dozens or even hundreds of alterations in gene expression. Secondly, much of the technological development associated with sequencing, such as fluorescent-tagged primers and automated gel reading, and capillary electrophoresis, is readily adaptable to RAP–PCR. Therefore, a large fraction of the genes expressed in many situations can be surveyed given existing technology and this capability will be greatly enhanced by further technological developments. There are, however, other intrinsic limitations that should be considered in the design of an experiment based on RNA arbitrarily-primed PCR. Primarily, the influence of the abundance of messenger RNA on the fingerprint must be considered.

## 3.2 An approximate kinetic model of RAP–PCR

We have developed a simple working model for each of the steps in RAP–PCR. If the initial priming event resulting in first strand synthesis is represented by the probability, $p_1$, and the second priming event is represented by the probability $p_2$, the exponential growth of that product in the subsequent PCR can be approximately modelled as:

$$m = p_1 p_2 c X^n$$

where $X$ is a number between 1 (no amplification) and 2 (perfect amplification), $c$ is the concentration of the template molecule, $n$ is the number of *effective* exponential growth cycles, and $m$ is the mass of the product.

By effective exponential growth cycles, we refer to only those cycles that occur before any of the reaction components become limiting. Usually the limiting component is the polymerase, which is eventually titrated by the primed sites to the extent that only linear amplification can occur. The mass of each of the products can, in principle, be characterized by specific values for these variables. Naturally, there will occur instances in which any one of the variables may dominate. For example, given a normal distribution for the product, $p_1 p_2 X^n$, there will occasionally be high values for $c$ that compensate for low values of $p_1 p_2 X^n$. Simply put, the fingerprint cannot be expected to be abundance-normalized. On the contrary, we expect fingerprints generated by this simple version of RNA arbitrarily primed PCR to represent differences in abundant RNAs more faithfully than differences in low abundance messages. The over-representation of abundant messages can be a serious problem when the genes of interest give rise to rare messages. On the other hand, our preliminary information, bearing on the quality of match between the primer and the RNA in the first strand synthesis reaction and between the primer and the first strand cDNA in the second strand synthesis, suggests that adequate matches may be found only once in a few million nucleotides. Furthermore, Northern analysis has indicated that RNA abundances from differentially expressed genes revealed by RNA arbitrarily primed PCR vary over about two orders of magnitude. None the less, it is known from hybridization kinetics that RNA abundance can vary over many orders of magnitude. We have developed a partial solution to this problem, termed 'nested RAP–PCR'. Although we have only recently developed this method and have not fully explored its parameters, we feel that it is based on sound reasoning and, so far, appears to behave as expected. Therefore, we present this reasoning and a procedure in Section 3.6.

### 3.3 How large must a RAP–PCR experiment be?

The scale of a RAP–PCR experiment is determined by two factors:

(a) the number of tissues, cell types, or more generally, RNA populations to be compared;

(b) the number of products one wishes to compare.

Each RNA is fingerprinted at two concentrations, providing information on reproducibility, and typically, 40–150 bands might be observed, depending on the choice of primer and quality of materials. Thus, using a 96-well format PCR machine and 100-well gels (see *Protocol 6*), it is possible to examine over 2000 messages in *pair-wise* comparison in a single experiment (80 bands per lane × 100 wells, two concentrations, two RNA types). It is wise to screen a large number of primers against one or a few RNA preparations to maximize the likelihood of choosing useful primers.

### 3.4 RNA purification

The guanidinium thiocyanate–caesium gradient method of RNA purification of Chirgwin *et al.* (27) has been used successfully, but other methods may also work. Most purification methods yield RNA that is not entirely free of contaminating genomic DNA. This can be a serious problem because the genome is more than 10 times as complex as the RNA population, resulting in better matches with the primer. We therefore routinely treat the RNA with RNase-free DNase I prior to fingerprinting. The RNA concentration is checked spectrophotometrically and equal aliquots are electrophoresed on a 1% agarose gel and ethidium bromide stained to compare large and small ribosomal RNAs, qualitatively.

### 3.5 Choice of primers

Primers are chosen with several criteria in mind. First, the primers should not have stable secondary structure. Second, the sequence should be chosen such that the 3′-end is not complementary to any other sequence in the primer. In particular, palindromes should be avoided. Third, primers of 10–20 nucleotides in length can be used. Longer primers have some small theoretical advantages. In particular, 20-mers can be used at fairly high stringency such that any DNA contamination that persists will not contribute to the fingerprint. Also, as the reaction proceeds, some internal priming on amplified PCR products at the low stringency might be anticipated. Aside from these features, longer primers can contain more sequence information designed to aid in subsequent steps in the experiment, such as cloning, sequencing, etc. Primers 10 nucleotides in length can be obtained in kits from several companies and readers are referred to ref. 28 for a protocol.

### 3.6 Nested RAP–PCR

Nested RAP–PCR is a method designed to partially normalize the fingerprint with respect to mRNA abundance (29, 30). A procedure is described in *Protocol 7*. The strategy is very similar to standard nested PCR methods, except that we do not know *a priori* the internal sequences of the amplified products. In this method, the fingerprinting protocol (*Protocol 6*) is applied to the RNA, except that only 10 cycles of PCR amplification is performed rather than the 40 cycles in *Protocol 6*, step 4. A small aliquot of the first reaction is then further amplified using a second nested primer having one,

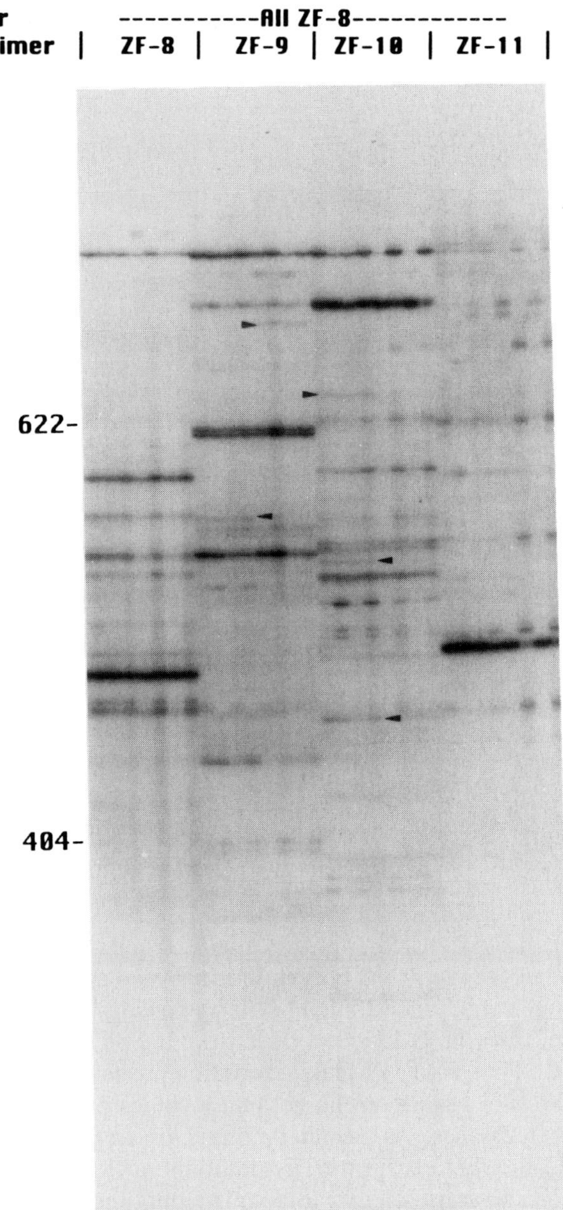


**Figure 3.** Nested RAP–PCR fingerprinting. Total RNA was prepared, in duplicate, from Mink lung epithelial cells, either untreated or treated with TGF-β. Then 40 ng and 10 ng of total RNA was fingerprinted using the ZF8 primer according to *Protocol 7*. Subsequently, ZF9, ZF10, and ZF11 were used for nested RAP. Arrows indicate differentially expressed RNAs.

two, or three additional *arbitrarily chosen* nucleotides at the 3′-end of the first primer sequence. *Figure 3* presents an example of the patterns obtained in such a RAP–PCR experiment.

**Protocol 7.** RNA arbitrarily primed PCR using nested primers

*Equipment and reagents*

- 96-well format thermocycler
- 0.5 ml reaction tubes
- A nested series of primers. We have used the series:
  - ZF8 CCAGAGAGAAACCCAGGA
  - ZF9 CAGAGAGAAACCCAGGAG
  - ZF10 AGAGAGAAACCCAGGAAGA
  - ZF11 GAGAGAAACCCAGGAAGAG
- Multichannel micropipette (see *Protocol 6*)
- First strand reaction mixture (see *Protocol 6*)
- Second strand reaction mixture (see *Protocol 6*)
- Nested primer reaction mixture: 10 mM Tris–HCl pH 8.3. 25 mM KCl, 2 mM $MgCl_2$, 0.2 mM each dNTP, 0.2 μCi/μl [α-$^{32}$P]dCTP, 1 μM nested primer, 0.1 U *Taq* polymerase (AmpliTaq, Cetus)
- Kodak AR-5 X-Omat X-Ray film

*Method*

**1–4.** Perform steps 1–4 of *Protocol 6* but perform the high stringency cycles of step 4 for only 10 cycles.

**5.** Transfer 3.5 μl of each reaction to 36.5 μl of nested primer reaction mixture.

**6.** Perform 40 high stringency PCR cycles (94°C, 1 min; 40°C, 1 min; 72°C, 2 min).

**7.** Add 2 μl of each reaction to 10 μl of denaturing loading buffer and electrophorese 2 μl on a 4% to 6% polyacrylamide sequencing-type gel containing 50% urea in 0.5 × Tris–borate–EDTA buffer.

**8.** Wrap the gel in plastic wrap (or dry the gel) and autoradiograph.

The nested RAP–PCR strategy partially abundance normalizes the sampling that occurs during RNA fingerprinting. If two mRNAs have equally good matches and equally good amplification efficiency but differ by 100-fold in abundance, then the products derived from them will differ by 100-fold after RAP–PCR. Thus RAP–PCR fingerprinting produces a background of products that are not visible on the gel that includes products derived from low abundance messages. A secondary round of amplification using a primer identical to the first except for an additional nucleotide at the 3′-end of the molecule can be expected to selectively amplify those molecules in the background that, by chance, share this additional nucleotide. The additional nucleotide will occur in 1/16 of the background molecules, accounting for both ends. There are many more molecules of low abundance in the RNA population than messages of high abundance, so most products produced by the high stringency nesting step should be sampled from the low complexity class. Each additional nucleotide at the 3′-end of the initial primer will contribute, in principle, a factor of 1/16 to the selectivity. In practice, the selectivity is probably somewhat less than this, because, although *Taq* polymerase is severely biased against extending a mismatch at the last nucleotide, it is more tolerant of mismatches at the second or third positions. None the less, our initial experiments are consistent with the interpretation that additional selectivity is achieved by this nested priming strategy.

Heteronuclear RNA (hnRNA) is thought to be about 10-fold more complex than mRNA. Complete normalization, therefore, can be expected to sample primarily hnRNA even if poly-$(A)^+$ selection of the RNA population is performed.

## References

1. Williams, J. G., Kubelik, A. R., Livak, K. J., Rafalski, J. A., and Tingey, S. V. (1990). *Nucleic Acids Res.*, **18**, 6531.
2. Nadeau J. H., Davisson M. T., Doolittle D. P., Grant P., Hillyard A. L., Kosowsky M. R., and Roderick, T. H. (1992). *Mammalian Genome*, **3**, 480.
3. Serikawa, T., Montagutelli, X., Simon-Chazottes, D., and Guenet, J.-L. (1992). *Mammalian Genome*, **3**, 65.
4. Woodward, S. R., Sudweeks, J., and Teuscher, C. (1993). *Mammalian Genome*, **3**, 73.
5. Welsh, J., Petersen, C., and McClelland, M. (1991). *Nucleic Acids Res.*, **19**, 303.
6. Al-Janabi, S. M., Honeycutt, R. J., McClelland, M., and Sobral, B. W. S. (1993). *Genetics*, **134**, 1249.
7. Neale, D. and Sederoff, R. (1991). *Probe*, **1**, 1.
8. Reiter, R. S., Williams, J. G., Feldmann, K. A., Rafalski, J. A., Tingey, S. V., and Scolnik, P. A. (1992). *Proc. Natl Acad. Sci. USA*, **89**, 1477.
9. Michelmore, R. W., Paran, I., and Kesseli, R. V. (1991). *Proc. Natl Acad. Sci. USA*, **88**, 9828.
10. Kubota, Y., Shimada, A., and Shima, A. (1992). *Mutation Research*, **283**, 263.
11. Chalmers, K. J., Waugh, R., Sprent, J. I., Simons, A. J., and Powell, W. (1992). *Heredity*, **69**, 465.
12. Welsh, J., Pretzman, C., Postic, D., Saint Girons, I., Baranton, G., and McClelland, M. (1992). *Int. J. Systematic Bacteriol.*, **42**, 370.
13. Welsh, J. and McClelland, M. (1993). In: *Diagnostic molecular microbiology* (ed. D. H. Persing, T. F. Smith, F. C. Tenover, and T. J. White), p. 595. ASM Press, Washington D.C.
14. Peinado, M. A., Malkhosyan, S., Velazquez, A., and Perucho, M. (1992). *Proc. Natl Acad. Sci. USA*, **89**, 10 065.
15. Ionov, Y., Peinado, M. A., Malkhosyan, S., Shibata, D., and Perucho, M. (1993). *Nature*, **363**, 558.
16. Liang, P. and Pardee, A. (1992). *Science*, **257**, 967.
17. Welsh, J. and McClelland, M. (1990). *Nucleic Acids Res.*, **18**, 7213.
18. Welsh, J. and McClelland, M. (1991). *Nucleic Acids Res.* **19**, 5275.
19. Welsh, J., McClelland, M., Honeycutt, R. J., and Sobral, B. W. S. (1991). *Theoretical and Applied Genetics*, **82**, 473.

20. Peinado, M. A., Fernandez-Renart, M., Capella, G., Wilson, L., and Perucho, M. (1993). *International J. Oncology*, **2**, 123.
21. Sambrook, J., Fitsch, E. F., and Maniatis, T. (ed.) (1989). *Molecular cloning: a laboratory manual* (2nd edn). Cold Spring Harbor Press, Cold Spring Harbor, New York.
22. Ruanto, G. and Kidd, K. K. (1991). *Proc. Natl Acad. Sci. USA*, **88**, 2815.
23. Welsh, J., Liu, J.-P., and Efstradiadis, A. (1990). *Genet. Anal. Tech. Appl.*, **7**, 5.
24. Welsh, J., Chada, K., Dahl, S. S., Ralph, D., Chang, R., and McClelland, M. (1992). *Nucleic Acids Res.*, **20**, 4965.
25. Wong, K. K., Mok, S.C-H., Welsh, J., McClelland, M., Tsao, S-W., and Berkowitz, R. S. (1993). *Internat. J. Oncology*, **3**, 13.
26. Wong, K. K. and McClelland, M. (1994). *Proc. Natl Acad. Sci. USA*, **91**, 639.
27. Chirgwin, J., Prezybyla, A., MacDonald, R., and Rutter, W. J. (1979). *Biochemistry*, **18**, 5294.
28. McClelland, M., and Welsh, J. (1994). *PCR Methods and Applications* **4**, S66.
29. McClelland, M., Chada, K., Welsh, J., and Ralph, D. (1993). In: *DNA fingerprinting: state of the science* (ed. S. D. Pena, R. Charkraborty, J. T. Epplen, and A. J. Jeffereys). Birkhauser Verlag, Basel.
30. Ralph, D., Welsh, J., and McClelland, M. (1993). *Proc. Natl Acad. Sci. USA*, **90**, 10710.

# 12

# Mutational analysis: known mutations

CLIVE R. NEWTON

## 1. Introduction

This chapter attempts to illustrate the various PCR options that are available to detect specific mutations and/or polymorphisms in genomic DNA. In so doing, the relative merits and drawbacks of each technique are addressed in the respective sections. For some methods of PCR-assisted mutation detection covered in this chapter, a protocol has not been included. In these instances the respective method is rarely used and is therefore discussed and the seminal reference cited. Protocols have, as far as possible, been described so that they can be applied in a generic fashion. However, detection of some mutations or polymorphisms may require deviations from the respective protocols as a result of adjacent genomic DNA sequence. This is more likely where detection of the specific DNA variation is dependent on oligonucleotide hybridization as opposed to primer/template annealing. To achieve accurate diagnoses, unequivocal results are required; therefore all precautions should be taken to avoid PCR contamination (see Chapter 1). In the case of an equivocal result from any of the methods described in this chapter, it is essential to perform a secondary analysis. The preferred method is direct sequencing of amplified DNA (see Chapter 5) as the result may be due to the presence of a rare or uncharacterized mutation, polymorphism, or even a substantial deletion or rearrangement in a region targeted either by amplification primers or hybridization oligonucleotides for the PCR product.

## 2. Substrate DNAs

Peripheral leucocytes are often used as a source of genomic DNA. A selection of methods for isolating large quantities of high quality DNA from blood are adequately documented elsewhere (1). Although DNA of the purity afforded by these methods works well in PCR-based assays, it is not necessary to achieve the quantity and quality of DNA that these protocols provide. Indeed, in some instances it has been found that if the substrate DNA concentration is particularly high then either the PCR reaction may be inhibited, specificity reduced, or streaking of PCR products will result during gel electrophoresis. If highly purified DNA is used, it should first be diluted to 5–50 μg/ml. *Protocols 1–4* generally allow the isolation of DNA of an acceptable quality for PCR analyses and can be performed simply and rapidly. The choice of protocol will usually be dictated by the reason the diagnosis is required and the current circumstances. Each protocol has different attributes. For example, DNA isolation from buccal cells is non-invasive, chorionic villus biopsy DNA allows pre-natal diagnoses (2), and Guthrie card isolated DNA allows retrospective analyses to be performed; in our laboratory DNA has been successfully amplified from Guthrie cards after 21 years of storage (3).

Whichever of the DNA isolation protocols are followed, it is essential to avoid cross-contamination of samples. Therefore rigorous attention should be paid to changing pipette tips between samples and between working solutions. The use of plugged pipette tips is also advocated to avoid aerosol contamination of pipette barrels. These considerations are particularly important in ARMS analyses where specific alleles are preferentially amplified.

### **Protocol 1.** DNA isolation from peripheral leucocytes

*Equipment and reagents*

- Microcentrifuge
- 5 ml EDTA blood tubes (e.g. Sarstedt 49.355.001)
- 1.5 ml screw-capped microcentrifuge tubes (e.g. Sarstedt 72.692)
- 170 mM $NH_4Cl$ (freshly-prepared)
- 10 mM NaCl, 10 mM EDTA (pH 7.5)
- 50 mM NaOH
- 1 M Tris–HCl pH 7.5

*Method*

1. Collect the blood sample into an EDTA tube and freeze the sample at −20°C.
2. Thaw the sample and transfer a 200 μl aliquot to 800 μl of freshly-prepared 170 mM $NH_4Cl$ in a screw-capped microcentrifuge tube. Rotate the tube for 20 min to mix the contents thoroughly.
3. Centrifuge the sample for 2 min in a microcentrifuge to obtain a white cell pellet and discard the supernatant.
4. Wash the cell pellet with 300 μl of 10 mM NaCl, 10 mM EDTA pH 7.5, and collect the pellet by centrifugation for 15 sec. Repeat this washing step three more times to remove all visible haem.
5. Resuspend the final cell pellet in 500 μl of 50 mM NaOH by vortexing for 10 sec.
6. Incubate the sample at 100°C for 5 min.
7. Neutralize the sample with 100 μl of 1 M Tris–HCl pH 7.5 and vortex the mixture for 5 sec.

8. Centrifuge the sample for 15 sec to remove cell debris and retain the supernatant containing the DNA.
9. 5 μl of DNA prepared in this way is sufficient for a single PCR reaction (2 × 5 μl per ARMS test). The DNA solution should be stored at −20°C until use.

**Protocol 2.** DNA isolation from buccal cells

*Equipment and reagents*

- Microcentrifuge
- Plastic universal tubes (sterile)
- 1.5 ml screw-capped microcentrifuge tubes (e.g. Sarstedt 72.692)
- 4% (w/v) sucrose
- 10 mM NaCl, 10 mM EDTA pH 7.5
- 50 mM NaOH
- 1 M Tris–HCl pH 7.5

*Method*

1. Agitate 10 ml of 4% sucrose in the mouth for 20 sec and collect the mouthwash in a sterile plastic universal tube.
2. Collect the buccal epithelial cells by centrifugation at 800 *g* for 10 min at room temperature and discard the supernatant.
3. Resuspend the pellet in 500 μl of 10 mM NaCl, 10 mM EDTA pH 7.5, and transfer the cell suspension to a screw-capped microcentrifuge tube.
4. Centrifuge the sample in a microcentrifuge for 15 sec and discard the supernatant.
5. Resuspend the cell pellet in 500 μl of 50 mM NaOH by vortexing for 10 sec.
6. Incubate the sample at 100°C for 5 min.
7. Neutralize the sample with 100 μl of 1 M Tris–HCl pH 7.5 and vortex the mixture for 5 sec.
8. Centrifuge the sample for 15 sec to remove cell debris and retain the supernatant containing the buccal cell DNA.
9. 5 μl of DNA prepared in this way is sufficient for a single PCR reaction (2 × 5 μl per ARMS test). Store the DNA solution at −20°C until use.

**Protocol 3.** DNA isolation from chorionic villus samples

*Equipment and reagents*

- Light microscope
- Microcentrifuge
- 1.5 ml screw-capped microcentrifuge tubes (e.g. Sarstedt 72.692)
- Dispase (Boehringer Mannheim, grade II, 2 U/ml)
- 0.85% NaCl

**Protocol 3.** *Continued*

*Method*

1. Chorionic villus biopsy material should be dissected with the aid of light microscopy to provide villi free of maternal tissue.[a]
2. Place the biopsy sample in a sterile Petri dish and add 500 μl dispase solution. Incubate it at ambient temperature for 5 min.
3. Wash the sample four times with separate 1 ml aliquots of 0.85% NaCl.
4. Transfer the sample (with minimal NaCl carryover) to a screw-capped microcentrifuge tube, add 500 μl of sterile water and then incubate at 100°C for 5 min.
5. Centrifuge at 13 000 *g* for 5 min in a microcentrifuge and retain the supernatant containing the DNA.
6. 20–50 μl of DNA prepared in this way is sufficient for a single PCR reaction (2 × 20–50 μl per ARMS test). Store the DNA at −20°C until use.

[a] In some chorionic villus biopsy cases, the fetal/maternal origin of the tissue may not be unequivocal. In such cases a paternal X chromosome contribution may be sought by using a PCR-based test using highly polymorphic X-linked microsatellites (4).

**Protocol 4.** DNA isolation from Guthrie cards

*Equipment and reagents*

- Microcentrifuge
- 1.5 ml screw-capped microcentrifuge tubes (e.g. Sarstedt 72.692)
- Paper hole punch
- 0.05% Tween 20

*Method*

1. Cut a 5 mm diameter disc from the blood spot using the hole punch and cut it into small pieces and place the pieces in a screw-capped microcentrifuge tube.
2. Add 100 μl 0.05% Tween-20 to the dissected blood spot and incubate at 100°C for 5 min.
3. Centrifuge the sample in the microcentrifuge at 13 000 *g* for 5 min.
4. Remove the supernatant into a fresh tube and centrifuge it at 13 000 *g* for a further 1 min.
5. Remove the supernatant containing the DNA into another fresh tube and use 5 μl per PCR reaction (2 × 5 μl per ARMS test). Store the DNA at −20°C until use.

# 3. Amplification refractory mutation system (ARMS)

ARMS has also been described in the literature as allele-specific PCR (ASP), PCR amplification of specific alleles (PASA), and allele-specific amplification (ASA).

ARMS (2, 5) has been successfully applied to the analysis of a wide range of polymorphisms, germ-line mutations, and somatic mutations. The utility of ARMS has thus been demonstrated in carrier detection and prenatal diagnosis of inherited disease and in the detection of residual disease during and after the treatment of malignancies. The ability of ARMS to detect such residual disease resides in the discrimination of the technique to selectively amplify a mutant allele within a vast background of normal alleles. Specificity and sensitivity of detection of one malignant cell in a background of $10^5$ normal cells have been reported (6). This selective amplification simplifies downstream analysis as, after ARMS, the background relative to amplified DNA becomes negligible. This compares with having to detect a mutant allele in a background of normal alleles at the original ratio if conventional PCR followed by hybridization is employed.

## 3.1 The ARMS reaction

A scheme of the ARMS reaction is shown in *Figure 1*.

A typical ARMS assay comprises two PCRs, each conducted using the same substrate DNA. Both reactions contain a common primer that anneals to an invariant DNA sequence to one side of the mutation to be detected. The 3′-terminus of the common primer is orientated towards the mutation. In one of the reactions the second primer is specific for one allele and in the other reaction the alternative allele-specific primer is included. The allele specificity of these primers is conferred by the 3′-nucleotide of the primer and the allele-specific primers are of the same sense as the strand primed by the common primer. It is also good practice to co-amplify a different region of the genome with a pair of internal control primers, to ensure that the PCR has been efficient thus helping to avoid a false negative result for the allele under investigation.

The ARMS reaction relies on the absence of a 3′ proof-reading activity of *Taq* DNA polymerase therefore some other thermostable DNA polymerases, such as Vent and *Pfu* DNA polymerases, are wholly unsuitable for this application. The reaction also requires the ability of the enzyme to initiate primed synthesis from a mismatch to be severely impaired such that amplification is essentially non-existent.

Inspection of the PCR products by agarose gel electrophoresis and ethidium bromide staining is all that is required to determine the zygosity of the

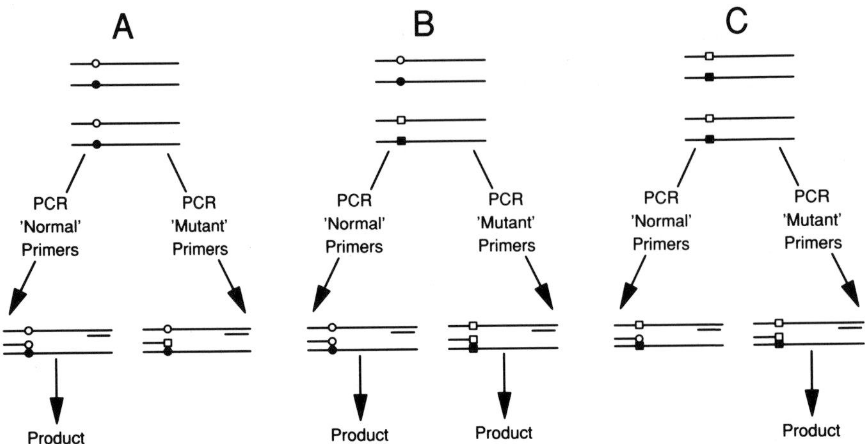


**Figure 1.** The ARMS assay. (A) normal homozygote DNA; (B) heterozygote DNA; (C) affected homozygote DNA. Circles represent the appropriate nucleotide of the normal allele and squares represent the equivalent nucleotide of the 'mutant' allele. Open and filled circles are complementary base pairs. Similarly, open and filled squares are complementary (i.e. a circle paired with a square is mismatched). (A) Only the normal allele is present therefore only the normal primer is incorporated into PCR product with this substrate DNA. (B) Both alleles are present, therefore in their respective PCR reactions, both ARMS primers give rise to PCR product. (C) With only the defective allele present, only the 'mutant' primer has a complementary substrate, is extended and generates amplified product.

DNA for the allele being examined. Therefore no additional enzymatic manipulations, hybridizations or the use of radioisotopes are needed.

## 3.2 Primer design

When designing primers for ARMS, as with other PCR techniques, it is necessary to consider genomic sequence as opposed to cDNA sequence. If the genomic DNA sequence is not available but the intron/exon boundaries within cDNA are known, it is still usually possible to position the common primer and ARMS primers so as to generate a suitable product for agarose gel electrophoresis. If this is not the case then intron sequence is required. A convenient method of acquiring this information is by the use of vectorette PCR (see Chapter 10).

Whilst an ARMS test can be designed using primers in the region of 20 nucleotides in length, it is useful to increase the length of the primers to about 30 residues. When longer primers are used it becomes more feasible to establish a generic set of conditions that are applicable to the loci to be examined. Furthermore, destabilizing the allele-specific primers by introducing deliberate mismatches close to the 3′-terminal nucleotide often improves specificity but this may reduce yield. When introducing additional mismatches

**Table 1.** Summary of the prediction for allele specificity[a,b]

| **Alleles** | **Mismatches** | **Specificity prediction** |
|---|---|---|
| A·T T·A | A·A and T·T | Excellent for A·A, good for T·T |
| C·G G·C | C·C and G·G | Excellent for both |
| C·G T·A | A·C and G·T | Good |
| | or | |
| | T·G and C·A | Good |
| G·C T·A | A·G and C·T | Excellent for A·G, poor for C·T |
| | or | |
| | T·C and G·A | Excellent for G·A, good for T·C |
| A·T G·C | C·A and T·G | Good |
| | or | |
| | G·T and A·C | Good |
| A·T C·G | G·A and T·C | Excellent for G·A, good for T·C |
| | or | |
| | C·T and A·G | Excellent for A·G, poor for C·T |

[a] Reproduced from ref. 7 with permission.
[b] All of the possible allele differences are shown on the left of the table. For some allelic differences, there are two different possible sets of primers depending upon which of the two strands of the DNA are used as template for extension of the allele-specific primers. A·T represents A(primer)·T(template); T·A represents T(primer)·A(template).

the type of mismatch, the position within the primer, the GC content of the five or six nucleotides preceding the 3′-nucleotide and the discriminatory 3′-nucleotide that is dictated by the difference between the alleles must all be considered. A recent study (7) has been carried out that compares all possible primer/template mismatches and a prediction derived for the ability of *Taq* DNA polymerase to extend these and their effect in allele-specific amplification. This prediction is reproduced in *Table 1*.

The closer to the 3′-terminus of the primer that an additional destabilizing mismatch is incorporated, the greater is the effect on destabilization. High GC content of the penultimate five 3′-nucleotides reduces the destabilizing effect. Qualitatively, additional mismatches can be ranked in the order CC > CT > GG = AA = AC > GT in their destabilizing effect. In our laboratory, we favour mismatch introduction at the 3′ penultimate nucleotide after considering all of the above factors. The effect of additional mismatches on the ARMS assay has to be determined empirically but it is worth testing the ARMS primers with alternative internal control primer pairs since the choice of control amplification fragment occasionally affects the specificity of the ARMS primers. In rare circumstances, absolute specificity is difficult to achieve, this can usually be overcome by designing a second set of primers that are complementary to the DNA strand primed by the first primer set.

If the ARMS test is carried out to detect a deletion or insertion mutation,

**A**

| | |
|---|---|
| Normal allele | AAATATCATCTTTGGTGTTTCCTATGATGAATATAGATACAGA-3′ |
| Normal ARMS primer | AAACCACAAAGGATACTACTTATATCTATG-5′ |
| Mutant ARMS primer | **t**AACCACAAAGGATACTACTTATATCTATG-5′ |

**B**

| | |
|---|---|
| Phe 508 deletion allele | AGAAAATATCATTGGTGTTTCCTATGATGAATATAGATACAGA-3′ |
| Normal ARMS primer | **a**AACCACAAAGGATACTACTTATATCTATG-5′ |
| Mutant ARMS primer | TAACCACAAAGGATACTACTTATATCTATG-5′ |

**C**

| | |
|---|---|
| Normal allele | AAATATCATCTTTGGTGTTTCCTATGATGAATATAGATACAGA-3′ |
| Normal ARMS primer | AGTAGAAACCACAAAGGATACTACTTATA-5′ |
| Mutant ARMS primer | **ta**TAG**t**AACCACAAAGGATACTACTTATA-5′ |

**D**

| | |
|---|---|
| Phe 508 deletion allele | AGAAAATATCATTGGTGTTTCCTATGATGAATATAGATACAGA-3′ |
| Normal ARMS primer | **ag**TAG**a**AACCACAAAGGATACTACTTATA-5′ |
| Mutant ARMS primer | TATAGTAACCACAAAGGATACTACTTATA-5′ |

**Figure 2.** Primer design. (A, B) The normal CF Phe 508 allele with the deleted codon of the mutant allele underlined (A) and the mutant allele (B); the poorly designed ARMS primers show (lower case and bold) the mismatched nucleotides. (C, D) As A and B but with well-designed primers substituted, showing three mismatches of each primer with the non-respective allele.

no further destabilization is required since the mutation can provide additional mismatching between one primer and the non-respective allele. An example of good and poor primer design for the cystic fibrosis (CF) *Phe* 508 deletion mutation is shown in *Figure 2*.

The choice of common primer is less demanding than for the allele-specific primers. The general rules are that the common primer should be selected for a region that has approximately 50% GC content, has no repeated or unusual sequences, provides an appropriate sized fragment for electrophoresis, and shares no 3′ complementarity with either allele-specific primer or internal control primers.

The experience in our laboratory has shown, furthermore, that primers for ARMS assays need not be purified after synthesis. Primers are routinely used subsequent to deprotection, removal of the ammonia using a rotary evaporator and resuspension in 1 ml of either 5% aqueous ethanol or sterile water.

This experience is, however, limited to oligonucleotides synthesized on Applied Biosystems DNA synthesizers. Synthesizers that may give lower coupling yields per nucleotide addition, or employ alternative chemistries, may give rise to the need for primer purification. In general, using crude primers allows approximately 1000 reactions to be performed per primer synthesis performed on a 0.2 μmol scale.

## 3.3 Specificity of ARMS primers

Several factors affect the specificity of ARMS primers. The effect on specificity attributable to these factors is reduced when additional destabilization of the primers has been included during primer design (see Section 3.2). The factors that have been noticed to most significantly reduce specificity are magnesium, enzyme, and primer concentrations. Indeed, magnesium concentrations above about 3 mM can completely abolish specificity. In general, the lower the concentration of each of these reagents, the higher is the specificity. However, if the concentrations of any of these reagents are too low they can adversely affect the efficiency of amplification. Specificity can also be adversely affected if the substrate DNA concentration is particularly high (see Section 2). Specific conditions can usually be found with the magnesium concentration at 1.2 mM, primer concentrations in the range 0.1–1 mM, and an enzyme concentration of 1 U per 100 μl reaction volume. A typical thermocycling regimen would comprise 30–40 cycles of 94 °C denaturation, 60 °C annealing, and 72 °C extension, with 1 min at each temperature. If, in the first instance, the primers are not specific, the reaction should be repeated with small temperature increments in the annealing step. A 'hot-start' procedure (see *Protocol 5*) often resolves specificity problems if they are not corrected by annealing temperature adjustment. The appropriate choice of destabilizing mismatches and primer lengths of 30 nucleotides however, will usually give success with the above conditions with minimal optimization required. A typical ARMS analysis is shown in *Figure 3*.

When several loci are co-amplified as in the case of multiplex ARMS (see Section 3.5) primer specificity is often improved. Here, yield may be affected and any additional destabilizing mismatches may need to be removed or altered to reduce their destabilizing effect.

**Protocol 5.** ARMS assay and optimization of PCR conditions

*Equipment and reagents*

- UV spectrophotometer
- Quartz UV spectrophotometry cells
- Thermocycler
- Agarose gel electrophoresis tank
- DC power supply
- 10 × reaction buffer: 100 mM Tris–HCl pH 8.3, 500 mM KCl, 12 mM $MgCl_2$, 0.1% gelatin
- 10 × dNTPs: most suppliers (e.g. Promega, BCL, Pharmacia) provide 2′-deoxynucleoside 5′-triphosphates (dNTPs) as 100 mM solutions. A 10 × dNTPs working solution (each dNTP 1 mM in combined aqueous mix) is conveniently prepared by adding 10 μl of each 100 mM dNTP to 960 μl of sterile water

**Protocol 5.** *Continued*

- *Taq* DNA polymerase (AmpliTaq, 5 U/μl; Applied Biosystems, a Division of Perkin–Elmer). Dilute this enzyme just before use with 1 × reaction buffer to 1 U/μl
- Oligonucleotide primers (common primer, 'normal' primer, 'mutant' primer, internal control forward primer, internal control reverse primer); 20 μM working solution of each primer in water[a]
- Light mineral oil

*Method*

1. Combine the primers for the ARMS assay in two tubes by mixing the following:

   *Tube 1*
   - common primer[a] (50 pmol) 2.5 μl
   - 'normal' primer[a] (50 pmol) 2.5 μl
   - internal control forward primer[a] (50 pmol) 2.5 μl
   - internal control reverse primer[a] (50 pmol) 2.5 μl

   *Tube 2*
   - common primer[a] (50 pmol) 2.5 μl
   - 'mutant' primer[a] (50 pmol) 2.5 μl
   - internal control forward primer[a] (50 pmol) 2.5 μl
   - internal control reverse primer[a] (50 pmol) 2.5 μl

2. Combine the following reagents in a separate tube:

   *Tube 3*
   - 10 × reaction buffer 10 μl
   - 10 × dNTPs 10 μl
   - DNA[b] 10 μl
   - water[b] 49 μl
   - *Taq* DNA polymerase[c] (1 U) 1 μl

3. Divide the contents of tube 3 equally between two fresh tubes ('normal' and 'mutant' tubes). Overlay the contents of each tube with light mineral oil and incubate them in the thermocycler at 94°C for 2 min.
4. Initiate 'hot-start' PCR by adding the contents of tube 1 to the 'normal' tube through the oil overlay. Likewise add the contents of tube 2 to the 'mutant' tube through the oil overlay.[d]
5. Run 30–40 cycles comprising 94°C, 1 min; 60°C, 1 min; 72°C, 1 min.
6. Run a final 72°C extension for up to 10 min.[e]
7. Remove 20–40 μl from below the oil overlay from each tube and analyse by agarose gel electrophoresis.[f]
8. Optimize the assay as appropriate by increasing the annealing tempera-

ture to increase specificity or lowering the annealing temperature and/or increasing the number of cycles to increase the yield (ideally, the cycle number should remain as low as possible). $Mg^{2+}$ may also be adjusted as discussed in Section 3.3.

[a] To calculate primer concentrations, first calculate their molar absorption coefficients at 260 nm. The molar absorption coefficient can be determined using the formula (15 200)A + (12 010)G + (7050)C + (8400)T where A, G, C, and T are the number of times that each respective residue occurs in the primer. The molar extinction coefficient calculated is equivalent to the $A_{260}$ of a 1 M solution of primer. Measure the absorbance of the primer solution at 260 nm (if the primer is crude rather than purified and prepared on 0.2 μmol scale (see Section 3.2), dilute a 20 μl aliquot to 1 ml, and multiply the $A_{260}$ of the dilution by 50 for the $A_{260}$ of the stock solution). Divide the $A_{260}$ of the primer solution by the molar absorption coefficient to derive the molar concentration of the primer. Dilute an aliquot of the stock solution to give a 20 μM working solution. In practice, it is often convenient when the same primer combinations are used routinely, to combine these primers in a mix where each primer is 20 μM within the mix.

[b] Use the appropriate volume of DNA according to the isolation protocol used and adjust the volume of water added to tube 3 to give a final volume of 80 μl.

[c] *Taq* DNA polymerase should be diluted to 1 U/μl with 1 × reaction buffer just prior to use.

[d] A 'hot start' comprises leaving an essential reaction component (e.g. primers, *Taq* polymerase) out of the reaction mixtures until the reaction mixtures have been denatured for the first time, then these are added before the mixtures have cooled to the annealing temperature. Adding a large volume through the paraffin oil overlay is easier than adding a small volume. Therefore each reaction can be set up conveniently in 0.8 × the final volume in 1.25 × reaction buffer with primers omitted. With all reactions held at denaturation temperature, the appropriate primer mixes are then added at 5 × the final required concentration through the oil overlay; thorough mixing is not necessary since this is achieved by convection within the reaction as thermal cycling commences. Products such as Ampliwax™ (Perkin-Elmer Cetus) can simplify the 'hot start' by providing a solid barrier to separate reaction components prior to thermal cycling. On heating, the wax melts, the reaction components mix, and the wax then fulfils the function of the oil overlay.

[e] The final prolonged extension is required only when the PCR products are particularly large and GC-rich.

[f] A typical ARMS analysis is shown in *Figure 3*.

## 3.4 Plasmid cassette system for primer and/or PCR optimization

Ideally, primer or reaction optimization should be performed using DNAs previously characterized with respect to the alleles to be detected. In some instances, suitable typed DNA may be scarce. In many cases, particularly when a number of mutations may give rise to the affected phenotype of a recessive disorder, homozygotes for specific mutations may not be available. ARMS primer specificity is essential in these situations (non-specificity of a 'mutant' allele primer may lead to misdiagnosis of an unaffected heterozygote as an affected compound heterozygote; in the case of more common mutations an affected homozygote could be mistyped as heterozygous if the 'normal' allele primer were non-specific).

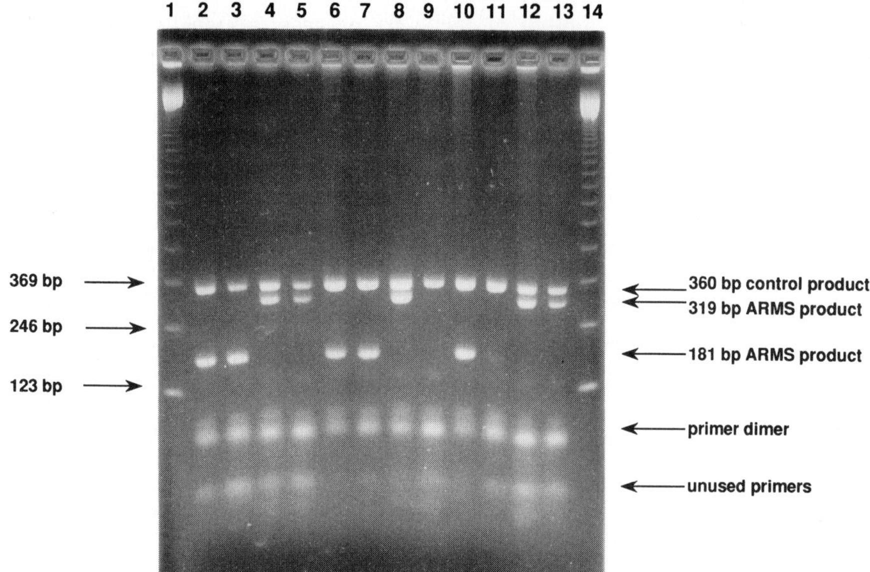


**Figure 3.** Ethidium bromide stained 1.5% agarose, 1.5% Nu-sieve agarose composite gel showing ARMS analysis of apolipoprotein (Apo) E genotypes. Lanes 1 and 14: 123 bp ladder. Lanes 2–5: DNA heterozygous for Apo E alleles Arg-158 and Cys-158 (lanes 2 and 3) and Cys-112 and Arg-112 (lanes 4 and 5); genotype $\epsilon_4\epsilon_2$. Lanes 6–9: DNA heterozygous for alleles Arg-158 and Cys-158 (lanes 6 and 7) and homozygous for allele Cys-112 (lanes 8 and 9); genotype $\epsilon_3\epsilon_2$. Lanes 10–13: DNA homozygous for allele Arg-158 (lanes 10 and 11) and heterozygous for alleles Cys-112 and Arg-112 (lanes 12 and 13); genotype $\epsilon_4\epsilon_3$. Figure reproduced from ref. 8.

The plasmid cassette system for primer optimization goes some way to allowing the testing of ARMS primers in the absence of clinical material or where typed homozygous DNA is not available. The system allows the duplication of the critical region of the 'normal' and 'mutated' alleles within a plasmid background. This permits the 3′-terminal mismatches and any additional mismatches of ARMS primers to be examined in the context of the approx. 60 bp that would flank them in genomic DNA. Having created 'artificial' alleles within a plasmid, then ARMS primers for either orientation can be assessed in conjunction with appropriate primers designed to the plasmid sequence. An alternative plasmid primer pair can also serve as PCR control. If synthetic genomic cassettes are routinely cloned between the same plasmid restriction sites, a generic primer set can be established for ARMS common primers, PCR control primers and sequencing primers (9). A scheme for plasmid cassette preparation is shown in *Figure 4*.

*Figure 5* shows two ARMS primer pairs tested using the plasmid cassette system during the development of an ARMS assay for $\alpha_1$-antitrypsin (AAT) deficiency and *Figure 6* shows the result of the use of both primer pairs each in conjunction with their respective plasmid-derived common primer.

### **Protocol 6.** Making a cassette and testing ARMS primers

*Reagents*

- 10 × T4 DNA ligase buffer: 0.66 M Tris–HCl pH 7.6, 66 mM $MgCl_2$, 100 mM dithiothreitol, 4 mM ATP
- L-broth containing 50 μg/ml ampicillin
- 'Cassette' oligonucleotides
- Plasmid pAT153 cut with *Eco*RI and *Bam*HI
- T4 DNA ligase (>1 U/μl)
- L-agar plates containing 50 μg/ml ampicillin
- L-agar plates containing 15 μg/ml tetracycline hydrochloride

*Method*

1. Combine the 'cassette' oligonucleotides (0.3 pmol each) and anneal them by adjusting them to 10 μl in 1 × T4 DNA ligase buffer. Briefly heat the mixture to 100 °C then cool this gradually to ambient temperature.
2. Purify the plasmid vector fragment (large fragment of pAT153 digested with *Eco*RI and *Bam*HI) by agarose gel electrophoresis.[a]
3. Add 0.3 pmol of the eluted vector fragment to the annealed 'cassette' oligonucleotides. Add 1 μl of 10 × T4 DNA ligase buffer and 0.1 U T4 DNA ligase, then adjust the volume to 20 μl with water. Incubate the reaction at 4°C for 1 h then at 15°C for 15 h.
4. Transform competent *E. coli* host cells with 1 μl from the ligation mixture and plate on to L-agar containing 50 μg/ml ampicillin. Incubate the plates overnight at 37 °C.[a]
5. Replica plate colonies from step 4 on to L-agar plates containing 50 μg/ml ampicillin and L-agar plates containing 15 μg/ml tetracycline hydrochloride. Incubate the plates overnight at 37 °C.[a]
6. Pick 12 ampicillin-resistant, tetracycline-sensitive colonies and grow 10 ml cultures in L-broth containing 50 μg/ml ampicillin. Incubate these overnight at 37 °C with shaking.[a]
7. Prepare plasmid mini-prep DNA from each culture.[a]
8. Re-transform competent *E. coli* cells with each plasmid mini-prep DNA and plate on to L-agar containing 50 μg/ml ampicillin. Incubate the plates overnight at 37 °C.[a]
9. Isolate plasmid DNA from 100 ml cultures of separate colonies from step 8 and sequence the cassette inserts using primer 3 (*Figure 4*).[a]

**Protocol 6.** *Continued*

**10.** Perform ARMS analyses according to *Protocol 5* using 0.1 ng of normal and mutant plasmids and a mixture of 0.05 ng of each plasmid to simulate heterozygous DNA using primers as orientated for the cassette primers as described in *Figure 4*.

[a] Detailed methods for restriction digestion, DNA purification from agarose gels, making and transforming competent *E. coli* cells, preparing media, plasmid DNA isolation, and double-stranded plasmid sequencing are adequately documented elsewhere (1).

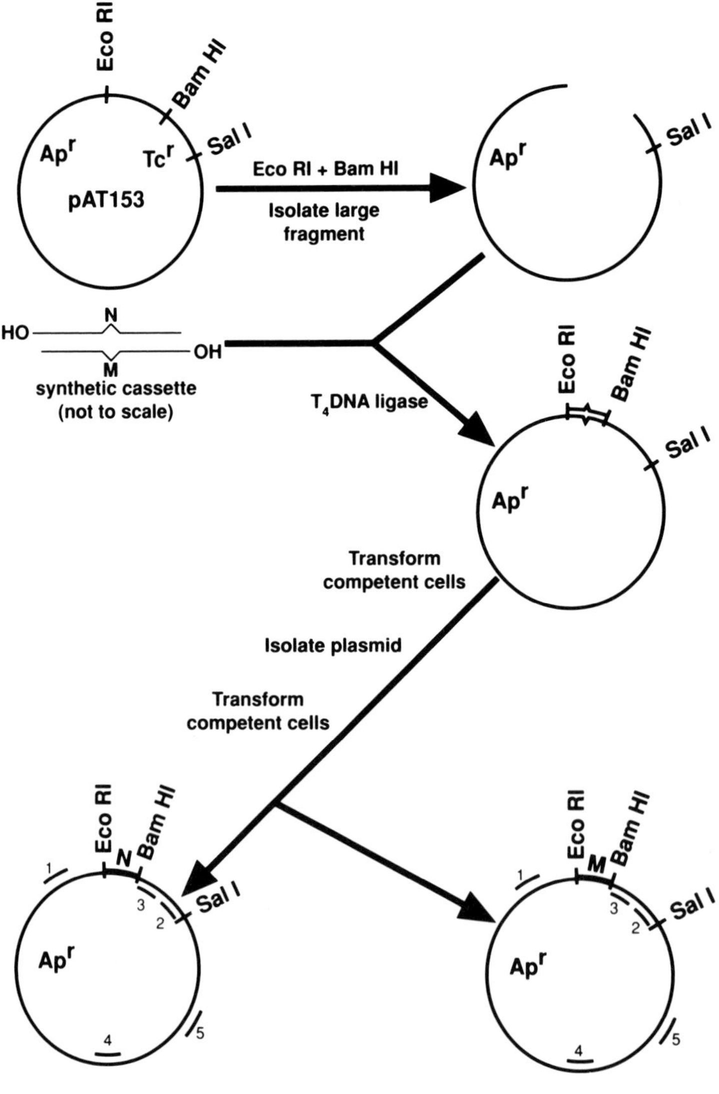

## 3.5 Recent advances: multiplex ARMS

It has recently been shown that several ARMS reactions can be combined in the same reaction tubes, even in the event of closely situated variant loci. Indeed, the multiplexing of ARMS can even be performed where individual PCR products overlap as recently demonstrated in the development of a multiplex ARMS test for the more common UK mutations of the cystic fibrosis transmembrane conductance regulator (CFTR) gene (10). The multiplex reaction is essentially the same as individual ARMS reactions. Two

```
  5'-CCGTGCATAAGGCTGTGCTGACCATCGCCA  Primer 7c
  5'-CCGTGCATAAGGCTGTGCTGACCATCGCCG  Primer 7b
AATTCCGTGCATAAGGCTGTGCTGACCATCGACgAGAAAGGGACTGAAGCTGCTGGGGCCATG
    GGCACGTATTCCGACACGACTGGTAGCTGtTCTTTCCCTGACTTCGACGACCCCGGTACCTAG
                     Primer 8b   CTCATTCCCTGACTTCGACGACCCCGGTAC-5'
                     Primer 8c   TTCATTCCCTGACTTCGACGACCCCGGTAC-5'
```

**Figure 5.** The synthetic ARMS duplex with the terminal *Eco*RI (left) and *Bam*HI (right) cohesive ends; the mismatched nucleotides equivalent to the AAT normal allele (top strand) and Z allele (bottom strand) are shown in lower case superscript (g) and lower case subscript (t) respectively. The ARMS primers 7b and 8b correspond to the normal allele, primers 7c and 8c correspond to the AAT Z variant. Underlined nucleotides depict the deliberately introduced additional mismatches to achieve specificity. The 7 series primers were used with common primer 2 to generate a 280-bp product; the 8 series primers were used with common primer 1 to give a 360-bp product. Reproduced from ref. 9 with permission from Eaton Publishing.

---

**Figure 4.** Scheme of the plasmid cassette system for ARMS primer optimization. Synthetic 'cassette' duplexes comprising one oligonucleotide equivalent to one strand of genomic DNA with 'normal' variant of polymorphic nucleotide (N) situated centrally hybridized to the complementary genomic DNA equivalent bearing the complement to the 'mutant' variant nucleotide (M). Each synthetic strand bears the appropriate restriction enzyme overhang for the cloning sites chosen (*Eco*RI and *Bam*HI are exemplified). The synthetic duplex is ligated between the *Eco*RI and *Bam*HI restriction sites of plasmid pAT153, inactivating the tetracycline resistance ($Tc^r$) gene. After transformation, plasmid isolation, and secondary transformation, ampicillin resistant ($Ap^r$), tetracycline sensitive ($Tc^s$) plasmids have inserts corresponding to the normal (N) or the mutated (M) allele. Primers 1 and 2 are universal primers for use with ARMS primers; primer 3 is a universal sequencing primer; primers 4 and 5 are universal amplification control primers. The nucleotide sequence of primer 1 is dTTACTTTCACCAGCGTTTCTGGGTGAGCAA; primer 2, dAAGCAGCCCAGTAGTAGGTTGAGGCCGTTG; primer 3, dGGTGATGCCGGCCACGATGC; primer 4, dCAGACCCCGTAGAAAAGATCAAAGGATCTT; primer 5, dACAGGACTATAAAGATACCAGGCGTTTCCC. Note: the secondary transformation is performed with isolated plasmid DNA since the primary transformants carry a mixture of both 'normal' and 'mutant' plasmids due to the mismatch from the cassette. Sequencing of the secondary transformants using primer 3 types individual secondary transformants. To simulate heterozygous DNA, equimolar quantities of 'normal' and 'mutant' plasmid are combined. Reproduced from ref. 9 with permission from Eaton Publishing.

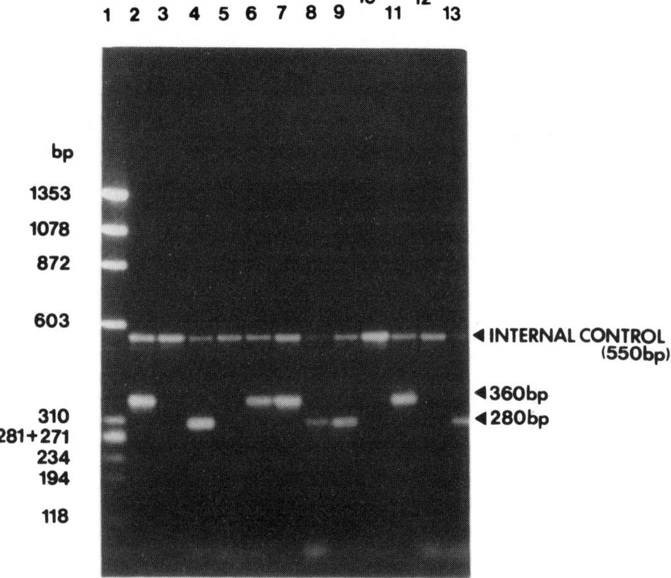


**Figure 6.** Ethidium bromide stained agarose gel analysis of the ARMS reactions using primers as described in *Figure 5*. Lane 1, size markers (ϕX174 *Hae*III digest); lane 2, primer 8b (normal plasmid substrate); lane 3, primer 8c (normal plasmid substrate); lane 4, primer 7b (normal plasmid substrate); lane 5, primer 7c (normal plasmid substrate); lane 6, primer 8b (mixed 'cassette' simulated heterozygote DNA); lane 7, primer 8c (mixed 'cassette' simulated heterozygote DNA); lane 8, primer 7b (mixed 'cassette' simulated heterozygote DNA); lane 9, primer 7c (mixed 'cassette' simulated heterozygote DNA); lane 10, primer 8b (mutant (AAT ZZ) plasmid substrate); lane 11, primer 8c (mutant plasmid substrate); lane 12, primer 7b (mutant plasmid substrate); lane 13, primer 7c (mutant plasmid substrate). The internal control band is the 550-bp amplification product of plasmid primers 4 and 5 (*Figure 4*). Reproduced from ref. 9 with permission from Eaton Publishing.

reactions are performed but each tube has a mix of ARMS primers for both normal alleles and mutant alleles. Given that compound heterozygotes will have more normal alleles than mutant alleles when five or more loci are multiplexed, this distributes the number of expected amplification products evenly between the two reactions. The requirement for an internal amplification control is also obviated if the number of combined reactions is four or more (10).

Standard ARMS reaction conditions apply when multiplexing individual ARMS reaction. However, it has been observed that, when individual primers that discriminate well in isolation are combined, there may be increased effects of any deliberately introduced additional mismatches as shown by reduced PCR product yields. In such situations it may be necessary to revise any deliberate additional primer mismatches to compensate for this. Naturally,

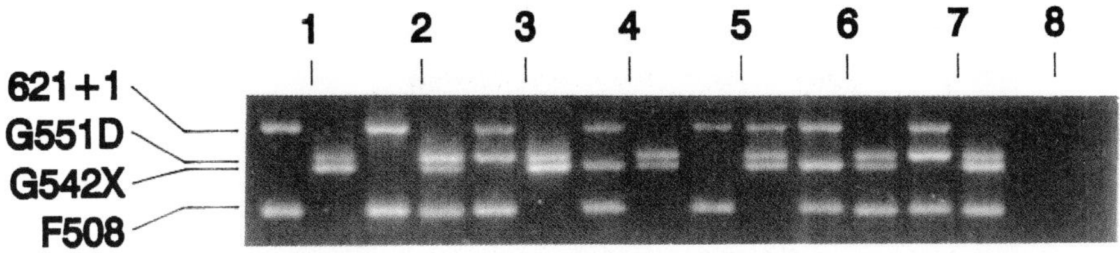


**Figure 7.** Ethidium bromide stained 3% agarose, (3:1 agarose:Nu-sieve agarose) composite gel showing multiplex ARMS. Lane pairs 1 to 7, multiplex ARMS analyses of seven separate individuals; lane pair 8, reagent blank (no DNA). The left of each lane pair is derived from a reaction containing the normal specific primers for the CFTR mutations 621+1 and ΔF508 and the mutant specific primers for the CFTR mutations G551D and G542X. Conversely, the right of each lane pair is derived from a reaction containing the mutant specific primers for mutations 621+1 and ΔF508 and the mutant specific primers for mutations G551D and G542X. The genotypes for the seven individuals are therefore: 1, normal/normal; 2, ΔF508/normal; 3, G551D/normal; 4, G542X/normal; 5, 621+1/normal; 6, G542X/ΔF508; 7, G551D/ΔF508.

specificity of the revised primers will need to be confirmed. An example of a CF multiplex ARMS analysis is shown in *Figure 7*.

The reliability and accuracy of multiplex ARMS has been demonstrated by the validation of the CF multiplex ARMS test exemplified in *Figure 7*. The validation study genotyped over 500 samples and the results were in complete agreement with the genotypes as determined by a variety of other methods (10). Thus, the ability to analyse accurately and simultaneously several mutations or polymorphisms in one simple test has led to multiplex ARMS becoming the method of choice in the analysis of known mutations in many genetic testing laboratories.

## 3.6 Recent advances: fluorescent ARMS using four-colour technology

The advent of phosphoramidite reagents that allow the conjugation of fluorophores to PCR primers during primer synthesis (see Chapter 3) has created the opportunity to adapt the ARMS technique further. Not only does the use of different fluorophores on the normal and mutant ARMS primers allow the two primers to be combined in one tube to amplify competitively but it also permits automated data capture and analysis. In theory, using the combination of multiplex ARMS (see Section 3.5 above) with the development of multiple fluorophores, fluorescent analysis has the potential to revolutionize genetic testing. *Figure* 8 shows the operating principle of the 362 Gene Scanner and 373A and 373 DNA sequencers (Applied Biosystems, Inc.) which allow automated ARMS data capture. *Figures 9* and *10* depict results of automated apolipoprotein E fluorescent ARMS genotyping which can be compared with *Figure 3* where equivalent non-fluorescent primers

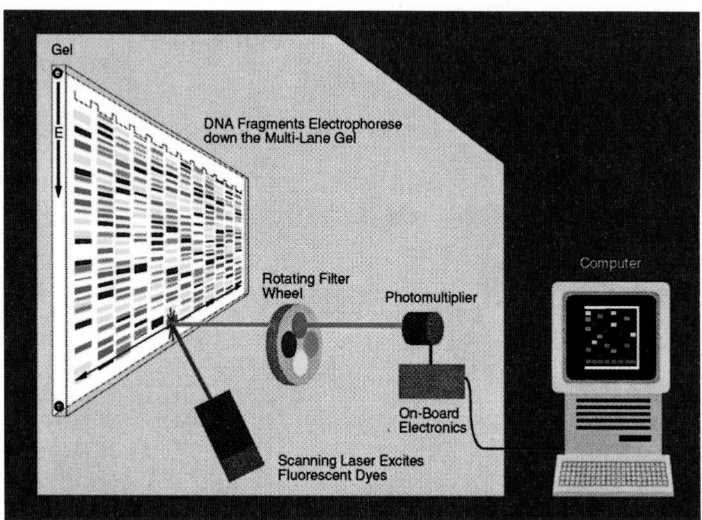


**Figure 8.** Fluorescent DNA analysis technology. The fluorescent DNA analysis technology implemented in the 362 Gene Scanner, 373 and 373A DNA sequencers uses multiple wavelength fluorescence detection for submarine agarose and polyacrylamide gel electrophoresis respectively. As fluorophore-tagged DNA fragments electrophorese past a scanning laser beam, fluorescence measurements at four wavelengths are taken and stored as a function of scan number (time). Separate band-pass filters are used to select which of four different colour fluorophores are detected by the photomultiplier. Stored data are analysed by the systems software to provide either primary DNA sequence information or, in the case of fluorescent fragment analysis, the sizes in base pairs, colours and concentrations of fluorescent molecules in each sample relative to a within-lane size standard. From Applied Biosystems Inc. with permission.

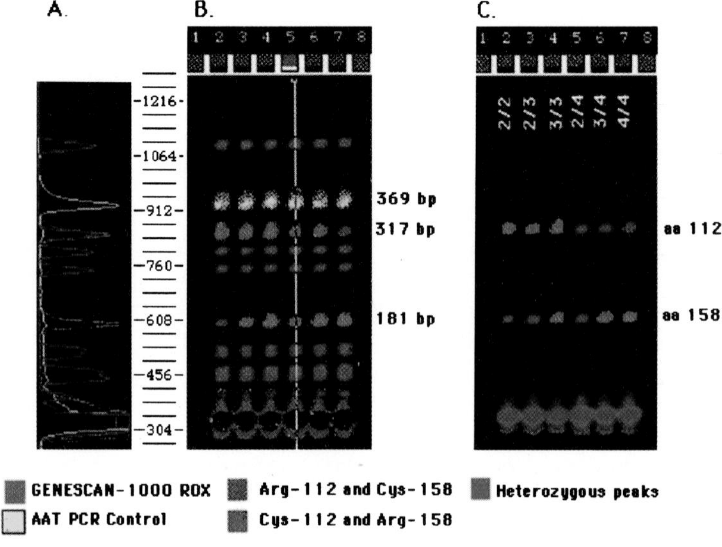

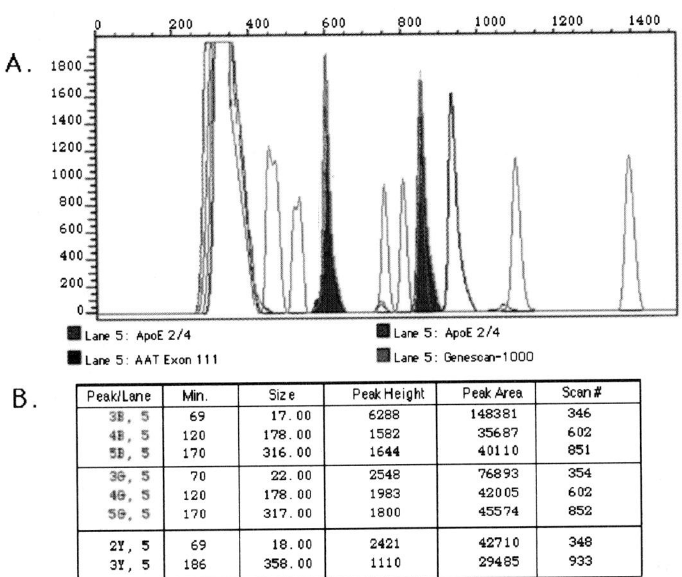


| Peak/Lane | Min. | Size | Peak Height | Peak Area | Scan # |
|---|---|---|---|---|---|
| 3B, 5 | 69 | 17.00 | 6288 | 148381 | 346 |
| 4B, 5 | 120 | 178.00 | 1582 | 35687 | 602 |
| 5B, 5 | 170 | 316.00 | 1644 | 40110 | 851 |
| 3G, 5 | 70 | 22.00 | 2548 | 76893 | 354 |
| 4G, 5 | 120 | 178.00 | 1983 | 42005 | 602 |
| 5G, 5 | 170 | 317.00 | 1800 | 45574 | 852 |
| 2Y, 5 | 69 | 18.00 | 2421 | 42710 | 348 |
| 3Y, 5 | 186 | 358.00 | 1110 | 29485 | 933 |

**Figure 10.** GENE SCANNER electrophoretogram for apo E 2/4 genotype. Genotype 2/4 apo E DNA is heterozygous for amino acid substitutions at both codon 112 and 158. This fluorescent fragment profile illustrates the 'lane view' of an ARMS assay of genotype 2/4 DNA from lane 5 of the gel in *Figure 9*. The profile consists of both blue and green fragments, sized by the GENE SCANNER as 178 and 316 bp, specific for codons 158 and 112, respectively. A 2μl aliquot of PCR product was electrophoresed for 3 h at 100 V in a 2% SeaPlaque agarose (FMC Inc.) gel and automatically analysed. The output consists of the four colour electrophoretogram (A) and an optional tabular display (B). Tabular data, displaying the detection time, fragment size, peak height, and area, as well as the scan number for each analysed fragment, may also be exported to other graphical or statistical programs for further analysis. From Applied Biosystems Inc with permission.

---

**Figure 9.** Automated genotyping of Apo E using fluorescence-based ARMS strategy. The 'gel view' of the apo E ARMS assay illustrates separation of DNA samples representing the six Apo E genotypes. 50 ng genomic DNA is subjected to 25 cycles of PCR amplification using competing allele-specific forward primers that are specifically tagged with blue (FAM) and green (JOE) dyes and a common unlabelled reverse primer. A 181-bp PCR product specific for codon 158 and a 317-bp product specific for codon 112 are amplified in separate PCR reactions and combined for analysis in one lane of the gel. The forward primers specific for Cys-112 and Arg-158 (amino acids encoded in the Apo E 3/3 genotype) are labelled with JOE (green) and primers specific for Arg-112 and Cys-158 are labelled with FAM (blue). A PCR control, labelled with TAMRA (yellow) co-amplifies a 360-bp region from the $\alpha_1$-antitrypsin gene and an internal lane standard, Genescan-1000 ROX (red), is added to allow automatic sizing and quantitation of PCR products. The 'gel view' consists of two analytical panels: (A) a lane scan showing the fluorescent fragment profile of a selected lane and (B) the four-colour gel image. To distinguish the six Apo E genotypes, only the blue-and green-labelled fragments are selected for viewing; heterozygous bands are distinguished as overlapping fragments in a third colour (pink) (C). From Applied Biosystems Inc. with permission.

have been used to determine some of the same genotypes in a standard ARMS assay (8).

## Protocol 7. ARMS assay using Applied Biosystems Inc. Gene Scanner

### *Equipment and reagents*

- 362 Gene Scanner or 373A DNA sequencer fitted with Gene Scanner software kit (Applied Biosystems, Inc.)
- Thermocycler
- 10 × Reaction buffer: 100 mM Tris–HCl pH 8.3, 500 mM KCl, 15 mM $MgCl_2$, 0.1% gelatin
- 12.5 × dNTPs: most suppliers (e.g. Promega, BCL, Pharmacia) provide 2′-deoxynucleoside-5′-triphosphates (dNTPs) as 100 mM solutions. A 12.5 × dNTPs working solution (each dNTP 2.5 mM in combined aqueous mix) is conveniently prepared by adding 25 μl of each 100 mM dNTP to 900 μl of sterile water
- *Taq* DNA polymerase (5 U/μl; Applied Biosystems, a Division of Perkin-Elmer). Dilute to 0.5 U/μl with double distilled water just before use
- Dimethylsulphoxide (DMSO)
- Primers labelled with fluorescent dyes:[a] allele 2-FAM primer, allele 3A-JOE primer, allele 4-FAM primer, allele 3B-JOE primer, AAT ExonIII-TAMRA primer, AAT ExonIII-TAMRA reverse primer (5 μM in water)
- Light mineral oil

### *Method*

**1.** Combine the reagents for the ARMS assay in two tubes by mixing the following:

*Tube 1*

- reverse primer[a] 3 μl
- allele 2-FAM primer[a] 3 μl
- allele 3A-JOE primer[a] 3 μl
- AAT ExonIII-TAMRA primer[a] 3 μl
- AAT ExonIII-TAMRA reverse primer[a] 3 μl
- 12.5 × dNTPs 4 μl
- 10 × reaction buffer 5 μl
- DMSO 5 μl
- DNA (20–80 ng) 1 μl
- water 18 μl

*Tube 2*

- reverse primer[a] 3 μl
- allele 4-FAM primer[a] 3 μl
- allele 3B-JOE primer[a] 3 μl
- AAT ExonIII-TAMRA primer[a] 3 μl
- AAT ExonIII-TAMRA reverse primer[a] 3 μl
- 12.5 × dNTPs 4 μl
- 10 × reaction buffer 5 μl

- DMSO 5 μl
- DNA (20–80 ng) 1 μl
- water 18 μl

2. Overlay the contents of each tube with light mineral oil and incubate them in the thermocycler at 96°C for 5 min.
3. Initiate a 'hot-start' using 1 unit (2 μl) diluted *Taq* DNA polymerase.
4. Run 25 cycles comprising 96°C for 30 sec; 60°C for 30 sec and 72°C, 1 min.
5. Run a final 72°C extension step for 10 min.[b]
6. Combine the PCR reaction products for each genotype and load 2 μl on a 2% agarose gel. Electrophorese to give a distance of 4 cm well to read,[c] for 3 h.

[a] These are the primers used for the assay shown in *Figures 9* and *10* and nucleotide sequences should be substituted appropriately for different loci. The fluorescent dyes are as described in Chapter 3 and the legend to *Figure 9*. The stock primer concentrations are 5 μM.

[b] This is important for the Apo E assay because Apo E is GC-rich and incomplete extension without this step is observed for this locus. Other loci may not require this extended final extension incubation.

[c] 'Well to read' is defined as the total migration distance between the wells and the scan region where the DNA is detected as it passes the laser; see *Figure 8*.

## 3.7 Recent advances: primers with non-amplifiable tails

Further developments in the chemistry of oligonucleotide synthesis have led to the ability to prepare non-nucleosidic phosphoramidite reagents (see Chapter 2). The incorporation of some of these types of reagents into PCR primers and consequently into PCR products results in the inability of *Taq* DNA polymerase to incorporate further nucleoside triphosphates when the non-nucleosidic moiety is reached. Thus, these non-nucleoside moieties form impassable barriers to the polymerase and result in single-stranded tails to the PCR product which may be any sequence of choice (11; see *Figure 11*).

Single-stranded tails can be used for either capture or signalling purposes. When an ARMS primer is made with a non-amplifiable tail which is exclusive to the allele it is to detect, the single-stranded tail can be used for capturing the product on a solid phase with an immobilized oligonucleotide complementary to the tail (see *Protocols 8* and *9*). A single-stranded tail similarly generated on the common primer to the ARMS reaction and common to all alleles and control PCR product can be used in a generic signalling system with a single complementary detection oligonucleotide. This may be for example, fluorescent as in *Section* 3.6 above or colorimetric as in *Protocol 9*.

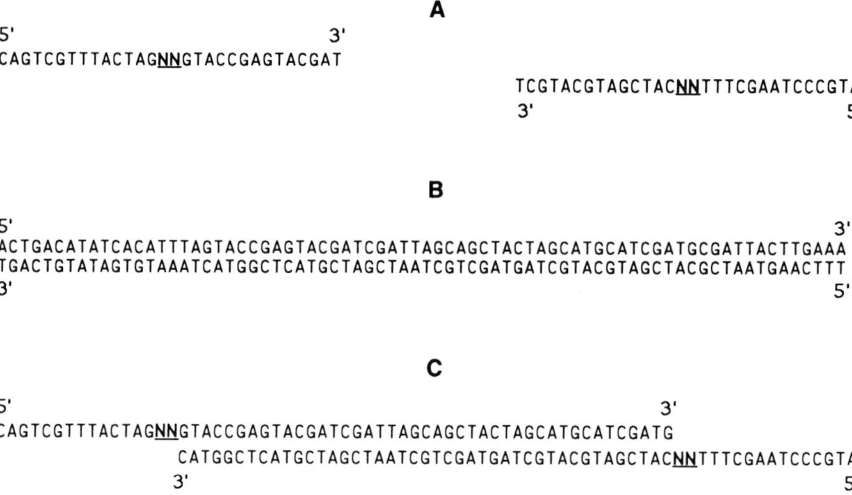


**Figure 11.** The principle of primers with non-amplifiable tails and the generation of tailed PCR products. (A) Two primers with non-nucleotidic inserts (*N*); note that the sequences 5′ of the inserts can be any sequence of choice. (B) Target DNA. (C) The PCR product.

## **Protocol 8.** Oligonucleotide immobilization to microtitre dishes[a]

### *Equipment and reagents*

- Nap-25 column prepacked gel filtration columns (Pharmacia)
- UV spectrophotometer
- Quartz UV spectrophotometry cells
- Clear polystyrene microtitre dishes, e.g. Nunc
- Laminar flow cabinet
- Oligonucleotide(s) for immobilization synthesized with 10 additional 5′-thymidine residues and as a 5′-aminoalkyl derivative (see Chapter 2)
- Thiolation buffer: 4 mg/ml iminothiolane in 0.2 M $Na_2CO_3/HCO_3$ pH 9.6
- Phosphate-buffered saline (PBS): 1 × PBS is 137 mM NaCl, 2.7 mM KCl, 8 mM $Na_2HPO_4$, 1.5 mM $KH_2PO_4$ pH 7.4
- Functionalization solution: poly(Phe-Lys), (Sigma) dissolved to 100 μg/ml in 50 mM $Na_2CO_3$ pH 9.6
- 0.05% Tween in PBS
- Conjugation buffer: 100 mM triethanolamine HCl, 1 mM $MgCl_2$ 1 mM $ZnSO_4$ pH 7.4
- Succinimidyl-4-(*N*-maleimidomethyl) cyclohexane-1-carboxylate (SMCC; Sigma)
- *N*,*N*-Dimethylformamide (DMF)
- Wash buffer: 5% lactose, 0.5% gelatin, 0.1% $NaN_3$, 6 mM PBS pH 7.5

### *Method*

1. Thiolate the oligonucleotide(s) by resuspending the product from 0.2 μmol syntheses in 200 μl water and mix with thiolation buffer (300 μl) then incubate at ambient temperature for 1 h.
2. Dilute the thiolation reaction(s) with 1 ml PBS and desalt by size exclusion chromatography using a Nap-25 column (Pharmacia) according

to the manufacturer's instructions. Wash the column with 2.2 ml PBS and collect the first 1.6 ml of eluate. Determine the oligonucleotide concentration by UV spectrometry (see *Protocol* 5, footnote *a*). Prepare a 10 μM solution in PBS and store at −20°C.

3. Activate the microtitre dish wells by adding 100 μl of functionalization solution and incubate overnight at 4°C.
4. Wash the wells three times with 0.05% Tween in PBS.
5. Prepare a solution of 8.7 mg/ml SMCC in DMF and dilute this solution 10 times with conjugation buffer. Add 100 μl of this solution to each well and incubate at ambient temperature for 1 h.
6. Wash each well five times with water.
7. Thaw the thiolated oligonucleotide(s) and add 100 μl per well. Incubate the microtitre dish overnight at 4°C.
8. Wash the wells twice with 0.05% Tween in PBS and twice with wash buffer.
9. Dry the coated dishes overnight in a laminar flow cabinet then store at 4°C.[b]

[a] As adapted from Running and Urdea (12).
[b] Microtitre dishes prepared in this way are stable for at least 1 year.

## Protocol 9. Capture and detection of PCR/ARMS products with single-stranded tails[a]

### *Equipment and reagents*

- Microcentrifuge
- Microtitre plate reader capable of measuring $A_{405}$ values (optional)
- Oligonucleotide functionalized microtitre dishes (see *Protocol 8*)
- Detection oligonucleotide synthesized as a 5′-aminoalkyl derivative (see Chapter 2)
- E-LINK™ oligonucleotide labelling kit (Cambridge Research Biochemicals) or equivalent
- Prehybridization solution: 0.6 M NaCl, 20 mM sodium phosphate buffer pH 7.5, 1 mM EDTA, 0.02% Ficoll 400, 0.02% polyvinylpyrrolidine 360 and 0.02% bovine serum albumin
- 20 × SSC stock solution: 3 M NaCl, 0.3 M trisodium citrate. Other SSC solutions are prepared from 20 × SSC by appropriate dilution with water
- Colour development solution. For 15 ml of this reagent, suspend 5 mg nitroblue tetrazolium (NBT) in 1.5 ml TNM buffer (0.1 M Tris–HCl pH 9.5, 0.1 M NaCl, 5 mM $MgCl_2$) in a microcentrifuge tube and vortex vigorously for 2 min. Centrifuge briefly in a microcentrifuge at full speed. Decant the supernatant into a polypropylene tube containing 10 ml TNM warmed to 37°C. Re-extract the NBT pellet twice with 1.5 ml TNM and combine the supernatants with the first solution. Rinse the microcentrifuge tube with 0.5 ml TNM and decant this also into the first solution. Dissolve 2.5 mg bromochloroindolyl phosphate (BCIP) in 50 μl *N,N*-dimethylformamide and add this solution dropwise with gentle mixing to the above NBT solution. This solution is stored at −20°C in convenient aliquots. Before use, prewarm the requisite number of aliquots at 37°C.

**Protocol 9.** *Continued*

*Method*

1. Conjugate the 5′-aminoalkyl oligonucleotide to alkaline phosphatase using the E-LINK™ oligonucleotide labelling kit as directed by the manufacturer. Dilute it to 1.5 μM.
2. Add 200 μl prehybridization solution to each well of a capture oligonucleotide functionalized microtitre dish and incubate at 37 °C with gentle shaking for 30 min. Aspirate off the prehybridization solution and replace it with 60 μl of 5 × SSC.
3. Add 20 μl of tailed PCR or ARMS product to the appropriate well(s).
4. Incubate the microtitre dish at 37 °C with gentle shaking for 1 h then aspirate off the hybridization solution and rinse the well(s) twice with 200 μl of 2 × SSC at 37 °C with gentle shaking. Aspirate off the 2 × SSC after each wash and then add 40 μl of 5 × SSC.
5. Add 10 μl of the alkaline phosphatase conjugated detection oligonucleotide (from step 1) and hybridize this to the captured PCR or ARMS products as described in step 4, then wash the wells also as described in step 4.
6. Add 200 μl of the colour development solution to the appropriate wells and incubate the microtitre dish at 37 °C for 1 h in darkness.
7. Assess bound PCR/ARMS products by eye or using a plate reader to read $A_{405}$ values.

[a] This protocol describes hybridization (immobilization and detection) for 20 residue oligonucleotides with equal A, G, C, and T content. Stringencies may need to be adjusted for different 'tail' lengths and base compositions.

# 4. Allele-specific oligonucleotide (ASO) analysis

## 4.1 Dot blots

Dot blots of PCR products using ASOs (13) is a technique that has essentially been superseded by others described herein. The reasons for this are the need for additional technical manipulations subsequent to the PCR reaction and for confirmation of amplification by agarose gel electrophoresis. A downstream signal generation system is also required and a typical test is restricted to the analysis of a single allelic variation. It is also impractical to establish a generic protocol for ASO hybridization since the hybridization conditions are always dependent on the melting temperature ($T_m$) of the probe/target duplex which will vary from one polymorphism to another. Nevertheless, with appropriately optimized ASOs and hybridization conditions, the test is accurate, and its choice may be dictated by the equipment available. Procedures for the immobilization of PCR products for dot blots or slot blots are described in *Protocol 10*.

**Protocol 10.** Immobilization of PCR products to nylon membranes

*Equipment and reagents*

- Thermocycler
- Agarose gel electrophoresis tank
- DC power supply
- DNA transilluminator
- Slot or dot blot manifold (e.g. from Schleicher and Schüll or BioRad)
- Nylon membrane (e.g. Hybond-N; Amersham)
- 20 × SSC (see *Protocol 9*)

*Method*

1. Perform the PCR reaction using primers that flank (and do not overlap) the polymorphism.
2. Electrophorese 10–20 μl of the PCR reaction product on an agarose gel against appropriate DNA size markers to confirm amplification of the target DNA.
3. Dilute two separate aliquots (2–5 μl each) of the PCR reaction product with 200 μl of 15 × SSC and incubate these at 100°C for 5 min.
4. Apply the diluted aliquots to a nylon membrane using the slot or dot blot manifold.
5. Immobilize the DNA on the membrane by UV irradiation on a transilluminator for 3 min.

The choice of either sense or antisense strand as target for the ASOs can be important; if the mutation or polymorphism is such that a G–T mismatch occurs between one allele and the ASO for the alternative allele then it may not be possible to achieve the discrimination between the alleles. In such situations the alternative strand should be the target for the ASOs where a C–A mismatch would be made.

ASO probes need to be labelled for signal generation. Advances in synthetic oligonucleotide chemistry have provided the capability of direct conjugation of biotin to oligonucleotides during their solid-phase synthesis (see Chapter 2). Biotinylated ASO probes can be conveniently used for non-isotopic signal generation which is described in Chapter 3. Alternatively, ASO probes may be radiolabelled. Both the radiolabelling and ASO hybridization protocols have been documented elsewhere (14).

## 4.2 Reverse dot blots

Like multiplex ARMS, reverse dot blots permit the simultaneous analysis of several mutations or polymorphisms in a single test. This method is similar to the dot blot probed with ASOs, but the format is inverted. ASOs are immobilized on to a membrane and hybridization is usually carried out using biotinylated or radiolabelled PCR product as the probe. The clear advantage of the format inversion is that many probes can be immobilized to detect specific alleles from a single PCR which can contain several primer pairs in a multiplex reaction. However, the principal limitations are as for the dot blot probed with ASOs. Specifically, because the $T_m$ for each immobilized ASO will be allele-dependent (for a constant length ASO), each filter can only have bound ASOs with a similar $T_m$. Optimal ASOs usually comprise approximately 20 nucleotides if they are around 50% GC content. Unfortunately, if attempts are made to alter the $T_m$ of an ASO by altering its length, the specificity of the probe will also be affected. A procedure for generating a reverse dot blot filter is described in *Protocol 11*.

Again, as with ASO probed dot blots, there is the additional need for hybridization analysis to be performed after the PCR. The other similarity with ASO-probed dot blots is the requirement for a signal generation system, which, as with dot blots, may be a radiolabel (if a [$^{32}$P]deoxynucleoside triphosphate is incorporated into the dNTP mix for the PCR) or colorimetric (if biotinylated PCR primers are used; see Chapter 3).

**Protocol 11.** Preparing a reverse dot blot filter

*Equipment and reagents*

- DNA transilluminator
- Slot or dot blot manifold (e.g. from Schleicher and Schüll or BioRad)
- Nylon membrane (e.g. Hybond-N; Amersham)
- Terminal deoxyribonucleotidyltransferase (this enzyme appears in many suppliers catalogues as 'terminal transferase')
- 5 × Terminal transferase buffer: 500 mM potassium cacodylate, 125 mM Tris–HCl pH 7.6, 5 mM $CoCl_2$, 1 mM dithiothreitol, 0.4 mM dTTP
- 10 mM EDTA
- 10 mM Tris–HCl pH 8.0, 0.1 mM EDTA
- 5 × SSPE 0.9 M NaCl, 50 mM $NaH_2PO_4$ 5 mM EDTA pH 7.2 containing 0.5% SDS
- ASOs

*Method*

1. Extend the ASOs to be immobilized with a poly(dT) tail by incubating each oligonucleotide (200 pmol) in 100 μl containing 20 μl of 5 × terminal transferase buffer and 60 U terminal deoxyribonucleotidyl-transferase.[a,b]
2. Incubate the reaction at 37 °C for 60 min then stop the reaction by adding 100 μl of 10 mM EDTA.[c] Dilute the tailed oligonucleotides into 100 μl 10 mM Tris–HCl pH 8.0, 0.1 mM EDTA.
3. Using a slot blot or dot blot manifold, apply the tailed oligonucleotides to a nylon membrane and immobilize by UV irradiation on a trans-illuminator for 3 min.

4. Wash the membranes in 5 × SSPE containing 0.5% SDS for 30 min at 55°C to remove unbound oligonucleotides.[d]

[a] Some suppliers market kits for this reaction, e.g. BCL.
[b] Some investigators have reported the immobilization of ASOs without a poly(dT) tail via a 5′-amino group (see Chapter 2) direct to carboxyl groups of alternative membranes with no apparent loss of specificity or sensitivity in the reverse dot blot analysis (15).
[c] These conditions should give a nominal tail length of 400 residues (16).
[d] If the filters are not to be used immediately, rinse them with water and air dry. The dried filters may be stored at room temperature.

The choice of protocol for hybridization and detection will depend on the signal generation system incorporated into the PCR reaction. If biotinylated PCR primers were employed, follow *Protocol 12*. If a [$^{32}$P]dNTP was incorporated into the PCR product(s), follow *Protocol 13*.

## **Protocol 12.** Hybridization and detection (colorimetric)[a]

*Equipment and reagents*

- Hybridization solution: 5 × SSPE (see *Protocol 11*) containing 0.5% SDS
- 2 × SSPE (0.36 M NaCl, 20 mM $NaH_2PO_4$, 2 mM EDTA pH 7.2) containing 0.1% SDS
- Streptavidin–horseradish peroxidase conjugate (Eastman Kodak) prepared as a 100 ng/ml solution in hybridization solution
- Red leuco dye (Eastman Kodak)
- 2 × PBS, see *Protocol 8*
- 400 mM NaOH, 10 mM EDTA
- PCR product

*Method*

1. Place each filter in a minimum volume of hybridization solution containing the streptavidin–horseradish peroxidase conjugate (~5 ml).
2. Denature 20 µl of PCR product with 20 µl 400 mM NaOH, 10 mM EDTA, and add this to the filter in the hybridization solution.
3. Incubate the filter at 55°C for 30 min.[b]
4. Rinse the filters twice in 2 × SSPE, 0.1% SDS at room temperature, then once at 55°C for 10 min.[b]
5. Briefly rinse the filters twice in 2 × PBS at room temperature.
6. Incubate the filters in 25–50 ml red leuco dye at room temperature for 5–10 min.
7. Photograph the blot if a permanent record is required.

[a] Adapted from ref. 16.
[b] The hybridization and washing temperatures may need to be optimized if specificity is not achieved using these conditions. Lower the hybridization and washing temperatures by a few degrees if subsequent colour generation is weak and allele-specific. If known homozygotes type as heterozygotes, the hybridization and washing temperature should be raised a few degrees.

**Protocol 13.** Hybridization and detection (isotopic)

*Equipment and reagents*

- Medical X-ray film e.g. Kodak X-OMat AR5
- Hybridization solution: 5 × SSPE (see *Protocol 11*) containing 0.5% SDS
- 2 × SSPE containing 0.1% SDS (see *Protocol 12*)
- 400 mM NaOH, 10 mM EDTA
- PCR product

*Method*

1. Place each filter in a minimum volume of hybridization solution (~5 ml).
2. Denature 20 μl of PCR product with 20 μl 400 mM NaOH, 10 mM EDTA, and add this to the hybridization solution.
3. Incubate the filter and probe at 55°C for 30 min.[a]
4. Rinse the filters twice in 2 × SSPE, 0.1% SDS at room temperature, then once at 55°C for 10 min.[a]
5. Air dry the filters and autoradiograph by exposure to X-ray film.

[a] The hybridization and washing temperatures may need to be optimized if specificity is not achieved using these conditions. Lower the hybridization and washing temperatures by a few degrees if subsequent autoradiograph signal is weak and allele-specific. If known homozygotes type as heterozygotes, the hybridization and washing temperature should be raised a few degrees.

## 5. Competitive oligonucleotide priming (COP)

Competitive oligonucleotide priming as first reported (17) employs allele-specific primers designed in the same manner as ASO probes. The allele specificity is derived from an appropriately variant nucleotide positioned roughly at the middle of each primer of a primer pair. Primer pairs are used in PCR with a common primer analogous to an ARMS reaction. However, the primers comprising the pair must be distinguishable, for example one can be radiolabelled as in ref 17. With this format, PCR is carried out and the products analysed by agarose gel electrophoresis. The expected result is an equivalent band in all samples analysed. To discriminate between alleles, the gel is dried and autoradiographed. Where a signal appears on the X-ray film, this corresponds to the incorporation of the labelled primer into the PCR product (see *Figure 12*). Thus, if the 'normal' primer were [$^{32}$P]-labelled, the originating genomic DNA would either be homozygous 'normal' or heterozygous (assuming specificity of the primers). The clear drawback using this format is the difficulty of distinguishing accurately between homozygotes and

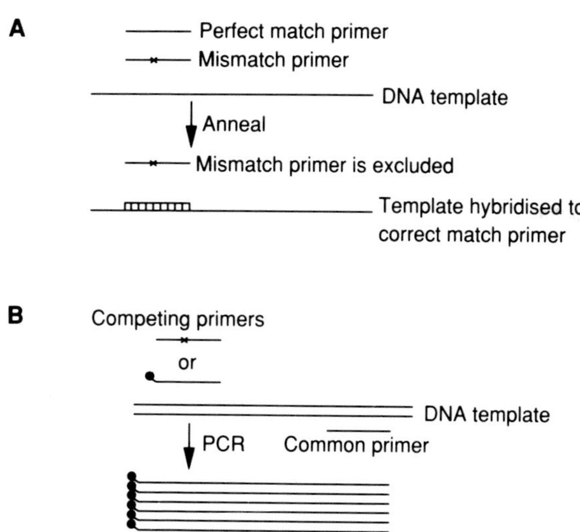


**Figure 12.** Scheme of competitive oligonucleotide priming. (A) Two oligonucleotide primers differing by one nucleotide (shown as x) are combined with a DNA template. The perfect match primer preferentially anneals to the template excluding the mismatched primer. (B) One of the primer pair, the correctly matched primer, is labelled (●). This labelled primer excludes the mismatched primer and when combined with a common primer in PCR is incorporated into product. Identification of the label incorporated into the product allows inference of the template sequence. (Redrawn from ref. 17, with permission.)

heterozygotes. To overcome this problem, the individual primers of the primer pair may be conjugated to different and distinguishable signal generation labels (18). For instance, each primer of a primer pair may be conjugated to a different fluorophore such as fluorescein and rhodamine (18). After PCR, agarose gel electrophoresis, and long-wavelength UV irradiation, the PCR product from heterozygous DNA will fluoresce yellow (rhodamine plus fluorescein incorporation). Homozygous DNA will exhibit green (fluorescein-labelled primer incorporation) or red (rhodamine-labelled primer incorporation) fluorescence for either allele, respectively. Alternatively, the electrophoresis gel can be scanned by laser during electrophoresis analogous to fluorescent ARMS using four colour technology (see Section 3.6).

COP has rarely been the method of choice for known mutation analysis given the difficulty of accurate heterozygote typing and the need for two detection systems (autoradiography in addition to gel electrophoresis) in the earlier format (17). The improved allelic discrimination contributed by positioning the allele-specific nucleotide at the 3′-terminus of primers as in ARMS (Section 3) permits ARMS to be performed as in the later COP format (18). Thus ARMS and COP combined have given rise to the fluorescent ARMS assay using four colour technology as described in Section 3.6.

## 6. Primer extension sequence test (PEST)

This method of mutation detection, like COP, relies on the extension or not of an allele-specific primer whose allele-specific nucleotide is within the primer. Thus the extension or non-extension of the primer is hybridization specific as opposed to enzyme catalytic specific. An ASO for the 'mutant' allele is used in PCR and the thermocycling conditions are tuned precisely such that the ASO is extended only on the correctly matched template. A further primer, upstream of the polymorphism/mutation, complementary to an invariant sequence and of the same sense as the ASO is also included (19). If the ASO does not hybridize, a PCR product is generated with the upstream primer (i.e. when the template DNA is 'normal'). If the ASO does hybridize, it generates the PCR product (i.e. when the template DNA is 'mutant'). Heterozygous DNA templates are characterized by incorporation of both primers into PCR products. Since one primer is upstream of the other, products of different sizes are generated (larger for the 'normal' allele, smaller for the 'mutant' allele). The scheme of the test is shown in *Figure 13*.

The principal drawbacks with this system are that only one mutation can be analysed per reaction and that each reaction must be tailored to the respective mutation under investigation since the $T_m$s for different ASOs will

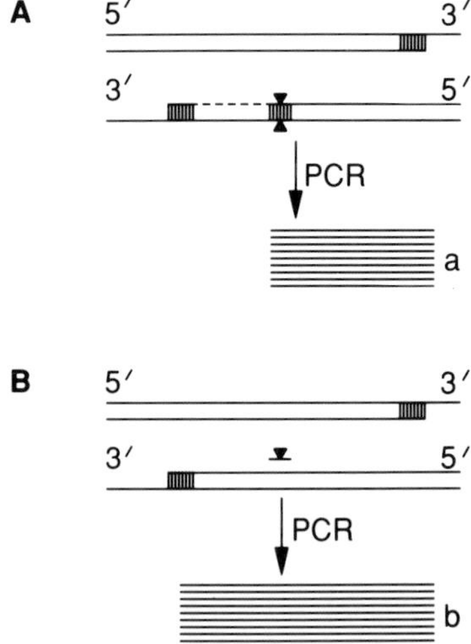


**Figure 13.** The primer extension sequence test. Closely-spaced vertical lines indicate the base-pairing of either mutant oligonucleotide (▼) or upstream primer (---). (A) The mutant oligonucleotide is complementary to the template DNA and can hybridize (blocking extension of the upstream primer) and be extended to give short PCR product 'a'. (B) The mutant oligonucleotide is non-complementary to the template, does not hybridize, and so a larger PCR product 'b' is derived from the primer upstream of the mutation. Reproduced from ref. 19 with permission from S. Karger AG, Basel.

vary. It is also likely that specificity problems will be encountered due to the non-competitive annealing of the ASO to template DNA.

## 7. Mutation detection using *Taq* 5′→3′ exonuclease activity

Whilst *Taq* DNA polymerase does not possess a 3′→5′ exonuclease activity, it does exhibit a 5′→3′ exonucleolytic property (20). This property has been exploited to demonstrate simultaneously the presence of target DNA as the PCR proceeds. This is achieved by including a further primer downstream of one of the conventional amplification primers. This additional primer is blocked for extension at the 3′-terminus and carries a label at the 5′-terminus (see *Figure 14*). Extension from the primer upstream of the additional primer results in liberation of the 5′-label by the *Taq* DNA polymerase as it reaches the downstream primer (21).

The disadvantage of utilizing such a method for mutation detection is the inability to multiplex the reaction and hence the requirement for a separate reaction for each allele for each DNA sample. Furthermore, such an application of the method would require that the allele specificity would need to reside within the additional, 3′-blocked, 5′-labelled primer which would necessitate this being designed as two variants in an ASO probe type manner. This situation would also bring with it the additional associated problems that are described for PEST (see Section 6 above).

## 8. Mutation detection by introduction of restriction sites

This method of detection of mutations or polymorphisms operates by incorporating a deliberate mismatch at or close to the 3′-terminus of a PCR primer adjacent to the variant nucleotide such that the variant nucleotide of one allele, after incorporation into PCR product, complements the preceding sequence of the primer to introduce a restriction enzyme recognition site.

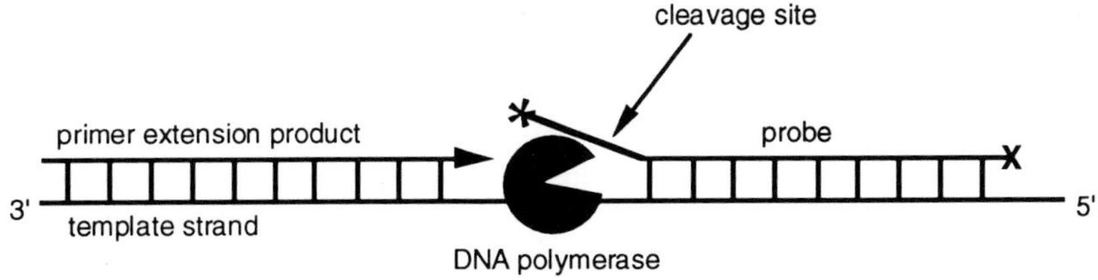


**Figure 14.** Diagram of 5′→3′ exonuclease cleavage of 5′-labelled oligonucleotide probe to generate labelled fragments in a PCR. The asterisk represents the 5′-label on the probe, the X represents the 3′-block on the probe. Modified from ref. 21.

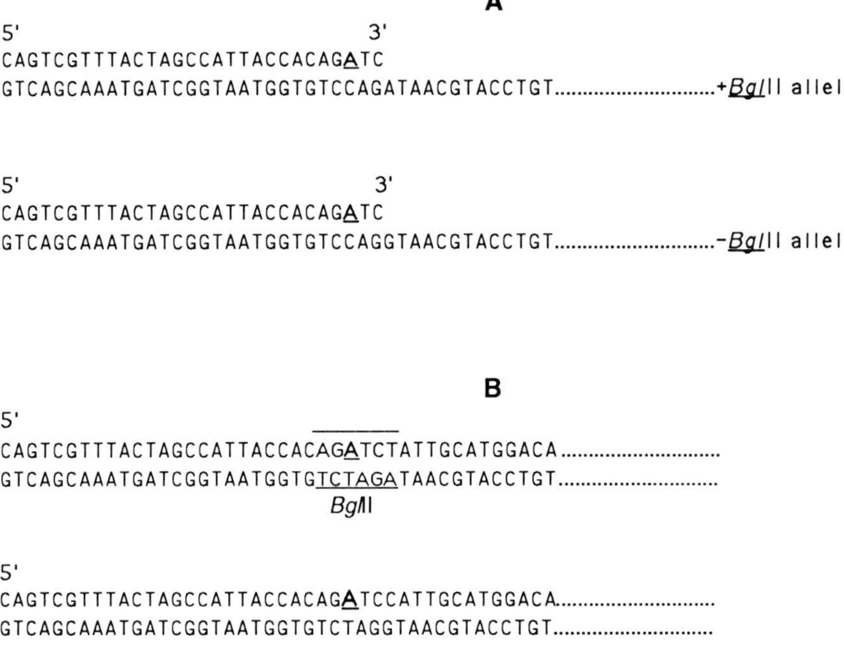


**Figure 15.** (A) PCR primer annealed to + *Bgl*II and − *Bgl*II alleles. The bold and underlined 'A' residue is deliberately changed to generate a partial *Bgl*II restriction recognition site. (B) Extension of the primer and incorporation into PCR product creates a *Bgl*II recognition site (under and overlined) from the + *Bgl*II allele, no restriction recognition site is created from priming the − *Bgl*II allele.

The corresponding variant nucleotide of the other allele does not complement the primer and the restriction site is not formed. Subsequent digestion of the PCR product with the restriction enzyme identifies which alleles were present in the template DNA (22). The principle of this method is shown in *Figure 15*.

The main limitation of the method involves the serendipity of the genomic DNA sequence that may or may not allow the design of the primer, as for some mutations there may not be a potential restriction site for incorporation. Furthermore, amplification of the target is likely to be inefficient since the 3′ modification of the primer sequence is analogous to the property of ARMS primers where 3′-mismatches are not extended (see Section 3). If this method is chosen, the subsequent detection system should be considered carefully since the restriction enzyme digestion of the PCR product will lead to only a small change in the size of the PCR product. Therefore if agarose gel electrophoresis is employed, a system that will resolve cut from uncut product must be used. If possible, it is useful to design the restriction primer such that the incorporation of a restriction site that would naturally occur elsewhere in the PCR product is incorporated. The 'natural' restriction site then serves as a

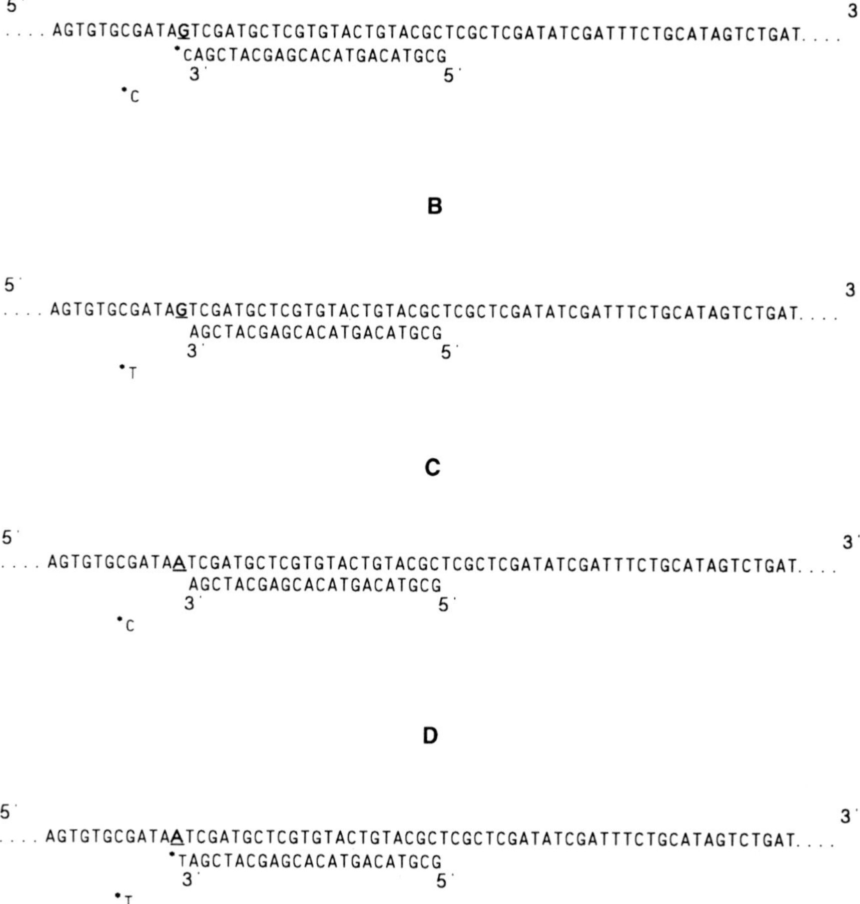


**Figure 16.** (A) SNuPE primer annealed to single strand from PCR product (G allele, underlined) and extended by incorporation of labelled dCTP (*C) due to complementarity with G of template. (B) SNuPE primer annealed to single strand from PCR product (G allele) and not extended by incorporation of mismatched, labelled dTTP (*T). (C) SNuPE primer annealed to single-strand from PCR product (A allele) and not extended by incorporation of mismatched, *C. (D) SNuPE primer annealed to single strand from PCR product (A allele) and extended by incorporation of complementary *T.

control for digestion. This is useful since a failed restriction digest would imply homozygosity for one allele which may be incorrect. A partial restriction digest could also be misdiagnosed as heterozygosity. If such design is not possible, the complete restriction enzyme recognition site may be incorporated into the other PCR primer. It is not practical to include a protocol for this method since each allele will have unique sequence attributes and many restriction endonucleases have different reaction requirements. Refs. 23–27 are therefore cited as examples of the potential of this method.

## 9. Single nucleotide primer extension (SNuPE) for mutation detection

This method (28) is a two stage test. The first stage comprises an initial PCR amplification of a DNA fragment that contains the polymorphic sequence. The amplified fragment is then purified from the remaining PCR reagents prior to the second stage. The second stage involves dividing the PCR product between two further reactions and the annealing of another primer, the SNuPE primer, whose 3′ end immediately precedes the polymorphism. In one reaction a single radiolabelled dNTP specific for one allele is included, the other reaction contains the single radiolabelled dNTP specific for the alternative allele. Incubation of these reactions with a DNA polymerase (typically *Taq* DNA polymerase) extends the SNuPE primer when the radiolabelled dNTP is complementary to the variant allelic nucleotide. The principle of SNuPE is shown in *Figure 16*.

SNuPE will detect the majority of mutations whether they be single base changes or frameshift. However, the method is cumbersome, requiring an initial PCR, a gel purification step, two further reactions followed by another electrophoretic step and autoradiography.

## Acknowledgements

*Figure 7, Table 1*, and *Protocols 1* and *2* are courtesy of Dr Steve Little, Zeneca Cellmark Diagnostics, Northwich, Cheshire, UK. *Figures 8, 9, 10*, and *Protocol 7* are courtesy of Elaine S. Mansfield, Applied Biosystems, Inc. Foster City, CA, USA: current address, Molecular Dynamics Inc. Sunnyvale, CA, USA. Finally, I thank Dr John Smith, Zeneca Pharmaceuticals, Alderley Park, Macclesfield, Cheshire, UK and Dr Steve Little for their critical reading of the manuscript.

## References

1. Sambrook, J., Fritsch, E. F., and Maniatis, T. (ed.) (1989). In *Molecular cloning: A laboratory manual* (2nd edn), Cold Spring Harbor Laboratory Press, Cold Spring Harbor, New York.
2. Newton, C. R., Heptinstall, L. E., Summers, C., Super, M., Schwarz, M., Anwar, R., Graham, A., Smith, J. C., and Markham, A. F. (1989). *The Lancet*, **2**, 1481.
3. Graham, A., Newton, C. R., Powell, S. J., Heptinstall, L. E., Summers, C., Brown, L., Anwar, A., Murray, K., Gammack, A., Kennedy, R., Kalsheker, N.,

and Markham, A. F. In Erlich, H., Gibbs, R., and Kazazian, H. Jr. (ed.) (1989), *Current communications in molecular biology: polymerase chain reaction*, p. 105. Cold Spring Harbor Laboratory Press, Cold Spring Harbor, New York.

4. Noble, J. S., Taylor, G. R., Stewart, A. D., Mueller, R. F., and Murday, V. A. (1991), *Disease Markers*, **9**, 301.
5. Newton, C. R., Graham, A., Heptinstall, L. E., Powell, S. J., Summers, C., Kalsheker, N., Smith, J. C., and Markham, A. F. (1989). *Nucleic Acids Res.*, **17**, 2503.
6. Billadeau, D., Blackstadt, M., Greipp, P., Kyle, R. A., Oken, M. M., Kay, N., and Van Ness, B. (1991). *Blood*, **78**, 3021.
7. Huang, M. M., Arnheim, N., and Goodman, M. F. (1992). *Nucleic Acids Res.*, **17**, 4567.
8. Wenham, P. R., Newton, C. R., and Price, W. H. (1991). *Clin. Chem.*, **37**, 241.
9. Newton, C. R., Summers, C., Heptinstall, L. E., Jenner, D. E., Graham, A., and Markham, A. F. (1991). *Biotechniques*, **10**, 582.
10. Ferrie, R. M., Schwarz, M. J., Robertson, N. H., Vaudin, S., Super, M., Malone, G., and Little, S. (1992). *Am. J. Hum. Genet.*, **51**, 251.
11. Newton, C. R., Holland, D., Heptinstall, L. E., Hodgson, I., Edge, M. D., Markham, A. F., and McLean, M. (1993). *Nucleic Acids Res.*, **21**, 1155.
12. Running, J. A. and Urdea, M. S. (1990). *Biotechniques*, **8**, 276.
13. Saiki, R. K., Bugawan, T. L., Horn, G. T., Mullis, K. B., and Erlich, H. A. (1986). *Nature*, **234**, 163.
14. Thein, S. L. and Wallace, R. B. (1986). In *Human genetic diseases: A practical approach* (ed. K. E. Davies), pp. 33. Oxford University Press, Oxford.
15. Chehab, F. F. and Wall, J. (1992). *Hum. Genet.*, **89**, 163.
16. Saiki, R. K., Walsh, P. S., Levenson, C. H., and Erlich, H. A. (1989). *Proc. Natl Acad. Sci. USA*, **86**, 6230.
17. Gibbs, R. A., Nguyen, P. N., and Caskey, C. T. (1989). *Nucleic Acids Res.*, **17**, 2437.
18. Chehab, F. F. and Kan, Y. W. (1990). *Lancet*, **335**, 15.
19. Efremov, D. G., Dimovski, A. J., Janovic, L., and Efremov, G. D. (1991). *Acta Haematol.*, **85**, 66.
20. Longley, M. J., Bennett, S. E., and Mosbaugh, D. W. (1990). *Nucleic Acids Res.*, **18**, 7317.
21. Holland, P. M., Abramson, R. D., Watson, R., and Gelfand, D. H. (1991). *Proc. Natl Acad. Sci. USA*, **88**, 7276.
22. Haliassos, A., Chomel, J. C., Tesson, L., Baudis, M., Kruh, J., Kaplan, J. C., and Kitzis, A. (1989). *Nucleic Acids Res.*, **17**, 3606.
23. Sorscher, E. J. and Huang, Z. (1991). *Lancet*, **337**, 1115.
24. Gregersen, N., Blakemore, A. I. F., Winter, V., Andresen, B., Kølvraa, S., Bolund, L., Curtis, D., and Engel, P. C. (1991). *Clin. Chim. Acta*, **203**, 23.
25. Stocks, J., Thorn, J. A., and Galton, D. J. (1992). *J. Lipid Res.*, **33**, 853.
26. Gasparini, P., Bonizzato, A., Dognini, M., and Pignatti, P. F. (1992). *Mol. Cell. Probes*, **6**, 1.
27. Bal, J., Rininsland, F., Osbourne, L., and Reiss, J. (1992). *Mol. Cell. Probes*, **6**, 9.
28. Kuppuswamy, M. N., Hoffmann, J. W., Kasper, C. K., Spitzer, S. G., Groce, S. L., and Bajaj, S. P. (1991). *Proc. Natl Acad. Sci. USA*, **88**, 1143.

# 13

# Mutational analysis: new mutations

K. MICHAELIDES, R. SCHWAAB, M. R. A. LALLOZ, W. SCHMIDT, and E. G. D. TUDDENHAM

## 1. Introduction

In the last few years the search for mutations and sequence polymorphisms has been dramatically accelerated by the use of PCR and subsequently by direct sequencing of PCR products (1, 2). In spite of these powerful new methods, direct sequencing is not always practicable in detecting mutations because they may be positioned anywhere in some of the very large genes which are being studied. For example, mutations have now been found in nearly every one of the 26 exons of the factor VIII gene. Thus a number of prescreening methods have been developed, such as discriminant oligonucleotide hybridization (3), enzymatic (4) and chemical cleavage methods (5), single stranded conformation polymorphism (SSCP) (6) and denaturing gradient gel electrophoresis (DGGE) (7). These methods are described and by way of example, some results of their application to screening the coding sequences and splice sites of the factor VIII gene for mutations amongst haemophilia A patients (8–10) are shown.

## 2. Genomic DNA amplification

The exon scanning techniques described in this chapter are all based on PCR. Genomic DNA is extracted from peripheral blood samples by conventional means (11, 12). The test DNA is then amplified by PCR (refs. 1, 13; *Protocol 1*) and the resulting amplification product used in the chosen mutation screening technique. The quality of the amplified DNA can determine the success of the method and the clarity of the results, especially for SSCP.

The selection of oligonucleotide primers for PCR is very important. Primer pairs with a similar dissociation temperature ($T_d$) will give a cleaner product since the annealing temperatures will be similar for both. As a rough guide, primers with a $T_d$ of 60 °C will anneal to their template at approximately 55 °C. Trial amplification with a control DNA is recommended to fine tune the reaction conditions. Non-specific background bands can often be eliminated by increasing the annealing temperature by 1–2 °C.

When very little genomic DNA is available and needs to be conserved (e.g. the patient is deceased), secondary PCR reactions can be seeded from the primary amplification. This enhances the yield of PCR product and leaves the genomic DNA in reserve. The same oligonucleotides can be used for both primary and secondary PCR reactions. Any mutation detected must be confirmed by going back to the genomic DNA sample and sequencing a freshly amplified PCR product. In this way possible infidelities introduced by the *Taq* polymerase enzyme may be excluded.

### **Protocol 1.** PCR amplification of genomic DNA

*Equipment and reagents*

- Thermal cycler
- Positive displacement pipettes, capillaries and pistons (Microman, Anachem)[a]
- 0.5 ml sterile microcentrifuge tubes
- 100 mM dNTPs in sterile distilled water (Pharmacia 27-2035-01)
- *Taq* DNA polymerase (5 U/μl)
- 1 × PCR buffer: 10 × buffer is supplied with the enzyme and should be diluted with sterile distilled water. Add 20 μl each of 100 mM dATP, dGTP, dCTP, dTTP for each 10 ml of 1 × buffer. The buffer will now contain 50 mM KCl, 10 mM Tris–HCl pH 8.8, 1.5 mM $MgCl_2$, 0.1% Triton X-100, and 200 μmol of each dNTP.[b] Aliquot in 1 ml amounts and store at −20°C
- Oligonucleotide primers 200–300 ng/μl
- Mineral oil (Sigma molecular biology grade M5904)
- Genomic DNA 400 ng/μl
- 10 × TBE buffer: 0.9 M Tris–HCl pH 8.0, 0.9 M boric acid, 20 mM EDTA

*Method*

1. Thaw all the reagents and keep on ice.
2. Prepare sufficient 1 × PCR master mix for the required number of reactions (100 μl per reaction). To do this, add 10 μl of each oligonucleotide primer and 4 μl *Taq* DNA polymerase per ml of 1 × PCR buffer. Mix well by inversion.
3. Aliquot 100 μl of the 1 × PCR master mix into each 0.5 ml sterile microcentrifuge tube.
4. Add 1 μl of genomic DNA and mix by tapping gently.
5. Overlay each reaction with 50 μl of mineral oil. The PCR reaction mixture will now contain
   - genomic DNA — 400 ng
   - oligonucleotide primers — 200–300 ng of each
   - *Taq* DNA polymerase — 2 U
6. Include a negative (no DNA) control with each batch of amplifications.
7. Programme the thermal cycler. The following conditions are a guide:

| | | | |
|---|---|---|---|
| Denature | 94°C | 5 min | |
| Denature | 94°C | 35 sec | 30 cycles |
| Anneal | 47–59°C | 1 min | |
| Extend | 72°C | 1 min[c] | |
| Final extension | 72°C | 10 min | |
| Soak | 10°C | | |

The annealing temperature will depend on the oligonucleotide primer sequences.

**8.** Check the amplification by running 10 μl of the PCR product on a 1–2% agarose gel in 1 × TBE.

[a] Filter tips are now also available for most pipettes to avoid aerosol contamination.

[b] $MgCl_2$ is sometimes supplied separately to the 10 × PCR buffer and should be added at this stage.

[c] An additional 3 sec/cycle extension time is often recommended for long amplification products but this is not always essential.

# 3. Single strand conformation polymorphism (SSCP)

The mobility of single-stranded nucleic acid molecules electrophoresed under non-denaturing conditions is determined by both their fragment length and their secondary structure which is sequence dependent (6, 14). A fragment may adopt several conformations for any given set of electrophoretic conditions and these are visualized as separate bands in the gel. Some or all of these bands may show a shift in a mutated sequence, a single base change being sufficient to alter secondary structure and hence mobility.

Screening genomic DNA for unknown mutations is now possible, provided that the gene under scrutiny has already been at least partially sequenced to enable oligonucleotide primers to be designed. Selected regions of very large genes, for example the factor VIII gene, consisting of 186 kb of genomic DNA and 26 exons, can now readily be screened. However, this method is equally applicable to any gene however large or small. Even when only cDNA sequence is available, it is possible in some cases to screen small genes at the genomic level by designing primers within the known coding sequence and amplify across introns. Fragments can then be directly sequenced, the only limit being intron size and the limits of the PCR reaction.

## 3.1 Standard SSCP

A wide variety of electrophoresis and gel conditions for SSCP have been cited in the literature (15); polyacrylamide concentrations varying from 4–6%, differing gel cross-linking ratios, the presence or absence of glycerol in the gel, and electrophoresis at either room temperature or 4 °C. The addition of glycerol allows gels to be run at room temperature but with a considerable increase in electrophoresis time. Cooling to 4 °C can produce erratic results when cooling fans are used, if a cold room is available carrying out electrophoresis at 4 °C is the best option.

The most important criteria for successful SSCP are as follows:

- DNA fragments should be less than 300 bp, since mobility shifts may not be detected in larger fragments.
- Polyacrylamide gels should be prepared with a low cross-linking ratio such as 39:1 acrylamide to bisacrylamide. *Protocol 2* describes a suitable procedure.
- A stable temperature environment must be maintained during the electrophoretic run, preferably in a cold room at 4 °C.
- Use [$^{33}$P]dATP which will give crisper bands during autoradiography than [$^{32}$P]dATP (9). The latter gives more diffuse bands because of its higher energy of emission. These diffuse bands can obscure minor shifts of mobility of fragments, e.g. 1–2 mm.
- Gels should be run for a sufficient length of time to enhance the possible shifts. It is also highly advisable to run samples for two different lengths of time (e.g. 3 h and 4 h) to minimize the risk of missing band shifts. The exact times chosen for electrophoresis depend on fragment length.
- Always run a double-stranded DNA control. This will enable identification of single- and double-stranded fragments, especially when samples have been digested with restriction enzymes and several fragments are visualized. Band shifts can frequently be seen with double-stranded fragments and sometimes are *only* seen with the double-stranded fragments.

*Protocol 3* describes a procedure for standard SSCP which satisfies all of these criteria. The conditions used in this protocol have produced >95% success in detecting mutated sequences in our laboratory. Results obtained from the standard SSCP protocol described here are illustrated in *Figures 1* and *2*.

### **Protocol 2.** Preparation of a polyacrylamide gel for SSCP

**WARNING**: Acrylamide and bisacrylamide are known neurotoxins and potential mutagens and teratogens in man. Always wear disposable plastic gloves, a dust mask, and safety glasses during handling of these reagents in powder form and plastic disposable gloves at least when handling their solutions.

*Equipment and reagents*[a]

- 40% acrylamide–bisacrylamide (39:1) stock solution.[b] Prepare by mixing 35.1 g acrylamide (Electran, Grade 1, BDH 44313 5B), 0.9 g *N,N'*-methylenebisacrylamide (Life Technologies 540–5516UB), and 90 ml distilled water. Dissolve at 65 °C for 10 min

- 4.5% acrylamide stock solution. Prepare by mixing 90 ml 40% acrylamide:bisacrylamide (39:1), 80 ml 10 × TBE buffer (see *Protocol 1*), and 630 ml distilled water. Store at 4°C
- 10% ammonium persulphate (Sigma A9164)
- TEMED (Sigma T7024)
- Polyacrylamide gel mould (or 5 mm thick float glass sequencing plates) using 0.4 mm spacers and 0.4 mm comb

*Method*

1. Filter the 4.5% acrylamide solution through a 0.45 μm filter (Sartorius) just before use. Degas the solution under vacuum.[c]
2. Mix:
   - 4.5% acrylamide solution 80 ml
   - 10% ammonium persulphate 460 μl
   - TEMED 60 μl
3. Pour the gel immediately and allow it to polymerize.

[a] It is important to use only the highest quality reagents.

[b] The 39:1 ratio of acrylamide to bisacrylamide is very important since the detection of mobility shifts will be decreased if lower ratios are used.

[c] Degassing prevents the formation of bubbles in the polymerized gel.

**Protocol 3.** Standard SSCP

*Equipment and reagents*

- High voltage power supply (e.g. Multidrive XL, Pharmacia)
- Vertical gel electrophoresis apparatus (e.g. Model S2, Life Technologies 580–1165SC or similar)
- Heated driblock
- [α-$^{33}$P]dATP (1000–3000 Ci/mmol) (Dupont NEN NEG-312H)
- Restriction enzymes as required
- Sample running buffer:
  For *single-stranded DNA samples*, this is prepared by mixing (per 1 ml buffer), 800 μl formamide, 100 μl of 1% Bromophenol blue, 100 μl of 1% xylene cyanol, 2 μl 0.5 M EDTA, 1 μl 10 M NaOH.[a]
  *Double-stranded DNA samples* require the same sample buffer but without NaOH
- 4.5% polyacrylamide gel (poured using 0.4 mm spacers and a 0.4 mm comb; see *Protocol 2*)
- Agarose gel (1–2%)

*Method*

1. Amplify the test DNA fragment (*Protocol 1*) in the presence of a radiolabelled nucleotide (e.g. 3 μCi [α-$^{33}$P]dATP per 100 μl reaction). This will label the PCR product during amplification. Some methods involve end-labelling the PCR primers by treating with polynucleotide kinase using [γ-$^{32}$P]ATP and so generate fragments which are only end-labelled. This is more laborious than the method described here and cannot be used where larger amplification products need to be digested, as internal digestion products are then not labelled.

**Protocol 3.** *Continued*

2. Confirm the amplification by running 10 μl of the PCR product on a 1–2% agarose gel.
3. Digest fragments larger than 300 bp with suitable restriction enzymes.
4. Take 5–10 μl of the labelled PCR product and make up to 25 μl with sample running buffer.
5. *For single-stranded DNA only,* denature the samples at 90–95°C for 5 min [a] and transfer immediately to ice.
6. After 5 min on ice, load 3–8 μl of each sample onto the 4.5% polyacrylamide gel. (Vary the sample volume according to the efficiency of amplification, whether the samples have been restricted or not, and the specific radioactivity of the label used.)
7. Electrophorese the gel in 1 × TBE at 4°C (preferably in a cold room) and at 40 W constant power for 2–6 h depending on the DNA fragment sizes. In a 4.5% polyacrylamide gel, xylene cyanol migrates at a position expected for single-stranded DNA fragments of 210 nt long.
8. After 1 h, load a second batch of the same samples as in step 6 above. By running for two different periods, the risk of missing shifted bands is reduced.
9. A lane of double-stranded DNA [a] should be run with each batch of samples to enable distinction between single-stranded and the double-stranded bands.
10. Transfer the gel to Whatman 3MM paper, cover with cling film, and dry under vacuum at 80°C for 30–45 min.
11. Expose the gel to X-ray film at room temperature without a screen. Develop the autoradiograph.

[a] Double-stranded DNA should be prepared in the same way as single-stranded DNA except that it is made up in sample running buffer without NaOH and the sample is not preheated before applying to the gel.

## 3.2 Consecutive SSCP

As stated earlier, large PCR fragments (e.g. 1 kb in length) have to be digested in order to detect a mobility shift by standard SSCP. However, the localization of a mobility shift can then prove to be difficult because multiple bands are visualized on the gel. Fortunately, the accurate assignment of a mobility shift can be achieved in such situations by applying consecutive SSCP (16).

The DNA fragment to be screened is amplified in the presence of an [$\alpha$-$^{33}$P]dNTP and using a biotinylated antisense primer (*Protocol 4*). The PCR product can then be bound by its biotinylated 3′-end to streptavidin-coated

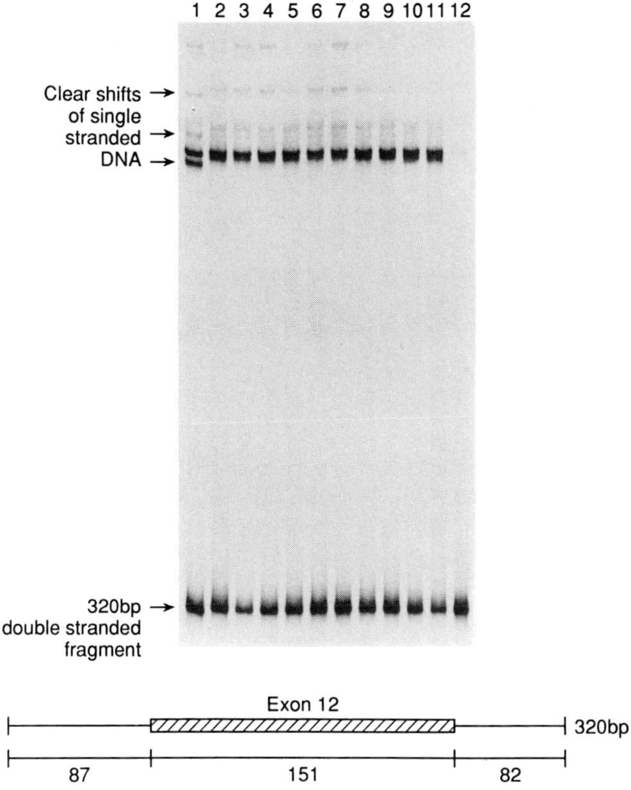


**Figure 1.** PCR–SSCP of exon 12 of the factor VIII gene. Clear shifts are seen in the main single-stranded doublet of lane 1. Fainter bands above the main doublet of single-stranded DNA are presumed to represent minor alternative conformations. Each of these is shifted in lane 1, confirming this interpretation. The line diagram at the base of the figure represents the PCR fragment analysed consisting of the whole of exon 12 (151 bp) and 87 bp and 82 bp of flanking intron sequence. Lanes 1–10, patient samples; lane 11, normal DNA control; lane 12, double-stranded non-denatured DNA control.

magnetic beads. A magnetic separator is used to remove the biotinylated product from the unincorporated primers, dNTPs, and radiolabelled nucleotides ready for restriction enzyme digestion. Careful selection of restriction enzymes allows sequential digestion of the biotinylated product and the individual collection of the restricted DNA fragments. The restriction enzymes are chosen so as to generate suitable DNA fragment sizes (100–300 bp) in a consecutive manner (see *Figure 3*). SSCP is then performed with each separate DNA fragment loaded individually onto the gel. In this way, previously observed shifts can be accurately assigned to a precise DNA fragment for subsequent direct sequencing. This adaptation of SSCP is therefore the approach of choice for rapid mutation screening of larger segments of DNA.

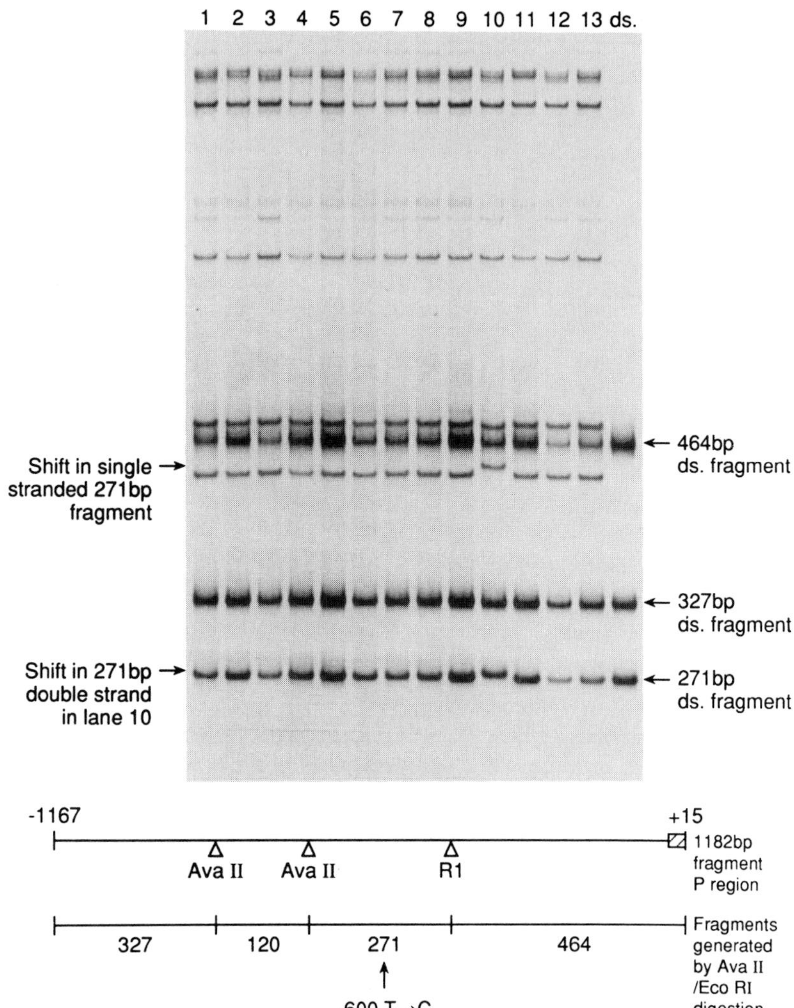


**Figure 2.** PCR–SSCP of a 1182 bp fragment of the 5′-untranslated region of factor VIII digested with *Ava*II/*Eco*RI. Four fragments are generated, three are shown here. There is a clear shift in a single-stranded fragment of lane 10 near the 464 bp double-stranded DNA. A shift in the 271 bp double-stranded DNA fragment is also seen. Consecutive SSCP subsequently confirmed this shift to be from the 271 bp single strands in this patient. Lanes 1–12, patient samples; lane 13, normal DNA control; ds, double-stranded non-denatured DNA control. The line diagram at the base of the figure illustrates the DNA fragments generated by restriction of the 1182 bp fragment with the enzymes *Ava*II and *Eco*RI. The small hatched area represents the first 15 bp of exon 1.

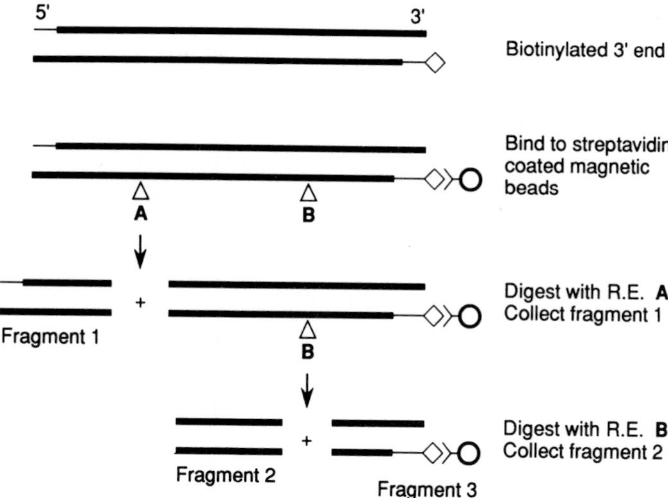


**Figure 3.** Schematic representation of consecutive digestion of biotinylated PCR products in preparation for consecutive SSCP. R.E. **A** refers to restriction enzyme **A**; R.E. **B** refers to restriction enzyme **B**. See the text for explanation.

## Protocol 4. Consecutive SSCP

### *Equipment and reagents*

- Streptavidin-coated magnetic beads (e.g. Dynabeads M-280 Streptavidin Dynal 112.05)
- Magnetic particle concentrator (e.g. Dynal MPC-E 120.04)
- PCR primers. The 3′-antisense primer must be biotinylated on its 5′-end
- Radiolabelled nucleotide, e.g. [$\alpha$-$^{33}$P]dATP (Dupont NEN)
- Selected restriction enzymes. Choose these so as to generate suitable DNA fragment sizes (100–300 bp) in a consecutive manner (see *Figure 3*).
- Sample running buffer (see *Protocol 3*)
- 2 × washing and binding buffer (WBB): 10 mM Tris–HCl pH 7.6, 1 mM EDTA, 2 M NaCl
- 4.5% non-denaturing polyacrylamide gel (see *Protocol 2*)
- Agarose gel (1–2%)
- 1 × TBE buffer (see *Protocol 1*)

### *Method*

1. Amplify the DNA (see *Protocol 3*) using 100–150 ng of the biotinylated primer per 100 μl reaction and 200 ng of the unbiotinylated primer. Remember to add 3 μCi of [$\alpha$-$^{33}$P]dATP to the 100 μl reaction mix.
2. Run 10 μl on a 1–2% agarose gel to confirm amplification.
3. Wash 40–50 μl of magnetic beads twice with 1 × WBB using the magnetic separator to remove the beads from the supernatant after each wash.
4. Resuspend the magnetic beads in 50 μl of 2 × WBB.

**Protocol 4.** *Continued*

5. Add 90 μl of the PCR mix to the beads and mix gently by pipetting. Do **not** vortex. Less PCR product can be used if the amplification is very efficient.
6. Incubate at 37°C for 10 min. This will bind the biotinylated PCR product to the magnetic beads. Incubation at room temperature is possible, but can be less efficient as the ambient temperature is variable.
7. Use the magnetic separator to remove the WBB from the magnetic beads. Pipette off the buffer completely.
8. Add 17 μl sterile distilled water to the beads. Mix gently by pipetting.
9. Add 2 μl of the appropriate 10 × restriction buffer and 1 μl of restriction enzyme. Mix by gentle pipetting.
10. Incubate at 37°C for 1 h.
11. Separate the magnetic beads from the supernatant using the magnetic separator. Transfer the supernatant to a clean 1.5 ml microcentrifuge tube and keep on ice. This contains fragment 1.
12. Wash the beads once with sterile distilled water and resuspend in 17 μl of sterile distilled water ready for the next digestion.
13. Repeat steps 9–12 as required until all the DNA fragments have been restricted, leaving the final fragment still attached to the beads.[a]
14. Take 5–10 μl of each restricted fragment including the final segment bound to the beads and make each one up to 20 μl in sample running buffer.
15. Denature the DNA samples at 90–95°C for 5 min. This will also cleave the final DNA fragment from the beads.
16. Immediately transfer to ice for 5 min.
17. Load 3–8 μl of each sample (single-stranded and double-stranded DNA) on to the gel.
18. Run the gel in 1 × TBE at 4°C (preferably in a cold room) at 40 W, constant power for 2–6 h depending on fragment length.
19. Continue as for *Protocol 3*, steps 10 and 11.
20. Examination of the developed autoradiograph will reveal the shift in one specific restricted fragment. This fragment can then be directly sequenced (see *Protocol 6*).

[a] 2 μl of each completed digest can be run on an agarose gel to confirm that the reactions have gone to completion. This information can be useful during interpretation of the consecutive SSCP gels.

**Table 1.** Multiplexed factor VIII exon groups

| Group | Proportion of essential coding sequence (bp) |
|---|---|
| 1 | 585 (exons 3,16,20,25) |
| 2 | 727 (exons 6,7,18,23) |
| 3 | 619 (exons 2,12,17,19) |
| 4 | 528 (exons 1,4,9) |
| 5 | 520 (exons 11,24,26) |
| 6 | 425 (exons 5,8,10) |
| 7 | 310 (exons 15,22) |
| 8 | 296 (exons 13, 21) |

## 3.3 PCR–SSCP multiplexed analysis: Application to the factor VIII gene

The aim of multiplexed analysis of very large genes like the factor VIII gene is to speed up SSCP analysis by amplifying several gene fragments simultaneously by the PCR method. All essential coding regions of the factor VIII gene (exons 1–3, 15–26) can be amplified as individual exons including intron/exon splice junctions. PCR fragment sizes range from 143 bp (exon 20) to 352 bp (exon 8). Exon flanking sequences range from 27–88 bp. Multiplexed PCR amplification was carried out in eight groups: three groups of four exons, three groups of three exons and two groups of two exons were multiplexed at an annealing temperature of 50 °C (*Table 1*). The exons within each group were separated by appropriate size differences and by large genomic map distances (the shortest distance between exon fragments being 5.63 kb) to avoid artefacts resulting from intronic sequence amplification.

The advantages of the method are fourfold:

- speed of analysis
- economy of reagents and sample
- simplicity of method
- sensitivity of detection

The whole of the essential coding region of the factor VIII gene is PCR amplified in only eight tubes. All primer pairs anneal at the same temperature (50 °C). Thus, a single patient DNA sample and a control (16 tubes) could have the essential coding region of the factor VIII gene amplified by the PCR-multiplexing system in about 4–5 h. All fragments are suitably small enough for SSCP gel analysis without prior restriction enzyme digestion. Thus, the preparation, running, and drying of the gel, and the exposure of the dry gel to X-ray film, determine the time taken before results are available. With appropriate forethought, results could be available within three days.

Multiplexed amplification is more economical on sample and reagents than individual PCRs performed for each exon; less *Taq* polymerase enzyme, radioactive isotope (making the procedure safer), nucleotides, and buffer are used, and considerably less patient genomic DNA sample is required. The multiplexed PCR–SSCP analysis described here (*Protocol 5*) is essentially a two-step procedure; each strand of the target DNA sequence is labelled as it is amplified in the thermal cycler and then transferred directly to electrophoresis on easily prepared gels. Since both strands of multiplexed PCR fragments are labelled, SSCP analysis may reveal a shift in either of the amplified complementary DNA strands after denaturation or in the double-stranded DNA product (*Figure 4*).

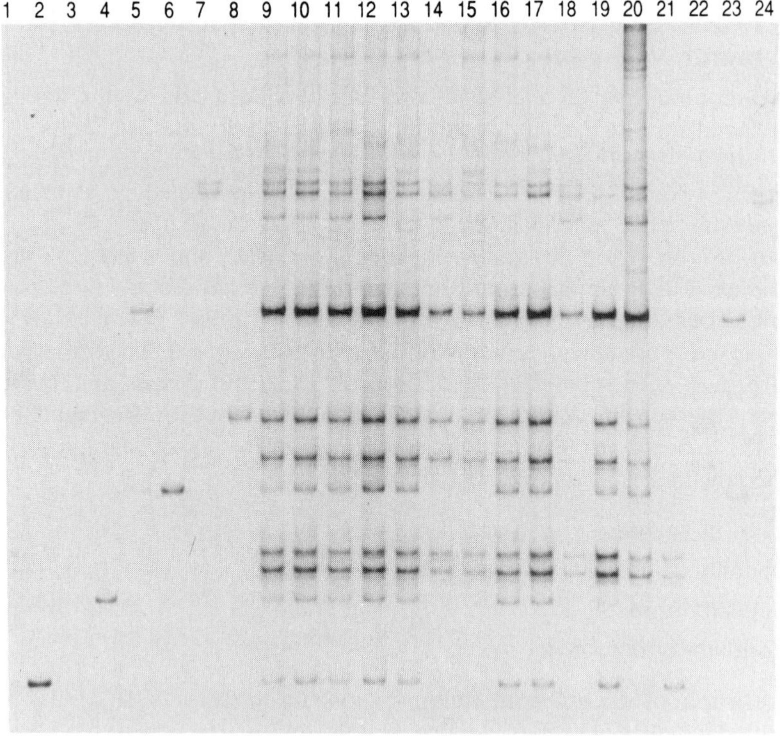


**Figure 4.** Multiplexed SSCP analysis of the factor VIII gene. Analysis of individual normal factor VIII exons as single-stranded DNA are shown in lanes 1 and 21 = exon 20, lanes 3 and 22 = exon 3, lanes 5 and 23 = exon 25 and lanes 7 and 24 = exon 16. The same normal exons 20, 3, 25, and 16 are shown as double-stranded DNA in lanes 2, 4, 6, and 8, respectively. Genomic DNA samples from 12 haemophilia A patients after group 1 multiplexed analysis (see *Table 1*), are shown in lanes 9–20. Despite the complexity of the multiplexed pattern, the bands are well-defined, migrate consistently, and can be individually assigned to exons of origin, enabling rapid localization of mutations.

**Protocol 5.** Multiplexed PCR

*Equipment and reagents*

- Genomic DNA
- Oligonucleotide primers for PCR
- *Taq* DNA polymerase
- [$\alpha$-$^{33}$P]dATP (1000–3000 Ci/mmol; 10 Ci/ml; NEN Radiochemicals)
- 1 × PCR buffer containing 200 μM each of dATP, dCTP, dGTP, dTTP (Pharmacia)
- Mineral oil (molecular biology grade; Sigma)
- Thermal cycler (e.g. Perkin-Elmer)
- 2% agarose gel

*Method*

1. Use the oligonucleotide primers directly after uncoupling and deprotection. Mix them in groups (as in *Table 1* for the factor VIII gene), dry down, and reconstitute such that 1 μl contains 200 ng of each primer.
2. For PCR, add 400–800 ng genomic DNA, 200 ng of each primer, 1.25 units *Taq* DNA polymerase/primer pair, and 0.25 μl of [$\alpha$-$^{33}$P]dATP/primer pair to 100 μl of PCR buffer containing 200 μM of dNTPs.
3. Overlay the reaction mixture with mineral oil.
4. Carry out PCR. The PCR conditions for a Perkin-Elmer thermal cycler are:
   (i) denature at 94°C for 5 min;
   (ii) perform 30 cycles of PCR as follows:
   - denaturation 94°C for 30 sec
   - annealing 50°C for 1 min
   - extension 72°C for 1 min with 3 sec additional elongation per cycle

   The final extension is 72°C for 10 min.
5. Check 10 μl of the PCR product on a 2% agarose gel to confirm amplification.
6. Continue as in *Protocol 3*, steps 4–11.

# 4. Direct sequencing of biotinylated PCR products using streptavidin-coated magnetic beads in a solid support system

Once a suspected mutation is detected using the exon scanning techniques described in this chapter, confirmation is required by DNA sequencing. The DNA fragments could be cloned and sequenced using traditional methods. Although easily achievable using T-vectors (17) or blunt-end ligation the cloning of PCR products is time-consuming. Furthermore, in order to exclude artefacts introduced by the infidelity of the *Taq* polymerase during amplification, several clones must be sequenced from each DNA sample being analysed.

This tedious approach can be avoided by direct sequencing of the PCR products. The DNA template must be clean and the excess primers and excess dNTPs must be removed. Various column, gel purification, and electroelution methods have been described but these are time-consuming and often good readable sequence is difficult to achieve.

A very elegant approach which avoids these problems utilizes streptavidin-coated magnetic beads (18, 19) in a solid phase system which is both rapid and reliable. Efficient concentration of the PCR product while simultaneously removing excess primers and nucleotides results in excellent readable sequence.

Amplified DNA template prepared using this solid phase system can be the initial step to incorporate into *Taq* cycle sequencing in a thermal cycler using *Taq* DNA polymerase (20), or T7 DNA sequencing (21) in a conventional dideoxy chain termination protocol (22). Both variations are easily adapted for automated DNA sequencing simply by substituting four fluorescently labelled nucleotides in place of a radiolabelled nucleotide (usually [$\alpha$-$^{35}$S]dATP).

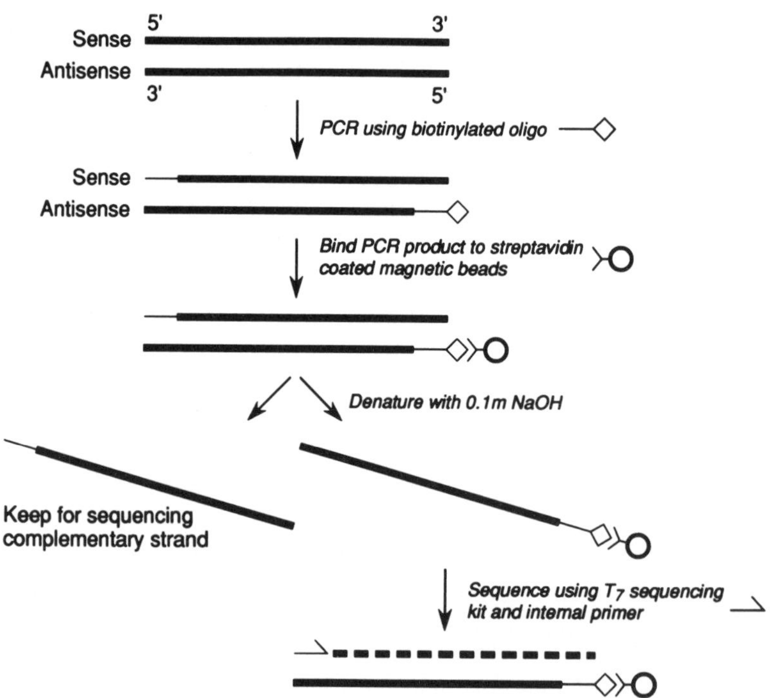


**Figure 5.** Schematic representation of the preparation of single-stranded DNA for direct sequencing using biotinylated oligonucleotide primers and streptavidin-coated magnetic beads. See the text for more details.

In *Protocol 6* we describe the direct sequencing of biotinylated PCR products by preparing single-stranded DNA templates and directly sequencing them using T7 DNA polymerase. Both DNA strands can be sequenced in this way. A schematic representation is shown in *Figure 5*.

**Protocol 6.** Preparation of single stranded DNA template and direct sequencing using streptavidin-coated magnetic beads and biotinylated PCR products

*Equipment and reagents*

- Genomic DNA
- PCR primers with the 3′ antisense primer biotinylated at its 5′-end
- Streptavidin-coated magnetic beads (e.g. Dynabeads M280-streptavidin, Dynal 112.05)
- Magnetic particle concentrator (MPC) (e.g. Dynal MPC-E, 120.04)[a]
- 2 × washing and binding buffer (WBB) (*see Protocol 4*)
- 0.1 M NaOH
- Internal primers for sequencing (20 ng/μl)
- T7 sequencing kit (e.g. Pharmacia 27–1682–01) and sequencing polyacrylamide gel (*Taq* cycle sequencing can also be used)
- Agarose gel (1–2%)

A. *Amplification*

1. Amplify the genomic DNA using only 100–150 ng of the biotinylated 3′ primer and 200 ng of the 5′ primer per 100 μl reaction. There should be little or no primer left after amplification.
2. Run 10 μl on a 1–2% agarose gel to check amplification.

B. *Binding of PCR product to the streptavidin-coated magnetic beads*

1. Pipette 50 μl of the beads into a 1.5 ml sterile microcentrifuge tube and remove the supernatant.[a]
2. Resuspend the beads in 50 μl of 1 × WBB. Mix gently.[b] Discard the supernatant.
3. Repeat the wash using 1 × WBB. Discard the supernatant.
4. Resuspend the beads in 50 μl of 2 × WBB. Vary the volume according to the quantity of PCR product. The final concentration of NaCl should be 1–2 M for efficient binding of PCR product to the beads.
5. Add an equal volume of PCR product, i.e. 50 μl. Mix gently.[b] Incubate at 37°C for 10 min.[c] Cool the tubes to room temperature.
6. Discard the supernatant and resuspend the beads in 50 μl of 1 × WBB. Mix gently. The PCR/magnetic bead complex can be stored for up to 1 week at 4°C prior to sequencing.

C. *Preparation of single-stranded DNA template*

1. Discard the 1 × WBB.
2. Add 50 μl of 0.1 M NaOH and incubate at room temperature for 10 min.

**Protocol 6.** *Continued*

3. Remove the supernatant[a] and **keep this**. This contains the DNA sense strand for subsequent sequencing (see below).
4. Wash the beads with 0.1 M NaOH. Discard the supernatant.
5. The streptavidin-coated magnetic beads are now bound to the biotinylated antisense DNA strand. Wash the beads once in 50 μl of 1 × WBB and then once in 50 μl $H_2O$.
6. Resuspend the beads in 10 μl $H_2O$ ready for sequencing.

D. *DNA sequencing*

1. Using an internal primer, sequence the antisense single-stranded DNA template using the T7 sequencing kit following the manufacturer's instructions for single-stranded DNA.[c] This will generate sense strand sequence.
2. Just prior to loading the polyacrylamide sequencing gel, heat the sequencing reactions at 72°C for 2–5 min. Load 3 μl per lane. If required, the magnetic beads can be kept to one side of the tube using the MPC to facilitate easier gel loading.
3. The sense strand template (see part C, step 3) can be sequenced after neutralizing with an equal volume of 0.1 M HCl and subsequent ethanol precipitation. Redissolve the DNA in 10 μl $H_2O$ and continue with the sequencing reactions.

[a] Use the MPC to isolate the beads from the supernatants between steps.
[b] Only mix by gentle pipetting.
[c] Agitate the beads during incubation steps to ensure contact of the beads with reagents.

## 5. Chemical cleavage detection of mismatched base pairs

This technique is based on the ability of hydroxylamine and osmium tetroxide ($OsO_4$) to modify single base mismatches (5, 23). This is used to identify point mutations at the genomic level by amplifying the DNA fragment of interest, creating a heteroduplex with the radiolabelled wild-type DNA, and subjecting the heteroduplex to chemical modification and cleavage.

Hydroxylamine at pH 6.0 modifies the C5═C6 double bond in cytosine and $OsO_4$ oxidizes the C5═C6 double bond in a thymine specific reaction. At the point of mismatch in the heteroduplex, there is effectively single-stranded DNA which is readily recognized and modified. Piperidine is subsequently used to cleave at the modified points of mismatch. Hydroxylamine modifies C–C, C–T, and C–A mismatches. $OsO_4$ modifies T–G, T–C, and

T–T mismatches. However by forming heteroduplexes between the mutant and wild-type sequences, all 12 possible point mutations and deletions give rise to at least one detectable mismatch or loop out.

After chemical modification and cleavage, the DNA heteroduplexes are run on a urea/polyacrylamide gel and mutations are identified by the presence of cleaved DNA fragments. Examples are shown in *Figure 6* where three mutations have been identified in exon 11 of the factor VIII gene. The size of the cleaved product gives positional information regarding the mutation which is confirmed by sequencing.

*Protocol 7* describes the preparation of 6% urea/acrylamide gels for chemical cleavage analysis and *Protocol 8* describes the cleavage detection procedure itself.

## **Protocol 7.** Preparation of 6% urea/acrylamide gels for chemical cleavage analysis

### *Equipment and reagents*

- 30% stock acrylamide solution. Prepare by mixing 28.5 g acrylamide, 1.5 g *N,N'*-methylene bisacrylamide and making up to 100 ml with sterile distilled water. Store at 4°C
- Urea buffer. Prepare by mixing 210 g urea and 50 ml 10 × TBE buffer (see *Protocol 1*). Make up to a total volume of 400 ml with sterile distilled water. Dissolve by heating at 50°C and filter through Whatman No. 1 filter paper. Store at 4°C.
- TEMED
- 10% ammonium persulphate
- Polyacrylamide gel mould with gradient spacers (0.4–1.2 mm thick) and a 0.4 mm comb

### *Method*

1. Prepare the 6% urea/acrylamide gel mixture[a] by mixing:
   - Urea buffer 96 ml
   - 30% acrylamide 24 ml
   - TEMED 54 μl
   - 10% ammonium persulphate 960 μl
2. Pour the gel immediately and allow it to polymerize.

[a] The quantity prepared here is for a gel of dimensions 394 × 300 mm.

## **Protocol 8.** Chemical cleavage detection of mismatched base pairs

### *Equipment and reagents*

- Precipitation solution: 0.3 M sodium acetate pH 5.2, 0.1 mM EDTA
- Ethanol (100% and 70%)
- [$\gamma^{32}$-P]ATP (4500 Ci/mmol; ICN Radiochemicals)
- T4 polynucleotide kinase (Pharmacia 27–0736–02)
- Kinase buffer: 500 mM imidazole-HCl, 100 mM $MgCl_2$, 1 mM EDTA, 50 mM DTT, 1 mM spermidine, 3 mM ADP

**Protocol 8.** *Continued*

- Nick columns (Pharmacia 17–0855–02 or similar Sephadex G50 columns)
- 5 × HET buffer: 1.5 M NaCl, 0.5 M Tris–HCl pH 8.0
- Hydroxylamine hydrochloride (Aldrich 25,558–0)
- Diethylamine
- 1 M Tris–HCl pH 8.0
- 500 mM EDTA
- Pyridine (Aldrich 18,452–7)
- 4% $OsO_4$ solution (Aldrich 25175–5)
- 10 mg/ml tRNA (from yeast; Sigma RO128)
- 1 M Piperidine (Sigma P5881, supplied as 10 M, dilute in sterile distilled water to 1 M just before use)
- 1 kb DNA size marker ladder (Gibco-BRL) and/or ΦX174 *Hae*III fragments
- 1 × TBE buffer (see *Protocol 1*)
- 1% agarose gel
- 6% urea/acrylamide gel (see *Protocol 7*)
- Stop solution: 80% formamide, 0.1% Bromophenol blue, 1 mM EDTA, 10 mM NaOH
- Methanol/Dry ice bath
- Speedvac (Life Sciences International)
- High voltage power supply
- Screw capped tubes (1.5 ml, Sarstedt)

A. *Amplification*

1. Design the oligonucleotide PCR primers within the intron sequences to enable analysis of whole exons and splice junctions. Allow 20–100 bp between the primer and the 5′/3′ end of each exon; otherwise the cleavage products may be too small to be detected.
2. Amplify the test DNAs (see *Protocol 1*) and include two normal DNA samples, one to be used as a labelled probe, the other as a negative control.

B. *Purification of the normal DNA fragment for use as the probe*

1. Run the total PCR product on a 1% agarose gel at a constant 30 V for 1–2 h, in 1 × TBE containing 1 μg/ml ethidium bromide.
2. View the DNA fragment using a UV transilluminator and cut out the gel slice containing the DNA fragment. Place the gel slice into a 1.5 ml sterile microcentrifuge tube.
3. Place the tube in a methanol/dry ice bath for 30 min.
4. Centrifuge in a microcentrifuge at 13 000 rpm for 30 min.
5. Transfer the supernatant to a clean 1.5 ml sterile microcentrifuge tube and concentrate in a Speedvac until the volume is reduced to 100 μl.
6. Add the following:
   - precipitation solution 200 μl
   - 100% ethanol 750 μl

   Place in a methanol/dry ice bath for 30 min.
7. Centrifuge in a microcentrifuge at 13 000 rpm for 30 min. Wash the pellet in 1 ml of 70% ethanol.
8. Dry the pellet in the Speedvac and store at −20°C until required.

C. *Labelling the probe*

1. Dissolve the DNA pellet in 10.5 μl sterile distilled water.
2. Add the following:
   - [$\gamma$-$^{32}$P]ATP (equivalent to 100 μCi) 10 μl
   - T4 polynucleotide kinase 2 μl
   - kinase buffer 2.5 μl

   Mix and incubate at 37°C for 35 min.
3. Remove the unincorporated radioactivity by passing the reaction through a Nick column.
4. Dry the probe in a Speedvac and then redissolve in 75 μl sterile distilled water. Count 1 μl in a liquid scintillation counter.
5. Use 3–5 $\times$ $10^5$ d.p.m./sample for the heteroduplex reaction (about 7 μl).

D. *Heteroduplex reaction*

1. Dry the PCR products in the Speedvac (all the test DNAs and the normal DNA).
2. Dissolve each DNA pellet in 25 μl sterile distilled water and then add:
   - 5 $\times$ HET buffer 8 μl
   - labelled probe 7 μl
     (adjust volume as necessary to give 3–5 $\times$ $10^5$ d.p.m.)
3. Incubate for 5 min at 95°C then allow the reactions to cool down overnight by floating the tubes in the switched off water bath.

E. *Hydroxylamine and $OsO_4$ modifications*

**NB.** The hydroxylamine and $OsO_4$ reactions should be performed in a fume cupboard.

1. For the hydroxylamine reaction, proceed as follows:
   (a) Mix 1.39 g hydroxylamine with 1.6 ml sterile distilled water. Adjust to pH 6.0 with diethylamine (about 1 ml).
   (b) Into a sterile 1.5 ml screw-capped tube add:
   - heteroduplex 7 μl
   - hydroxylamine solution 20 μl

   Incubate for 2 h at 37°C.
2. For the $OsO_4$ modification, proceed as follows:
   (a) Prepare a fresh $OsO_4$ solution in a 1.5 ml screw capped tube by adding:
   - 1 M Tris–HCl pH 8.0 6.9 μl
   - 500 mM EDTA 1.4 μl

**Protocol 8.** *Continued*

- pyridine 41.6 μl
- 4% osmium tetroxide 86.8 μl
- sterile distilled water 863 μl

Keep on ice.

(b) Vortex to mix. The solution should be lemon yellow in colour and turbid.

(c) Into a clean 1.5 ml screw-capped tube add:

- heteroduplex 7 μl
- osmium tetroxide solution 18 μl

Mix and incubate on ice for 2 h.

**3.** Precipitate the products of the hydroxylamine and $OsO_4$ modifications as follows:

(a) Add to each reaction in order:

- tRNA 1 μl
- precipitation solution 200 μl
- 100% ethanol 750 μl

(b) Mix and leave in a methanol/dry ice bath for 15 min. Then spin in a microcentrifuge for 15 min at 13 000 r.p.m.

(c) Wash the pellet in 1 ml 70% ethanol, spin in a microcentrifuge for 15 min at 13 000 r.p.m. and Speedvac dry.

F. *Piperidine cleavage*

**1.** Prepare a fresh 1 M piperidine solution in sterile distilled water.

**2.** Add 50 μl of 1 M piperidine to each sample, vortex, and briefly spin in a microcentrifuge.

**3.** Incubate at 90°C for 25 min. Dry in the Speedvac.

**4.** Add 200 μl of sterile distilled water, mix, and dry in the Speedvac. Repeat the wash.

**5.** Store the pellets at −20°C overnight.

G. *Labelling of the 1 kb ladder or Hae*III *digest of ΦX174*

**1.** Into a 1.5 ml sterile screw-capped tube add:

- 1 kb ladder or ΦX174 digest 5 μl
- kinase buffer 2.5 μl
- T4 polynucleotide kinase 2 μl
- [$\gamma^{32}$P]ATP (equivalent to 50 μCi) 5 μl
- Sterile distilled water 25.5 μl

Mix and incubate at 37°C for 35 min.

2. Pass the reaction through a Nick column.
3. Dry the labelled product in the Speedvac and redissolve in 25 μl sterile distilled water. The labelled size markers (1 kb ladder or ΦX174 digest) are usable for several weeks.

H. *Electrophoresis of the reactions*

1. Add 20 μl of stop solution to each reaction.
2. Vortex and then denature the samples at 95°C for 5 min and immediately snap cool on ice.
3. Load 2–10 μl on to a 6% urea/acrylamide gel depending on the radioactivity of the sample.
4. Also load 1 μl of labelled 1 kb ladder or ΦX174 fragments as size markers.
5. Run the gel at 2000 V, 60 mA, 60 W, constant power, until the Bromophenol blue reaches the bottom.
6. Transfer the gel to 3MM Whatman paper, cover with cling film and dry under vacuum at 80°C for 60 min.
7. Expose to Kodak XAR film at −70°C with an intensifying screen overnight.
8. Develop the autoradiograph.

## 6. Denaturing gradient gel electrophoresis (DGGE)

The DGGE method detects mutations (small deletions and insertions, point mutations) by separating PCR amplified DNA fragments, which differ from wild-type DNA in their melting behaviour, on a denaturing gradient gel.

The amplification products are composed of two different melting regions: a very high melting domain which is arbitrarily synthesized on the 5′-end of one primer (40 bases of GCs called the GC-clamp) and a lower melting domain which could vary in length between 25 bases and about 350 bases (24). The lower melting domain is the domain of interest which is to be screened for mutations. When a fragment migrating within the gel reaches a discrete denaturing concentration, the double-stranded DNA of the lower melting domain is melted into single-strands with the effect that the amplification product will cease migrating at this position in the gel. Both strands are still connected by the higher melting domain. The position where the melting takes place within the gel depends on the base pair sequence and the base pair composition of the lower melting domain. Even when two equal double-stranded DNA fragments differ by only one base pair, they melt at different denaturing positions. A schematic diagram of this method is given in *Figure 7*.

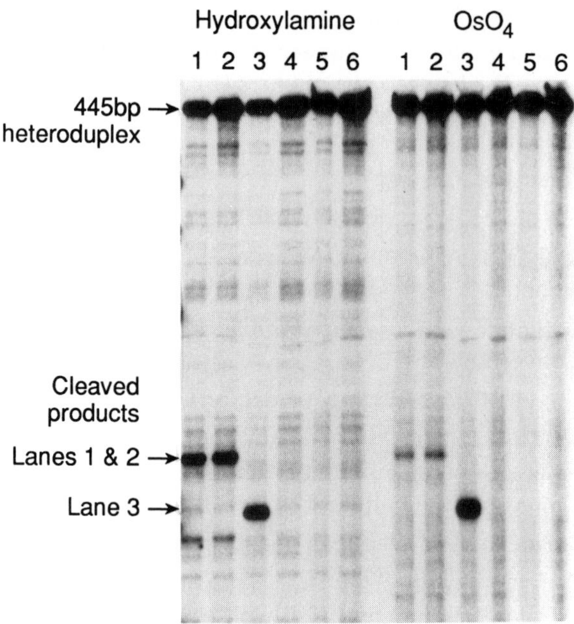


**Figure 6.** Chemical cleavage of exon 11 of the factor VIII gene in six patients, 1–6. Cleaved fragments are seen in lanes 1, 2, and 3. The cleaved products in lanes 1 and 2 are identical, consistent with the same single base substitution in these patients. The cleaved product in lane 3 was due to a different mis-sense mutation. All three mutations were detected by both hydroxylamine and osmium tetroxide ($OsO_4$). No cleavage products were detected in the samples run in lanes 4–6.

The separated amplification products can be made visible on an UV transilluminator after incubating the gel with an ethidium bromide solution.

Although DGGE is a powerful method for screening DNA regions for mutations or sequence polymorphisms, substantial theoretical and practical work is required to optimize it for each DNA segment under analysis.

## 6.1 Preliminary theoretical work

To identify all mutations within a DNA fragment by means of the DGGE method, it is first necessary to find a unit melting domain of the region. The computer programs MELT87 and SQHTX (25,26) have been designed for this purpose. MELT87 is very helpful for finding the ideal unit melting domain of a DNA fragment of interest (*Figure 8*). Defining such a melting domain is also influenced by the position of the artificial GC clamp, which could be attached to either end of the amplification product. SQHTX calculates the running distance on a gradient gel between a wild-type and a mutated DNA fragment. The displacement is given for a helix defect for each

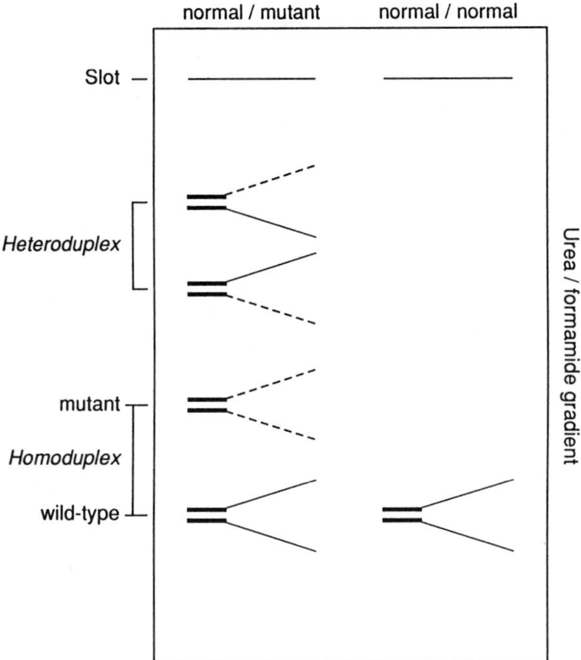


**Figure 7.** Diagram representing a denaturing gradient gel showing the separation of two mixed wild-type DNAs (homoduplexes) and a wild-type DNA mixed with a mutant DNA (heteroduplexes). While the wild-type homoduplexes melt at one position, the heteroduplexes show four bands (two different heteroduplexes and the original homoduplexes) which differ in their melting behaviour. The position of the two heteroduplexes always lies above the melting point of the homoduplexes. The mutated original homoduplex may be positioned above or below the wild-type homoduplexes. At their melting point the single strands are still held together by the GC clamp, shown as the thick line.

pair in the fragment as a function of gel running time (*Figure 8*). Further explanation of the operation and interpretation of the two programs is supplied with the software.

### 6.1.1 MELT87

When the unit melting domain is determined with the computer program MELT87, the positions of the amplification primers are automatically assigned. However, in most cases an optimal melting domain is not optimal for amplification primers. Therefore, since the composition of the amplification primers is very important for good PCR results, it is often necessary to find a compromise between a useful amplification primer and a unit melting domain of an amplification product. The melting temperature ($T_m$) of this unit melting domain should not deviate more than 0.5 °C up or down from the median temperature of its domain.

In our experience the following rules are helpful guidelines:

- an effective primer is 21–26 bp long (without the GC clamp), having a well-balanced composition of bases.
- Parts of each primer alone and parts of both primers together should not be able to hybridize against one another more than four base pairs in series.
- The two external bases of one primer's end should have no more than one base complementary to the bases at the end of the other primer.
- The $T_m$ of one primer should lie between 56 °C and 60 °C.

Programs are available to determine the $T_m$ of the amplification primers.

Besides its effect in helping to create a unit melting domain of a DNA fragment, the GC clamp is absolutely necessary for the gel run. When the DNA fragment has reached the point where the denaturing conditions, produced by the urea gradient, corresponds to its $T_m$, then the unit melting domain will melt at once into single strands. Only the GC clamps, by keeping the single strands together as double strands in this region, fix them at this level. This short section of double-stranded DNA also allows easy visualization of the migrated DNA fragment by ethidium bromide staining.

### 6.1.2 SQHTX

The values given by the program are temperature differences ($\Delta T$ °C) which exist between the wild-type DNA fragment and a fragment with any kind of mutation (*Figure 8*). The relationship between these temperature values and the denaturing concentration is $\Delta 1$ °C = $\Delta$ 3% denaturing concentration (approx. equivalent to 1 cm distance within a 20% urea gradient gel). For a given gel running time, the $\Delta T$ (°C) temperature values for all possible mutations within the melting domain should be higher than 0.1, otherwise it is not possible to separate the wild-type homoduplex fragment from the mutated one.

## 6.2 Optimizing the conditions for DGGE

### 6.2.1 Amplification

We have found that the same basic amplification conditions can be used for each different amplification reaction. To optimize the reaction for each amplification, it is necessary to vary only the annealing temperature. This is also the only value which has to be changed when using a different thermal cycler. When using a primer with an additional sequence (GC-clamp) on its 5′- or 3′-end, better results are obtained when the time of each annealing step is longer than in an ordinary amplification reaction. In our hands, 2 min annealing is sufficient.

### 6.2.2 DGGE

Although computer analysis is essential before one can start the practical procedure, it still takes some practical work to find the optimal running conditions. This optimization is carried out by running several ‘travel schedule

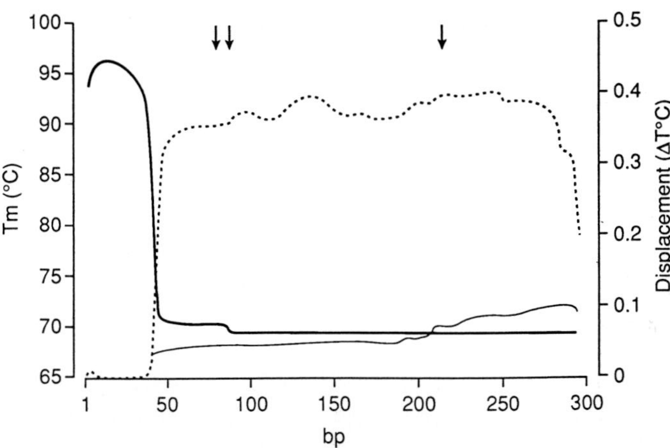


**Figure 8.** Graphic demonstration of the melt map (MELT87) and the displacement map (SQHTX-program) of the factor VIII gene exon 16. The thin solid line represents the melting map, $T_m$(°C), of the amplification product (exon 16 and flanking sequences) without the GC clamp. The *x*-axis indicates the base pairs of the exon 16 PCR product and the *y*-axis on the left side gives the melting temperature of each base pair within this amplification product. The thin line shows that the melting behaviour of the exon 16 product without the GC clamp is very heterogenous. The thicker solid line represents the effect of an attached GC clamp on the exon 16 PCR-product. Starting from a melting temperature of about 95 °C, the $T_m$ drops after 40 bp to a unit melting temperature of about 69 °C and keeps constant over the whole amplification product. The higher melting domain in the first 40 bases represents the GC clamp, the lower melting domain represents exon 16 and flanking sequences which can now be screened for mutations because of its unit melting behaviour. The dashed line demonstrates the displacement temperature map of factor VIII exon 16 complete with GC clamp. The *y*-axis on the right side gives the different temperature values ($\Delta T$°C) of each base pair within the PCR-product. These values arise if the SQHTX program compares the temperature differences between a possible mutation at each base pair position within a mutated fragment to the same position in a wild-type DNA duplex. The figure demonstrates that an average temperature difference ($\Delta T$°C) of about 0.35–0.40 exists over the whole of exon 16. Since we know under practical conditions that the lower limit is $\Delta T = 0.1$ °C for detecting mutations ($\Delta 0.1$ °C = $\Delta 0.3$% denaturing concentration = 0.1 cm running distance in the gel), we can calculate that each possible mutation is detectable within exon 16. The arrows show the position of some mutations which were detected in an analytical gel.

gels' which differ in the electrophoresis conditions (variation of the urea gradient and the voltage). Experience in finding the right running conditions for several different amplification products makes it easier in future work to select running conditions directly from the theoretical values since a strong correlation exists between theoretical and practical values. The following points should be borne in mind:

- Use a chosen constant voltage and a given denaturing gradient (it seems reasonable to use gradient gels which differ by 20%) to determine the running time.

- If the results are not satisfactory (no distinct band(s), or retardation) choose another voltage and/or denaturing gradient. However, to find the right running conditions, change only one variable at a time.
- In our experience, more distinct bands are obtained when the running time is longer, rather than the voltage higher.
- In our laboratory, best results are obtained using a voltage range between 25–60 V.
- All three variables (voltage, running time, and gradient) are dependent on the size and base content of the amplification products.

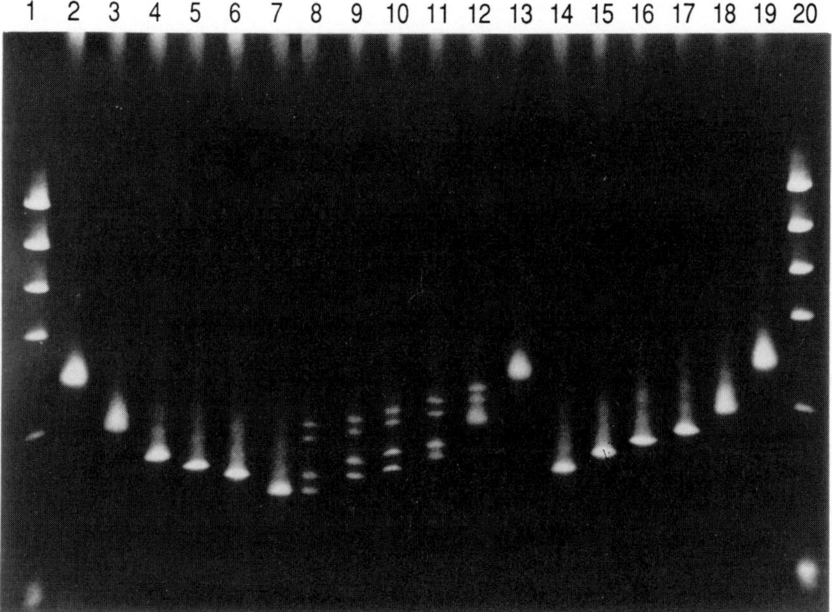


**Figure 9.** Travel schedule gel for the factor VIII gene exon 16. Lanes 2–7 represent a normal wild-type PCR product which is normally used to find the right running time for the analytical gel using a chosen voltage and denaturing gradient. Samples of the same amplification product have been loaded successively from the right to the left every 2 h during the first 10 h running time (total running time = 24 h). The optimal running time for the exon 16 PCR product is 20–22 h, represented by lanes 5 and 6. In these lanes retardation has already taken place (bands at the same position) and the bands are well focused compared to those in lanes 2 and 3. To demonstrate the correct running conditions for an analytical gel, a mutant exon 16 amplification product has been loaded successively (lanes 14–19) at the same time as the normal PCR-product from the left to the right. The different retardation compared to the wild-type fragment can be observed best after 20–22 h (lanes 5, 6 and 15, 16). Lanes 8–13 show the advantage given by analysing heteroduplex formation between a wild-type and a mutant DNA fragment. These amplification products were also loaded from the left to the right. Besides the two different homoduplexes which are the two bands below, two additional heteroduplex bands appear, thus allowing a better mutation identification in critical cases. For further explanation see the text.

At the optimum running time, the bands are well separated. In contrast to the slower migrating material a distinct pattern develops due to focusing in the gradient. In the example shown in *Figure 9*, the optimum running time is 20–22 h (see the bands in lanes 5–6, 9–10, and 15–16).

Having established the running conditions for a DNA fragment spanning a specific gene region, these conditions can now be used to run analytical gels for mutation screening.

## 6.3 Practical procedures

Full descriptions of the amplification and DGGE procedures are given in *Protocol 9*. The assembled equipment for DGGE is shown in *Figure 10*.

### Protocol 9. DGGE

*Equipment and reagents*

- 10 × TAE buffer pH 7.4: 0.4 M Tris–HCl, 0.2 M sodium acetate, 10 mM EDTA
- 40% formamide deionized with a mixed bed resin and filtered through a solution filter (Schleicher and Schüll)
- 1 × loading buffer: 20% Ficoll 400, 10 mM Tris–HCl pH 7.8, 1 mM EDTA, pH 8.0, 0.1% orange G
- 40% polyacrylamide stock solution (acrylamide:bisacrylamide = 37.5:1)
- 100% denaturing solution: 7.0% polyacrylamide, 7.0 M urea, 40% formamide, in 1 × TAE buffer pH 7.4.[a] Prepare this using the 40% polyacrylamide stock
- 0% denaturing solution: 7.0% polyacrylamide in 1 × TAE buffer pH 7.4.[a] Prepare this using the 40% polyacrylamide stock
- 10% ammonium persulphate stock solution
- TEMED
- Ethidium bromide stock solution (10 mg/ml)
- Electrophoresis apparatus (Hoefer SE 600); (glass plates 180 mm × 160 mm; spacer thickness 0.75 mm)
- Power supply
- Heating water bath with circulating pump to hold constant gel temperature (62°C) during gel run
- Gradient mixer in combination with a magnetic stirrer and a butterfly intravenous infusion set (a needle of outside diameter 1.9 mm, connected to a tube) for pouring a denaturing gradient gel
- Microsyringe for loading samples on to the gel.
- Peristaltic pump for circulating buffer between buffer tanks to prevent significant increase in pH during electrophoresis
- UV transilluminator
- Gel data documentation system (e.g. polaroid camera)
- Personal computer loaded with Lerman's software SQHTX and MELT87 (see Section 6.1)

A. *Amplification*

1. Mix the following compounds for the amplification reaction (total volume of 50 μl):

| | |
|---|---|
| • 10 × PCR buffer | 5 μl |
| • dNTP mix | 8 μl |
| • DNA | 500 ng |
| • oligonucleotide primer | 300 ng |
| • oligonucleotide primer with GC-clamp | 900 ng |
| • *Taq* DNA polymerase | 2.5 U |
| • Make up to final volume with water | 50 μl |

**Protocol 9.** *Continued*

2. Amplify according to the following conditions:
   - denaturing 94°C for 420 sec
   - denaturing 94°C for 35 sec
   - annealing 50–57°C for 120 sec
   - extension 72°C for 60 sec + 3 secs/cycle

   (denaturing, annealing and extension steps above: 40 cycles)
   - extension 72°C for 600 sec
3. Check the amplification by running 1/10th of the PCR product on a 1–2% agarose gel.

B. *Travel schedule gels*

Denaturing gradient gel electrophoresis can be carried out very successfully in a Hoefer SE 600 electrophoresis unit attached to a circulating water bath set to a constant temperature of 62°C. All electrophoresis is performed at this temperature.

1. Using the 0% denaturing and 100% denaturing stock solutions, prepare the two denaturing solutions (10 ml each) which are needed for the gradient gel (For example, for a denaturing gradient ranging between 30% and 50%, prepare a 30% and a 50% denaturing solution).
2. Clean the glass plates of the gel apparatus scrupulously first with warm water, then with *iso*-propanol and finally with distilled water.
3. Fix the glass plates within the equipment by slightly greasing the bottom of the glass plates.
4. Add 5.0 μl TEMED and 100 μl of 10% ammonium persulphate solution to the prepared solutions for the gradient gel and pour the solutions into the appropriate chambers of a gradient mixer. Make sure that the higher concentration solution is placed in the anterior chamber and the lower concentrated solution is placed in the posterior chamber. Stand the gradient mixer on a magnetic stirrer with a stirring bar inside the anterior chamber. Stir the anterior chamber fast but without producing any air bubbles. To avoid the use of a peristaltic pump, it is very important that the gradient mixer is positioned on a higher level than the glass plates within the fixation equipment. The connection between gradient mixer and glass plates is made with a 'butterfly' intravenous infusion set (a needle of outside diameter 1.9 mm, connected to a tube). The free end of the tube is connected to the output channel of the gradient mixer and the butterfly needle is put between the glass plates.
5. Pour the travel schedule gel by opening the channel between both mixer chambers first and by opening the output channel second. Pour the gel slowly (about 5 min), as this helps to exclude air bubbles from the gel and to achieve a gradient without waves. To start the solution running within the tube, it is sometimes necessary to tap the tube.

6. Put the sample comb in position in the gel and allow the gel to polymerize. Usually this takes about 2 h.
7. In the meantime, mix the amplification products with the loading buffer.
8. When the gel has polymerized, remove the sample comb. Load a wild-type DNA sample on to the gel. Use 1 × TAE pH 7.4 as running buffer.
9. Begin the electrophoresis (e.g. 25 V for a 25–45% urea/formamide gradient gel).
10. Load additional DNA samples consecutively five times with a 2 h gap between each loading.
11. Stop the gel run after 24 h running time.
12. Stain the gel with ethidium bromide solution and evaluate the results on a UV transilluminator. (Caution: ethidium bromide is a known carcinogen. Wear gloves at all times during handling of the solution. Safety glasses are recommended when large volumes or high concentrations are used.)
13. Select the optimum running time for the analytical gel.

C. *Analytical gel*

1. Select the same gel running conditions (denaturing gradient, voltage and running time) established by the above analysis using the travel schedule gel.
2. Pour the gel in the same manner as explained for the travel schedule gel above.
3. Form heteroduplexes[b] before loading the amplification products as follows:
   (a) Mix 10 μl (80 ng) of a wild-type PCR-amplified DNA with 10 μl (80 ng) of a patient DNA or preferably mix 10 μl of two different patient DNAs.
   (b) Add sample buffer equal to one-third of this volume.
   (c) Heat the mixture for 5 min at 96°C in a closed water bath, switch the power off, and let it cool down to room temperature overnight.
4. Load the DNA samples on to the analytical gel and start the gel run.
5. After the gel run is completed, stain the gel with EtBr for 1 h and evaluate the results on a UV transilluminator

[a] Both solutions (the 0% denaturing and the 100% denaturing solution) should be prepared in large amounts, because every new preparation has a slightly different urea concentration, which changes the practical conditions of the gel run. It is necessary to reoptimize running parameters when new solutions are prepared.

[b] Heteroduplexes can also be formed using a thermal cycler as follows: denature at 96°C for 10 min, then leave at 55°C for 10 min.

## 6.4 Comments on the DGGE protocol

Heteroduplex formation improves the detection of mutations for two reasons:

- It allows the simultaneous analysis of two amplification products of two different patients.
- It produces two additional bands and therefore allows the detection of mutations that do not change melting behaviour of the homoduplex.

Sometimes a point mutation results in the same base pairing, except that it is reversed (e.g. A/T→T/A). In such cases, the running behaviour of the mutated and the normal DNA fragment cannot be distinguished. This problem is solved by mixing a normal and a mutated DNA sample. The formation of two additional heteroduplexes, which present now a T/T and a A/A mismatch base pair, differ much more in their melting behaviour than the original complete double-stranded DNA, thus resulting in a different running distance.

An example of the different melting behaviour of amplification products, which represents the same DNA region, but differ only in one base pair

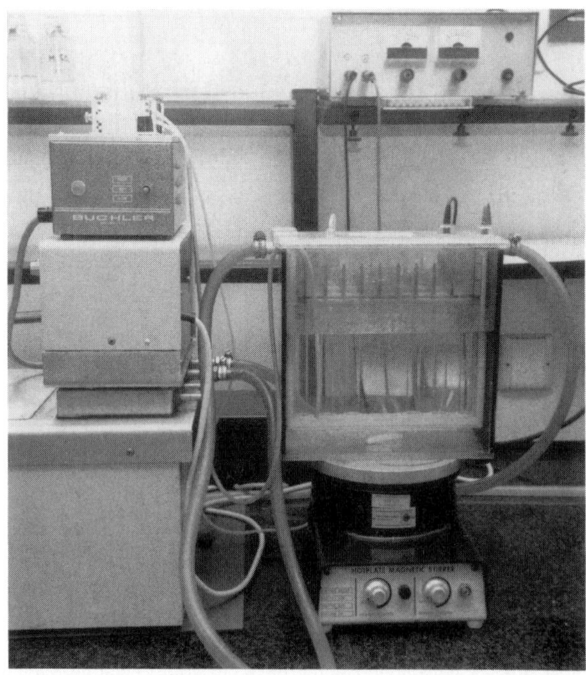


**Figure 10.** Assembled equipment for DGGE. Top left, pump circulating buffer from lower to upper reservoirs of the gel electrophoresis apparatus. Top right, power supply. Centre right, gel electrophoresis apparatus. Bottom right, magnetic stirrer. Bottom left, water bath set at 62 °C. Centre left, pump circulating water from bath through electrophoresis tank.

(point mutation), can be seen in *Figure 11*. If the base change of one DNA sample results in a completely different base pairing, we are able to identify four different bands: the original homoduplexes of the wild-type DNA and the patient DNA, and two different heteroduplexes.

Examples of a three-band pattern, where the wild-type and the mutated homoduplexes cannot be distinguished, can be seen in lanes 5 and 9 of *Figure 11*. If two different patient samples are used for the heteroduplex formation and a mutation pattern is found, it is not possible to decide which patient shows the mutation. In this case, the choice is to sequence both amplification products directly or to identify the mutated DNA by mixing them each with a definitive wild-type DNA before loading on to a new analytical gel.

## 7. Choice of method: DGGE, SSCP, or chemical cleavage

The three powerful screening techniques described in this chapter are able to detect all or virtually all mutations. The choice between these methods depends on a number of factors.

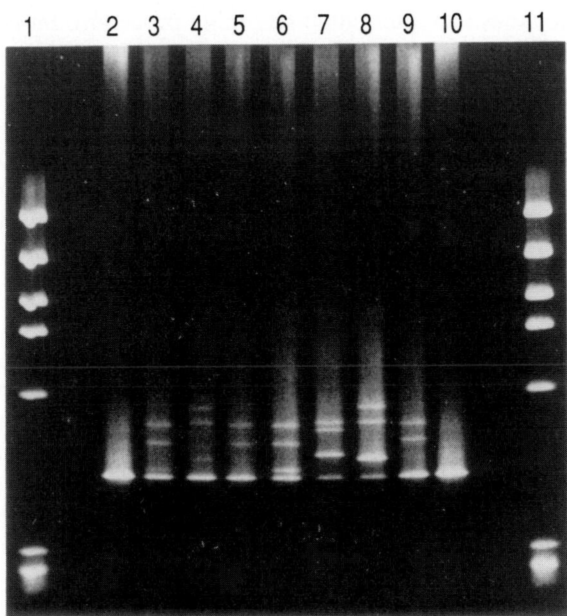


**Figure 11.** Analytical gel of the factor VIII gene exon 16 after heteroduplex formation of patient DNA samples with a normal DNA. Mutations can be seen by identification of additional bands above the normal band (see lanes 3–9). Controls consisting of heteroduplexes formed between two normal DNAs are in lanes 2 and 10. Lanes 1 and 11 contain a *Hin*dIII digest of ΦX174 DNA as size markers.

### 7.1 DGGE

DGGE is useful for screening a large number of individual cases for diverse mutations in a large gene. Once the conditions have been optimized it can be rapidly used to work through very large numbers of DNA segments as was successfully demonstrated for haemophilia A (8). However, the DGGE method is not the method of choice in every case of mutation or sequence polymorphism screening since:

(a) it takes substantial effort and time to find the optimal melting domain in connection with the primer determination and to transpose the theory into practice.
(b) DNA fragments which are about 350 bp in size give good results. However, mutation screening within fragments which are larger than 400 bp seems to be less sensitive and fails to identify every mutation, depending on the different migration behaviour of the mutant fragments within the gradient gel. As a consequence of this, exons (or amplification products), which are more than 400 bp in size should be divided into two separate smaller overlapping amplifications.
(c) Some genes are resistant to the DGGE method due to high G+C content, and for these an alternative screening procedure must be selected.

### 7.2 SSCP

SSCP has the advantage of simplicity in operation although conditions need to be carefully defined to obtain optimal results. This may explain why some workers have reported rather lower sensitivity than we have found in careful comparative studies (>95%). The size of fragment screened should not exceed 300 bp. By multiplexing it is possible to increase throughout to approach that achievable by DGGE.

### 7.3 Chemical cleavage

Chemical cleavage remains the standard by which other screening methods should be assessed as sensitivity approaches 100% and large fragments can be screened. However, the method is very laborious and technically demanding with requirements for toxic chemicals and a high energy emitting radiolabel. The combination of chemical cleavage with reverse transcribed transcript analysis allows large complex genes such as factor VIII or dystrophin to be screened.

## 8. Discussion and conclusions

However well these screening methods are performed, there remains a number of cases where no mutation is detected. Possible reasons for failure include the following:

- The wrong gene is being screened due to mis-diagnosis at the phenotype level.
- The gene has been rearranged in a subtle way not detectable by exon screening. For example, about half of all cases of severe haemophilia A are due to an inversion event that leaves two halves of the gene intact but separated by several megabases (27).
- The mutation causing the disease lies in an unscreened region of the gene.

Despite these difficulties and the technical demands of screening, the evident advantages of mutation specific diagnosis will probably lead to this becoming the preferred approach to genetic diagnosis in future. Large databases of mutations are accumulating for several disorders based on the results of gene screening. These have already contributed greatly to our understanding of the mechanisms of mutagenesis and the correlation between genotype and phenotype, as well as contributing to the analysis of structure and function of individual gene products. Another eventual benefit that may arise from precise identification of disease causing mutations would be the ability to repair genes through homologous recombination, as the ideal means of gene therapy.

## References

1. Saiki, R. K., Gelfand, D. H., Stoffel, S., Scharf, S. J., Higuchi, J., Horn, G. T., Mullis, K. B., and Erlich, H. A. (1988). *Science*, **239**, 487.
2. Wong, C., Dowling, C. E., Saiki, R. K., Higuchi, R. G., Ehrlich, H. A., and Kazazian, H. H. (1987). *Nature*, **330**, 384.
3. Pattinson, J. K., McVey, J. H., Boon, M., Ajani, A., and Tuddenham, E. G. D. (1990). *Br. J. Haematol.*, **75**, 73.
4. Meyers, R. M., Sheffield, V., and Cox, D. R. (1988). In *Genome analysis: a practical approach* (ed. K. Davies), p. 95. IRL, Oxford.
5. Cotton, R. G. H., Rodrigues, N. R., and Campbell, R. D. (1988). *Proc. Natl Acad. Sci. USA*, **85**, 4397.
6. Orita, M., Iwahana, H., Kanazawa, H., Hayashi, K., and Sekiya, T. (1989). *Proc. Natl Acad. Sci. USA*, **86**, 2766.
7. Meyers, R. M., Maniatis, T., and Lerman, L. S. (1987). In *Methods in enzymology* (ed. R. Wu), Vol. 155, p. 501. Academic Press, London.
8. Higuchi, M., Antonarakis, S. E., Kasch, L., Oldenburg, J., Economou-Peterson, E., Olek, K., Arai, M., Inaba, H., and Kazazian, H. H. Jr. (1991). *Proc. Natl Acad. Sci. USA*, **88**, 8307.
9. Michaelides, K., David, D., Schwaab, R., Lalloz, M. R. A., McVey, J. H., Brackmann, H., and Tuddenham, E. G. D. (1992). *24th Congress of the International Society of Haematology*. 506 Abstract.
10. Lalloz, M. R. A., Tuddenham, E. G. D., Schwaab, R., and David, D. (1993). In *Excerpta Medica International Congress Series* (ed. E. E. Polli), Vol. 1029, p. 101. Elsevier Science Publishers BV, Amsterdam.

11. Kunkel, L. M., Smith, K. D., Boyer, S. H., Borgaonkar, D. S., Wachtel, S. S., Miller, O. J., Breg, W. R., Jones Jr. H. W., and Rary, J. M. (1977). *Proc. Natl Acad. Sci. USA*, **74**, 1245.
12. Blin, N. and Stafford, D. W. (1976). *Nucleic Acids Res.*, **3**, 2303.
13. Saiki, R. K., Scharf, S., Faloona, F., Mullis, K. B., Horn, G. T., Erlich, H. A., and Arnheim, N. (1985). *Science*, **230**, 1350.
14. Orita, M., Suzuki, Y., Sekiya, T., and Hayashi, K. (1989). *Genomics*, **5**, 874.
15. Hayashi, K. (1991). *PCR Methods Applic.*, **1**, 34.
16. Wright, S. D., Michaelides, K., Johnson, D. J. D., West, N. C., and Tuddenham, E. G. D. (1993). *Blood*, **81**, 2339.
17. Marchuck, D. (1990). *Nucleic Acids Res.*, **19**, 1154.
18. Hultman, T., Stähl, S., Hornes, E., and Uhlén, M. (1989). *Nucleic Acids Res.*, **17**, 4937.
19. Thein, S. L. and Hinton, J. (1991). *Br. J. Haematol.*, **79**, 113.
20. Innis, M. A., Myambo, K. B., Gelfand, D. H., and Brow, M. A. D. (1988). *Proc. Natl Acad. Sci. USA*, **85**, 9436.
21. Tabor, S. and Richardson, C. C. (1987). *Proc. Natl Acad. Sci. USA*, **84**, 4767.
22. Sanger, F., Nicklen, S., and Coulson, A. R. (1977). *Proc. Natl Acad. Sci. USA*, **74**, 5463.
23. Montandon, A. J., Green, P. M., Giannelli, F., and Bentley, D. R. (1989). *Nucleic Acids Res.*, **17**, 3347.
24. Sheffield, V. C., Cox, D. R., Lerman, L. S., and Meyers, R. M. (1989). *Proc. Natl Acad. Sci. USA*, **86**, 232.
25. Lerman, L. and Silverstein, K. (1987). In *Methods in enzymology* (ed. R. Wu), Vol. 155, p. 482. Academic Press, London.
26. Abrams, E. S., Murdaugh, S. E., and Lerman, L. S. (1990). *Genomics*, **7**, 463.
27. Lakich, D., Kazazian, H. H., Antonarakis, S. E., and Gitschier, J. (1993). *Nat. Genet.*, **5**, 236.

# 14

# Linear amplification for the *in vivo* study of ligand/DNA interactions

HANS PETER SALUZ and JEAN-PIERRE JOST

## 1. Introduction

Genomic footprinting is a tool to study specific ligand/DNA interactions *in vivo* (1). The DNA in the region where proteins bind shows altered reactivity towards footprinting reagents and therefore usually escapes modification and/or cleavage. The cleavage products of the DNA are visualized by either electroblotting and indirect end-labelling (1, 2), linear amplification (3, 4), or via primer extension with labelled primers (5). Protected areas appear as a 'window' in the sequence ladder, when compared with treated protein-free control DNA.

Investigations by footprinting *in vitro* are less complicated than *in vivo*. *In vitro* interaction studies are usually performed by incubating cloned end-labelled DNA with a protein mixture to form specific protein/DNA complexes. In the *in vivo* situation, however, the genomic DNA is packaged into chromatin where different levels of organization can be considered: nucleosome structure, nucleosome positioning on the DNA, and higher order structure of the chromatin fibre.

The nucleosome structure has been well studied by means of X-ray crystallography. It was shown that the core nucleosome is cylinder-like (6) and a DNA fragment, forming 1.8 turns, is wrapped (146 bp) around the core histone octamer. Thereby the DNA helix becomes severely deformed and variations in widths in both the minor and the major groove can be observed. At 10 bp from the mid-point of the nucleosome core, a distortion of the DNA occurs due to a sharp bend. The size of the minor groove varies between 7 Å (inner face of the helix) and 13 Å (outer face). Variations in the major groove are between 11 Å (inner face) and 20 Å (outer face). In crystals, helical twists have 10 bp per turn instead of 10.4 bp which corresponds to the situation in solution. Such variations in topology may drastically affect the binding of proteins and suggest that the *in vitro* situation can be entirely different from the one *in vivo*, especially if one considers that the linker DNA varies from 4–104 bp. The nucleosome positioning on the DNA complicates the situation even more because the nucleosomes can be arranged in either a random, uniformly spaced, or positioned manner. The higher order structure of the chromatin fibre is still not fully known, especially how the nucleosomes are organized in the solenoid. It is expected that the degree of compaction varies depending on whether the chromatin fibre is associated with histone H1 or not and whether the histones are modified (acetylated or phosphorylated) or not.

For the above reasons, different problems may occur when *in vivo* footprinting techniques are applied. Superimposition of the chromatin footprint with specific protein footprints makes interpretation of the data more complex. Even more complexity may be encountered if one considers that the

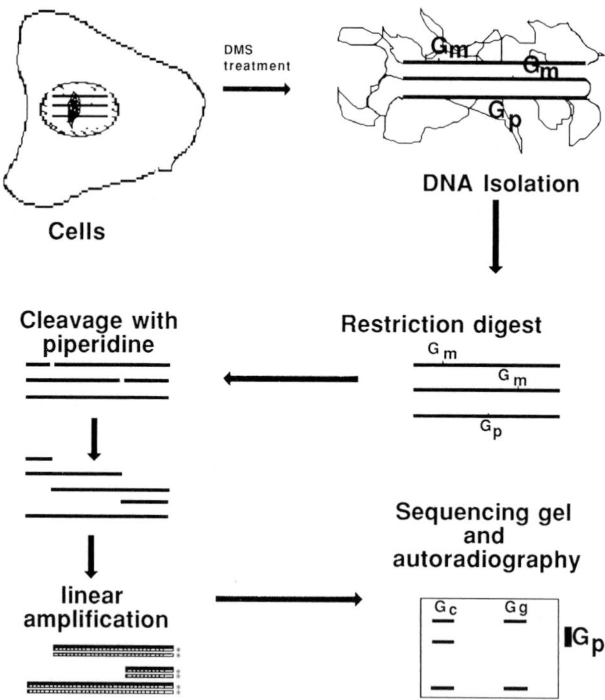


**Figure 1.** DNA footprinting *in vivo* by linear amplification. Cells in suspension are treated with dimethylsulphate (DMS) resulting in a modification of unprotected guanosine ($G_m$) residues whereas the protected guanosines ($G_p$) are not attacked by the footprinting probe. After DNA isolation, a restriction digestion is performed to reduce the viscosity of the DNA. Piperidine treatment is then applied to cleave the DNA at the modified sites ($G_m$). The DNA subfragments are linearly amplified (striped pattern) using a radioactively-labelled primer (▭). After purification of the amplification products, the genomic DNA fragments ($G_g$) are separated on a sequencing gel in parallel with a guanosine control reaction ($G_c$) allowing the detection of protected sites ($G_p$; footprint) upon exposure of the dried gels to an X-ray film.

reactivity of the footprinting probe can be increased or reduced depending on the chromatin structure. Because of the differences in accessibility of the target sequence and the differences in specific reactivity of the probe, different chemicals or enzymes should be applied to study *in vivo* protein/DNA interactions (see ref. 7 and references therein).

Several different genomic footprinting techniques are currently available (7). This chapter describes approaches involving linear amplification. This technique (3, 4) involves fewer steps than all other currently existing methods (7) and is suitable for experiments where the source of DNA target is not limiting. However, for the study of single-copy genes in organisms with less than 0.1 target copies per human genome (approx. $5 \times 10^9$ bp/haploid genome) the reader is recommended to use the procedure of Mueller *et al.* (8) which is described elsewhere in this volume (Chapter 15).

For *in vivo* genomic footprinting (*Figure 1*) organs or tissues are first treated with enzymes (collagenase, hyalurodinase) to produce a cell suspension. The cells in suspension are then exposed to the modifying agent. Often, after the isolation and purification of the now modified genomic DNA, a restriction digest is performed to reduce the viscosity of the DNA or to create DNA target fragments with specific start sites (7). The restriction enzyme is chosen such that the target fragments remain intact. When the footprinting probes modify the DNA without strand cleavage, a second reaction has to be performed to cut the DNA at all modified sites. The detection of the cleaved DNA fragments is achieved by the linear amplification with a radiolabelled primer and *Taq*-DNA polymerase or another thermostable DNA polymerase. The binding of a protein to DNA leads to the appearance of a 'window' in the sequence ladder, when compared to cloned control DNA (*Figure 1*). The amplified labelled DNA fragments are separated on a sequencing gel and the sequence can directly be read after exposure of the gel to an X-ray film.

## 2. Chemical and enzymatic probes for genomic footprinting

Numerous reagents have been applied for *in vivo* footprinting (see ref. 7 and references therein). These include exonuclease III, DNase I, micrococcal nuclease, dimethyl sulphate, diethylpyrocarbonate, bromoacetaldehyde, potassium permanganate, osmium tetroxide, methidiumpropyl-EDTA-iron(II), formaldehyde, 1,10-phenanthroline–copper, UV light, and others. Because these probes attack DNA in different ways, they provide quite detailed structural information about protein/DNA interactions. Within the limited scope of this chapter we describe the most common procedure for genomic footprinting, that using dimethylsulphate (DMS).

# 3. Genomic footprinting with dimethyl sulphate (DMS)

DMS is a very strong methylating agent that attacks nucleic acids but also proteins. A treatment of polynucleotides with DMS results in the methylation of unprotected deoxyguanosine residues at position N7 (major groove), adenine residues at positions N3 (minor groove) and N7 (major groove), and cytidine at position N3 (single-stranded DNA). Thymidine and uridine can also be modified with DMS but not under conditions which would be suitable for *in vivo* experiments (see ref. 7 and references therein). After isolation of the modified DNA, a piperidine reaction (β-elimination) had to be performed to cleave the DNA at the modified positions. Where a protein protects the DNA from the modifying agent, a 'window' in the sequencing ladder appears when compared with the unprotected control DNA. This reaction is adequate for the study of the interactions of proteins with the major grooves.

## 3.1. Preparation of cell suspensions

The main aim of genomic footprinting is to gain information about the *in vivo* state of a native gene. Therefore, it is important to use systems which are as close as possible to the natural situation. If the footprinting probe easily penetrates cell membranes, as does DMS, intact cells prepared from organs or tissues should be used. By this means, such chemicals penetrate all the cells in a minimum of time without forming concentration gradients. A cell suspension is obtained by a gentle treatment of small pieces of the target organs with collagenase and/or hyalurodinase (see *Protocol 1*). The number of cells, the incubation time, and the temperature are kept constant, whereas the concentration of the footprinting probe can be varied (see Section 3.3).

**Protocol 1.** Preparation of cell suspensions

*Equipment and reagents*

- Two nylon grid funnels; 1 mm mesh size
- Buffered saline: 0.15 M NaCl, 20 mM Hepes pH 8, 5 mM EDTA
- Dulbecco's medium
- Digestion buffer: 1 mg/ml hyalurodinase (Boehringer) and 0.5 mg/ml collagenase (Boehringer) in 0.15 M NaCl, 20 mM Hepes pH 8, 1 mM $CaCl_2$, 0.1 mM $ZnCl_2$, 0.1 mM $CoCl_2$. The buffer is sterilized by filtration and kept frozen in aliquots. The two enzymes are freshly just dissolved before use.
- 0.5 M EDTA pH 8.0

*Method*

1. Add 10–20 g of tissue to 50 ml of digestion buffer and mince it with scissors into the smallest possible pieces (the smaller the pieces, the better will be the yield of cells).
2. Incubate the tissue suspension in a conical (Erlenmeyer) flask in a rotatory water bath shaker for 30–40 min at 37°C.

3. Chill the suspension on ice and add 500 μl of 0.5 M EDTA to inhibit the enzymes and mix well.
4. Filter the suspension and debris through a nylon grid funnel over a 100 ml glass beaker and rinse the funnel with buffered saline.
5. Pour the filtered cell suspension into 30 ml centrifuge tubes and centrifuge at 1000 *g* at 4°C for 5 min.
6. Decant the supernatant and resuspend the cells in buffered saline by gently vortexing. Filter the suspension through a nylon grid funnel.
7. Centrifuge as described in step 5.
8. Resuspend the cells in 3 ml Dulbecco's medium. Adjust the cell density to $10^8$ cells/ml and use them immediately for the footprinting reaction (see Section 3.3).

## 3.2. Preparation of nuclei

The preparation of nuclei fullfils two different aims. First, high quality nuclei can be treated with footprinting probes, such as exonucleases, that do not penetrate cell membranes. Second, we have observed that the use of nuclei for the preparation of genomic DNA gives excellent results.

When nuclei are to be treated with footprinting probes, one has to bear in mind that the salt concentration used during the isolation of the nuclei can greatly influence the structure of chromatin (2) and therefore also the result of the footprint. Similar effects have been observed when using certain detergents, such as Triton X-100 (2). The procedure we describe here (*Protocol 2*) has been successfully applied for the isolation of nuclei from cells in tissue culture, chicken liver, chicken erythrocytes, kidney, and oviduct. However, it should also be suitable for other organs or tissues, provided that the sucrose concentration in buffers 2 and 3 is adapted for each case.

**Protocol 2.** The preparation of nuclei

*Equipment and reagents*

Only some of the following are required, depending on whether method A or B is used:

- A loose-fitting glass-Teflon homogenizer
- A tight-fitting Dounce glass–glass homogenizer
- Beckman SW28 rotor and ultracentrifuge or their equivalents
- Buffer 1: 80 mM NaCl, 1 mM EDTA, 20 mM HEPES pH 7.5. Add 2 mM DTT, 0.5 mM spermidine, 0.15 mM spermine just before use
- Buffer 2: 1.9 M sucrose in buffer 1
- Buffer 3: 0.35 M sucrose in buffer 1
- 0.15 M NaCl

A. *Isolation of nuclei from cells in tissue culture*

1. Resuspend the cell pellet in 5–7 vol. (v/w) buffer 3.

**Protocol 2.** *Continued*

2. Homogenize on ice with 20 strokes in a tight-fitting Dounce glass–glass homogenizer.
3. Sediment the nuclei at 800–1000 *g* for 10 min at 4°C.
4. Decant the supernatant.
5. Resuspend the nuclei in buffer 3.
6. Repeat steps 3 and 4.
7. Resuspend the nuclei either in the buffer used for footprinting with enzymes or in the DNA isolation buffer if the nuclei are to be used immediately for DNA preparation (see Section 3.4).

B. *Preparation of nuclei from organs (e.g. liver)*

1. Keep the livers on ice and perfuse them with ice-cold 0.15 M NaCl.
2. Put the liver into an ice-cold beaker and add 4 vol. (v/w) buffer 2.
3. Mince the livers with scissors and homogenize on ice with 4–5 strokes at 800 r.p.m. in a loose-fitting glass-Teflon homogenizer. The homogenate should have a final sucrose concentration of 1.57–1.66 M.
4. Overlay the homogenate with 10 ml buffer 2 in a SW-28 polyallomer centrifuge tube (Beckman), and centrifuge for 40 min at 30 000 *g* in a Beckman SW-28 rotor at 4°C.
5. Remove the thick top layer (fat and membranes) with a spatula. Decant the supernatant, wipe the sidewall of the tube clean with a tissue, and put the tubes on ice.
6. The pellet of nuclei is either resuspended in the buffer used for footprinting with enzymes or in the DNA isolation buffer if the nuclei are to be used immediately for DNA preparation (see Section 3.4).

## 3.3 Treatment *in vivo* with DMS

*Protocol 3* describes the *in vivo* footprinting procedures with DMS. For detailed studies of protein/DNA interactions with DMS, the reader is recommended to use different concentrations of this reagent while keeping temperature, substrate concentration, and time constant (*Figure 2*). This is because, for kinetic reasons (7), a chemical probe such as DMS can visualize either a major conformational state of the DNA or a related conformation in rapid equilibrium. At high DMS concentrations, the initial rate of attack is independent of the concentration of DMS and only the rate of access to the distorted reactive intermediate can be measured. At low concentrations, however, the initial rate of attack reflects the deformability of the protein/DNA complex. Consider the kinetics of chemical attack by DMS:

$$\mathrm{G} \underset{k_2}{\overset{k_1}{\rightleftharpoons}} \mathrm{G_i} \xrightarrow{k_3} \mathrm{G_m}$$

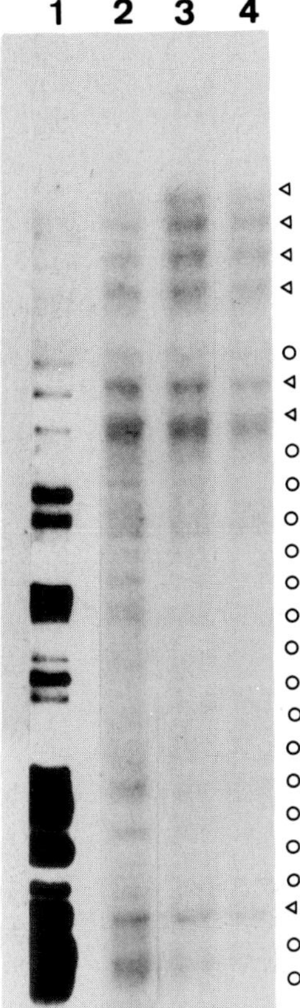


**Figure 2.** Different amounts of DMS for genomic footprinting. The naked control DNA (lane 1) was treated with 0.5% DMS. For the footprinting reactions (lanes 2–4), the cells in suspension ($10^8$ cells/ml) were treated with different amounts of DMS: 0.5% (lane 2), 0.05% (lane 3), 0.005% (lane 4). (◁) Sites which reacted to the same extent for all the different DMS concentrations used; (○) G residues which were stronger protected by proteins (bands disappear or get weaker).

where G is guanine, $G_i$ is interacting guanine in a reactive state and $G_m$ is modified guanine. At low DMS concentrations, a specific protein/DNA complex can be visualized via a footprint only if $k_1/k_2$ is greater than $k_3$.

**Protocol 3.** Treatment of cells in suspension with DMS

*Reagents*

- Cell suspension: $10^8$ cells/ml in Dulbecco's medium
- DMS (Aldrich)
- DMS stop buffer: phosphate-buffered saline (PBS) containing 1% BSA and 100 mM 2-mercaptoethanol

**Protocol 3.** *Continued*

- 15 ml Corex tubes
- Nuclei buffer: 0.3 M sucrose, 60 mM KCl, 15 mM NaCl, 1 mM EDTA, 0.5 mM EGTA, 15 mM Hepes pH 7.5, 1 mM spermidine, 0.3 mM spermine. Add the spermine and spermidine just before use
- Nuclei buffer containing 1% Nonidet P40

*Method*

1. Precool six 15-ml Corex tubes in ice and add 1 ml of cell suspension ($10^8$ cells/ml of Dulbecco's medium) to each tube.
2. Add various volumes of DMS (final concentrations of DMS: 0.5% for tube number 1; 0.05% for 2; 0.005% for 3; 0.0005% for 4; 0.00005% for 5, no DMS for tube number 6), while gently vortexing.
3. Incubate for 5 min at 20°C. Stop the reaction by adding 10 ml of ice-cold DMS stop buffer.
4. Vortex gently and centrifuge the cells for 5 min at 1000 *g* (4°C).
5. Discard the supernatant and resuspend the cells in 10 ml DMS stop buffer. Centrifuge as described in step 4.
6. Discard the supernatant and resuspend the cells in 1.5 ml of nuclei buffer containing 1% Nonidet NP40.
7. Keep the suspension for 5 min on ice. Centrifuge the nuclei at 3000 *g* (Sorvall HB-4 rotor or equivalent) at 4°C for 10 min.
8. Decant the supernatant. Resuspend the nuclei in 1.0–1.5 ml of nuclei buffer without Nonidet NP-40 and start immediately with the DNA isolation as described in Section 3.4.

## 3.4. Preparation of DNA

The quality of the DNA is of paramount importance for *in vivo* footprinting experiments. The better the quality of the DNA, the better will be the final result. The DNA has to be free of binding proteins and it has to be as intact as possible. Proteins bound to the DNA decrease the efficiency of the polymerase used for linear amplification and each incomplete copy of the target DNA sequence contributes to the final background. Similar comments can be made to nicks within the target sequence. Wherever a nick occurs, the polymerase will stop and DNA fragments shorter than the gel exclusion size (usually approx. 300 nucleotides) will enter the gel and contribute to the background as well. The best quality of DNA has always been obtained from isolated nuclei (see Section 3.2). However, it is also possible to extract high molecular weight DNA directly from fresh or frozen tissues, cells in tissue culture or from protoplasts. *Protocol 4* describes a method for DNA isolation which can be used for many different types of cells, different kind of tissues and also for isolated nuclei.

The quality of the isolated DNA can easily be tested by determination of the absorbance at 260 nm and at 280 nm. The ratio $A_{260}/A_{280}$ should be within the range of 1.9–2.0. In addition, a Southern blot experiment should be performed, with appropriate restriction endonucleases, to test the quality of the target sequence. If the target DNA is degraded, it should not be used for genomic footprinting experiments.

## **Protocol 4.** Isolation of DNA

*Reagents*

- Nuclei resuspension buffer: 0.15 M NaCl, 5 mM EDTA, 10 mM Tris–HCl pH 8
- DNA isolation buffer: 0.15 M NaCl, 50 mM EDTA, 5 mM EGTA, 10 mM Tris–HCl pH 8, 2% 2-mercaptoethanol, 1% SDS
- Proteinase K (1 mg/ml in DNA isolation buffer). Dissolve the proteinase K just before use.
- Pancreatic ribonuclease A (20 mg/ml water)
- Redistilled phenol containing 0.1% hydroxyquinoline, saturated with water
- Chloroform
- Dialysis tubing. Prepare this as follows. Boil the tubing for 30 min in 4% $NaHCO_3$, 30 min in 5 mM EDTA pH 7.5 and finally for 10 min in distilled water. Autoclave the tubing in 10 mM Tris–HCl pH 8, 1 mM EDTA and store it at 4°C.
- Dialysis buffer; 10 mM Tris–HCl pH 8, 0.5 mM EDTA

*Method*

1. Resuspend the cells or pellet of nuclei in 2.5–3.5 vol. nuclei suspension buffer.
2. Add an equal volume of 1 mg/ml proteinase K in DNA isolation buffer.
3. Incubate the lysed nuclei overnight at 37°C on a rocking platform (slow movements).
4. Add pancreatic ribonuclease A to a final concentration of 50–100 μg/ml.
5. Incubate for approx. 2 h at 37°C.
6. Transfer the viscous mass to a conical (Erlenmeyer) flask and add 1 vol. phenol.
7. Extract for 5 min at room temperature on a gyratory shaker; adjust the speed such that the two phases are well mixed.
8. Add one volume of chloroform. Mix well and transfer the contents to a Corex tube.
9. Separate the phases by centrifugation for a few minutes in a bench centrifuge. Recover the aqueous phase.
10. Repeat steps 6–9 at least six times.
11. Add 1 vol. of chloroform, mix well and centrifuge at 7000 *g* for 5 min.

**Protocol 4.** *Continued*

12. Using a broad-tipped pipette, transfer the DNA into dialysis tubing. Dialyse at 4°C for 2–3 days in a large volume of 10 mM Tris–HCl pH 8, 0.5 mM EDTA. Change the dialysis buffer several times during this period.

## 3.5. Restriction digestion of the genomic DNA

For reasons of easier handling, a restriction digestion of the genomic DNA should be carried out following its isolation in order to reduce its viscosity. The restriction enzyme should not cleave within the target sequence, so as not to shorten the length of the readable sequence. After the restriction digestion, perform at least one phenol/chloroform extraction of the DNA, followed by an ethanol precipitation.

## 3.6 Cleavage reaction of the chemically modified genomic DNA and sequencing reactions of the control DNA

DMS only modifies specific unprotected bases in a random manner and does not cleave the DNA. In order to cut the DNA strands at the modified position a second chemical reaction (β-elimination) has to be performed. We use piperidine (*Protocol 5*). As traces of piperidine may interfere with the polymerase reaction and, in addition, lead to poor electrophoretic resolution, piperidine has to be eliminated completely from the cleavage reaction products, before using the samples for linear amplification. The elimination of piperidine is achieved either by several lyophilization steps or by repeated ethanol precipitation.

Chemical sequencing (G,A,T,C, reactions) has to be performed with the naked control DNA which is used for the precise localization of the protected areas (*Protocol 6*). Only the control DNA has to be sequenced.

**Protocol 5.** Cleavage of the DMS-treated genomic DNA with piperidine

*Equipment and reagents*

- 1 M piperidine: 100 μl of piperidine (10 M piperidine, Fisher Scientific Company) in 900 μl water. This reagent must be freshly prepared. 1 ml 1 M piperidine is sufficient for 10 reactions.
- DNA modified with DMS *in vivo* (*Protocol 3*), isolated (*Protocol 4*) and restricted (Section 3.5) then dried in a Speedvac (Savant)
- Microcentrifuge tubes (can be siliconized)

*Method*

1. Add 100 μl of 1 M piperidine to 50 μg dried DNA.
2. Mix gently and seal the tubes. Incubate at 90–95°C for 30 min.

3. Freeze the samples in dry ice, open the tubes, and lyophilize under a good vacuum.
4. Repeat step 3 at least six times.
5. Redissolve the DNA in water (50 μg/69 μl). Store it at −70°C if it is not immediately used for the linear amplification.

## **Protocol 6.** Chemical sequencing of the control DNA

### *Equipment and reagents*

- Vacuum centrifuge (Speedvac; Savant)
- Cloned target DNA: recombinant plasmid containing the target insert DNA and digested with the same restriction enzyme as the genomic DNA. Approximately 4 μg per sequencing reaction (for a 9 kb PBR322 clone including insert) is sufficient for 300 sequencing lanes
- Sonicated *E. coli* DNA; 46 μg per reaction
- Dimethylsulphate (DMS) (Aldrich)
- DMS buffer: 50 mM sodium cacodylate pH 8.0, 1 mM EDTA
- DMS stop buffer: 1.5 M sodium acetate, pH 7.0, 1.0 M 2-mercaptoethanol
- Formic acid (p.a.; Fluka)
- Hydrazine (HZ) (Aldrich)
- Hydrazine stop buffer (HZ stop buffer): 0.3 M sodium acetate pH 7.5; 0.1 mM EDTA
- 0.3 M sodium acetate, 0.5 mM EDTA pH 5
- Piperidine (Fisher Scientific Company)
- Sample dye: 94% formamide, 10 mM EDTA pH 7.2, 0.05% xylene cyanol (XCFF), 0.05% bromophenol blue

### *Methods*

Set up four sequencing reactions

A. *G-reaction*

1. To a microcentrifuge tube, add 2 μl (4 μg) of digested recombinant plasmid DNA and 12.5 μl (46 μg) of sonicated *E. coli* DNA. Dry in a Speedvac.
2. Dissolve the DNA pellet in 6 μl of water and 200 μl DMS buffer and chill the sample on ice.
3. Add 1 μl of DMS and mix by tapping the tube.
4. Centrifuge for a few seconds in a microcentrifuge at 4°C.
5. Incubate the sample for 10 min at 20°C.
6. Add 50 μl of ice-cold DMS stop buffer, mix, and add 750 μl of pre-cooled (−20°C) ethanol.
7. Mix and chill for 15 min in solid $CO_2$.
8. Centrifuge the tubes for 10 min at 30 000 *g* in a Sorvall centrifuge (or equivalent) at 0°C.
9. Carefully pour off the supernatant and resuspend the pellet in 250 μl 0.3 M sodium acetate, 0.5 mM EDTA pH 5.0.

**Protocol 6.** *Continued*

10. Precipitate the DNA with 750 μl ethanol for 15 min on solid $CO_2$ and centrifuge as described in step 8.
11. Wash the pellet with 1 ml 70% ethanol/water. Centrifuge for 5 min in a microcentrifuge and carefully pour off the supernatant.
12. Add 100 μl 1 M piperidine to the dried DNA.
13. Mix gently and seal the tubes. Incubate at 90–95°C for 30 min.
14. Freeze the samples in dry ice, open the tubes, and lyophilize under a good vacuum.
15. Repeat step 3 at least six times.
16. Redissolve the DNA in 300 μl water. Store it in small aliquots at −70°C. 1 μl of this DNA is mixed with 68 μl water and used for the linear amplification reaction (see Section 4).

*(G+A)-reaction*

1. To a microcentrifuge tube, add 4 μl (8 μg) digested recombinant plasmid DNA and 12.5 μl (46 μg) sonicated *E. coli* DNA. Dry in a Speedvac.
2. Dissolve the DNA pellet in 11 μl water. Chill on ice.
3. Add 25 μl concentrated formic acid.
4. Mix by tapping the tube. Centrifuge for a few seconds in a microcentrifuge.
5. Incubate at 20°C for 4.5 min.
6. Add consecutively 200 μl of hydrazine stop buffer and 750 μl cold ethanol.
7. Mix well and chill for 15 min on solid $CO_2$.
8. Centrifuge the tubes for 10 min at 30 000 *g* in a Sorvall centrifuge (or equivalent) at 0°C.
9. Pour off the supernatant carefully and resuspend the pellet in 250 μl 0.3 M sodium acetate, 0.5 mM EDTA pH 5.0.
10. Precipitate the DNA with 750 μl of ethanol for 15 min on solid $CO_2$ and centrifuge as in step 8 above.
11. Wash the pellet with 1 ml of 70% ethanol/water. Centrifuge for 5 min in a microcentrifuge and pour off the supernatant. Dry the DNA pellet in a Speedvac.
12. Add 100 μl 1 M piperidine to dried DNA.
13. Mix gently and seal the tubes. Incubate at 90–95°C for 30 min.
14. Freeze the samples in solid $CO_2$, open the tubes, and lyophilize under a good vacuum.

15. Repeat step 3 at least six times.
16. Redissolve the DNA in 300 μl water. Store it in small aliquots at −70°C. 1 μl of this DNA is mixed with 68 μl water and used for the linear amplification reaction (see Section 4).

*(T+C)-reaction*

1. To a microcentrifuge tube, add 4 μl (8 μg) digested recombinant plasmid DNA 12.5 μl (46 μg) of sonicated *E. coli* DNA. Dry in a Speedvac.
2. Dissolve the DNA pellet in 21 μl water. Chill on ice.
3. Add 30 μl hydrazine and mix by tapping.
4. Centrifuge for a few seconds in a microcentrifuge.
5. Incubate the sample at 20°C for 10 min.
6. Add consecutively 200 μl of HZ stop buffer and 750 μl cold ethanol and mix well
7. Chill for 15 min on solid $CO_2$.
8. Centrifuge the tubes for 10 min at 30 000 *g* in a Sorvall centrifuge (or equivalent) at 0°C.
9. Pour off the supernatant carefully.
10. Resuspend the DNA pellet in 250 μl 0.3 M sodium acetate, 0.5 mM EDTA pH 5.0.
11. Precipitate the DNA with 750 μl ethanol for 15 min on solid $CO_2$ and centrifuge as described in step 8.
12. Wash the pellet with 1 ml 70% ethanol/water. Centrifuge for 5 min in the microcentrifuge and carefully pour off the supernatant.
13. Dry the pellet in the Speedvac.
14. Add 100 μl of 1 M piperidine to dried DNA.
15. Mix gently and seal the tubes. Incubate at 90–95°C for 30 min.
16. Freeze the samples in solid $CO_2$, open the tubes, and lyophilize under a good vacuum.
17. Repeat step 3 at least six times.
18. Redissolve the DNA in 300 μl water. Store it in small aliquots at −70°C. 1 μl of this DNA is mixed with 68 μl water and used for the linear amplification reaction (see Section 4).

*C-reaction*

1. To a microcentrifuge tube, add 2 μl (4 μg) digested recombinant plasmid DNA and 12.5 μl (46 μg) sonicated *E. coli* DNA. Dry in a Speedvac.

**Protocol 6.** *Continued*

2. Dissolve the DNA pellet in 5 μl water and add 15 μl 5 M NaCl.
3. Mix by tapping the tube and chill on ice.
4. Add 30 μl of hydrazine, mix by tapping the tube, and centrifuge for a few seconds in a microcentrifuge.
5. Incubate the sample at 20°C for 10 min.
6. Add consecutively 200 μl hydrazine stop buffer and 750 μl cold ethanol.
7. Mix well by inversion and chill for 15 min on solid $CO_2$.
8. Centrifuge the tubes for 10 min in a Sorvall centrifuge (or equivalent) at 30 000 *g* at 0°C.
9. Pour off the supernatant carefully and resuspend the pellet in 250 μl 0.3 M sodium acetate, 0.5 mM EDTA pH 5.0.
10. Precipitate the DNA with 750 μl ethanol for 20 min on solid $CO_2$ and centrifuge as described in step 8.
11. Wash the pellet with 1 ml 70% ethanol/water. Centrifuge for 5 min in the microcentrifuge. Carefully pour off the supernatant.
12. Dry the DNA pellet in the Speedvac.
13. Add 100 μl 1 M piperidine to the dried DNA.
14. Mix gently and seal the tubes. Incubate at 90–95°C for 30 min.
15. Freeze the samples in solid $CO_2$, open the tubes, and lyophilize under a good vacuum.
16. Repeat step 3 at least six times.
17. Redissolve the DNA in 300 μl water. Store it in small aliquots at −70°C. 1 μl of this DNA is mixed with 68 μl water and used for the linear amplification reaction (see Section 4).

## 4. Linear amplification

The purpose of linear amplification of the cleaved DNA with a thermostable DNA polymerase and a radiolabelled primer is to increase the number of each target sequence in order to obtain stronger final signals.

### 4.1 Selection of the oligonucleotide primer

Since DNA polymerases elongate a strand in $5' \rightarrow 3'$ direction, a primer used for linear amplification has to be complementary to the 3′-end of the target sequence. Under the gel electrophoresis conditions described in this chapter for analysing the linear amplification products (Section 4.6), the primer

Filling-in with Sequenase (version 2)

**Figure 3.** Synthesis of the radiolabelled primer by a filling-in reaction. A 6–9-mer segment (B) of the primer is annealed to a 33-mer template (A). The segment is annealed such that a short tail at 3′-end of the 33-mer template is produced. This tail is used for a fast strand separation on a short sequencing gel. The filling-in reaction is performed with Sequenase (version 2) free of exonuclease activity. The filled-in DNA stretch is indicated by a black box.

should be selected to bind to the target DNA approximately 90–180 bp away from the start site of the sequence of interest. Primers approximately 30 nt long work well. The primer should be chosen such that its average (G+C) content is similar or greater than the rest of the sequence to be amplified.

## 4.2 Labelling of the oligonucleotide primer

For genomes up to the size of yeast genome, we usually carried out a 5′-end-labelling of the primer using T4 polynucleotide kinase (5 U/μl). Linear amplification of DNA from larger genomes, however, requires primers of a much higher specific radioactivity to achieve sufficiently strong final signals. This can be obtained by the 'filling-in' reaction. A synthetic 6–9-mer segment of the primer is annealed, approximately 3 nt away from the 3′-end of a 33-mer oligonucleotide complementary to the primer (*Figure 3*). This strategy allows both high incorporation of radioactivity and subsequent rapid strand separation on a short sequencing gel. The labelled primer is recovered by elution from the gel.

For the filling-in reaction, it is important to chose a polymerase such as Sequenase (version 2), free of any 3′ → 5′ exonuclease activity, because otherwise the 3′ → 5′ exonuclease activity digests the 3′-tail which is used for strand separation. This 'filling-in' reaction and subsequent gel purification of the radiolabelled primer is described in *Protocol 7*. Due to the fact that the primer is so highly radioactive, it should be used without delay for linear amplification.

**Protocol 7.** 'Filling-in' reaction and purification of the primer

*Reagents*

- 5 × Sequenase buffer: 200 mM Tris–HCl pH 7.5, 100 mM $MgCl_2$, 250 mM NaCl
- 9-mer primer: 0.14 μg/ml
- 33-mer primer: 0.5 μg/ml
- dNTP mixture: 10 mM each of dCTP, dTTP, dGTP (sequencing grade)
- [α-$^{32}$P]dATP (3000 Ci/mmol; 10 μCi/μl)
- Sequenase, version 2 (13 U/μl)
- Sample dye: 94% formamide, 10 mM EDTA pH 7.2, 0.05% bromophenol blue, 0.05% xylene cyanol

**Protocol 7.** *Continued*

- Short 15% sequencing gel
- Elution buffer: 0.5 M ammonium acetate, 0.5 mM EDTA pH 7.5
- Ultrafree MC membrane (0.45 μm; Millipore)
- *E. coli* carrier DNA

*Method*

1. To a sterile microcentrifuge tube, add consecutively 7 μl of 5 × Sequenase buffer, 1 μl of the 9-mer primer, and 1 μl of the 33-mer oligonucleotide.
2. Mix and spin for a few seconds in a microcentrifuge. Incubate at 75°C for 3 min.
3. Cool down the sample to room temperature within 10–15 min.
4. Spin the sample briefly in the microcentrifuge. Place it on ice and add consecutively 3 μl of the dNTP mixture minus dATP, 22 μl of [$\alpha$-$^{32}$P]dATP (220 μCi), 1 μl of Sequenase, version 2 (13 U/μl).
5. Mix, spin briefly in a microcentrifuge, and incubate at 23°C for 20 min.
6. Add 40 μl of formamide sample dye.
7. Heat denature the sample for a few min at 90°C and then quick chill in ice/water. Load the sample on to the short sequencing gel (15%) and run the gel until xylene cyanol dye has migrated 10 cm.
8. Expose the gel for a few seconds to an X-ray film using radioactive ink for orientation. Cut out the gel fragment containing the labelled primer.
9. Elute the primer in an appropriate amount of elution buffer by shaking for 2 h (e.g. in an Eppendorf shaker) at 40–50°C (this can also be done at room temperature with lower efficiency).
10. Filter the eluate using an Ultrafree MC membrane.
11. Add 3–5 mg of *E. coli* carrier DNA and 3 vol. ethanol. Mix well and place the sample for 20 min on solid $CO_2$.
12. Centrifuge at 30 000 *g* in a Sorvall centrifuge (or equivalent). Discard the supernatant.
13. Dry the pellet. Add an appropriate amount of water (sufficient for approximately 10 reactions; 17 μl/reaction).

## 4.3. Determination of the melting temperature

In order to avoid non-specific amplified fragments which contribute to the background, it is important to use very stringent temperatures for the annealing of the primer to the genomic DNA target. The optimal temperature for annealing is that at which the lowest background is obtained. Good results have been obtained with an annealing temperature of 2–4 °C below the melting temperature. The melting temperature of the sequencing primer has to be determined experimentally. This is done either by using a thermo-programmable spectrophotometer (measuring the $T_m$ in *Taq* polymerase buffer without BSA and 2-mercaptoethanol) or by increasing the annealing temperature stepwise by 5 °C, starting at approximately 60 °C for 27-mer primers, and following the change in the $A_{260}$.

## 4.4 Suitable enzymes for linear amplification

Another parameter which influences the final results is the quality of the thermostable DNA polymerase. Since the quality is not always consistent, only good batches of enzyme should be used for genomic footprinting experiments. We routinely use AmpliTaq or the Stoffel fragment from Perkin-Elmer Cetus and sometimes also *Pfu* DNA polymerase (Stratagene). Other thermostable polymerases such as Vent polymerase (New England Biolabs) might also be acceptable.

Polymerases often do not work efficiently on genomic target sequences which have a very high G+C content (over 65%). In such cases, one should use amplification buffers containing 10% DMSO. With such buffers it is possible to properly amplify sequences with a G+C-content of over 80%.

## 4.5 The linear amplification reaction

The procedure is described in *Protocol 8*.

### **Protocol 8.** The linear amplification reaction

*Equipment and reagents*

- Thermocycler
- Mineral oil
- Footprinted genomic DNA (50 μg/69 μl of $H_2O$)
- Radioactively labelled primer (approx. 10 ng/17 μl)
- 10 × *Taq* buffer (166 mM $(NH_4)_2SO_4$, 67 mM $MgCl_2$, 100 mM β-mercaptoethanol, 2 mg BSA/ml, 670 mM Tris–HCl pH 8.8 at 25°C)
- dNTP mixture: 10 mM each of all four deoxynucleotides dATP, dGTP, dTTP, dCTP (sequencing grade).
- Thermostable polymerase (e.g. AmpliTaq, see Section 4.4)

*Method*

1. To a microcentrifuge tube, add 69 μl sequenced genomic DNA, 17 μl of the radioactively labelled primer, 10 μl of 10 × *Taq* buffer.
2. Mix gently and spin for a few seconds. Incubate the samples in the thermocycler at 95°C for 5 min.
3. Chill the samples in ice-water for 1 min (do not interrupt the denaturing program of the thermocycler in the meantime).
4. Spin the samples briefly. Place them immediately back on ice-water and add 3 μl of dNTP mixture and 2–5 U of thermostable polymerase. Mix briefly and put the samples back on ice-water.

**Protocol 8.** *Continued*

5. Add 100 μl mineral oil to each sample and incubate them immediately at 95°C in the thermocycler for another minute.
6. Start the amplification program: 25–30 cycles, each with 1 min denaturation at 94°C, 2 min at the annealing temperature,[a] and 3 min chain elongation at 72°C.
7. Put the samples on ice-water without delay and begin the purification of the amplified DNA fragments (Section 4.6).

[a] The annealing temperature has to be determined experimentally (see Section 4.3).

## 4.6 Purification and electrophoresis of the amplification products

Prior to gel electrophoresis, the amplified DNA targets have to be purified by a precipitation procedure with *N*-cetyl-*N*,*N*,*N*-trimethylammonium bromide (CTAB) in the presence of ammonium sulphate (9). This is carried out to avoid smearing of the bands during gel electrophoresis due to the presence of contaminants.

Due to the relatively large amounts of DNA, we use gels of 1 mm thickness. To obtain a good resolution, the gel length should be approx. 60 cm (3). The methods for purification of the amplification products and their analysis by gel electrophoresis are described in *Protocol 9*.

**Protocol 9.** Purification and electrophoresis of amplification products

*Equipment and reagents*

- Drawn-out microcapillaries
- Ultrafree MC membranes (0.45 mm; Millipore)
- 1% *N*-cetyl-*N*,*N*,*N*,-trimethylammonium bromide (CTAB) in water
- 0.5 M ammonium acetate
- 3 M sodium acetate, 5 mM EDTA pH 5.0
- Loading buffer 1: 100 mM NaOH, 1 mM EDTA
- Loading buffer 2: 8 M urea, 0.04% bromophenol blue, 0.04% xylenecyanol
- 8% sequencing gel: 7–8 M urea; acrylamide:bisacrylamide = 29:1). The gel is poured and allowed to polymerize overnight. The gel dimensions are 60 cm × (20–40 cm) × 0.1 cm
- Fixing solution: 10% acetic acid, 10% methanol and 80% water

*Method*

1. Using a drawn-out microcapillary, transfer the aqueous phase from the linear amplification reaction mixture (*Protocol 8*, step 7) on to an Ultrafree MC membrane for filtration. Avoid any mineral oil contamination.

2. Centrifuge for a few min in a bench-top centrifuge.
3. To the filtrate, add 10 μl 1% CTAB. Mix gently and place the sample for 20 min on ice.
4. Centrifuge for 15 min at 30 000 *g* in a Sorvall centrifuge (or equivalent) at 4°C.
5. Remove the supernatant and redissolve the pellet in 100–200 μl of 0.5 M ammonium acetate.
6. Add 0.1 volume of 3 M sodium acetate, 5 mM EDTA pH 5.0, and 3 vol. ethanol. Mix gently and place at −70°C for 15 min.
7. Centrifuge at 30 000 *g* (Sorvall or equivalent) and 4°C for 20 min.
8. Pour off the supernatant. Spin this for approx. 1 min in a microcentrifuge. Take off the residual supernatant with a drawn out microcapillary and dry the pellet.
9. Precipitate the DNA once more as described in steps 5–8.
10. To the DNA pellet, add 4 μl loading buffer 1 and 4 μl loading buffer 2. Redissolve the pellet and then heat the samples for approx. 15 sec at 94°C.
11. Load the hot samples on to a prerun and preheated 8% sequencing gel. Electrophorese at 60 mA (constant power).
12. After electrophoresis, cut the gel laterally into 30 cm pieces. Fix them for 30 min each in one litre of freshly prepared fixing solution. Dry the gel pieces on Whatman paper (gel drying grade) for 60–120 min in a vacuum gel dryer at 80°C.
13. Expose the gels to X-ray films with intensifying screens.

# References


1. Church, G. M. and Gilbert, W. (1984). *Proc. Natl Acad. Sci. USA*, **81**, 1991.
2. Saluz, H. P. and Jost, J. P. (1990). In *A laboratory guide for in vivo studies of DNA methylation and protein/DNA interaction*, p. 129. Birkhauser, Basel–Boston.
3. Saluz, H. P. and Jost, J. P. (1989). *Nature*, **338**, 277.
4. Saluz, H. P. and Jost, J. P. (1989). *Proc. Natl Acad. Sci. USA*, **86**, 2602.
5. Becker, M. M. and Wang, J. C. (1984). *Nature*, **309**, 682.
6. Van Holde, K. (1989). *Chromatin*. Springer-Verlag, Berlin.
7. Saluz, H. P. and Jost, J. P. (1993). *Critical Reviews in Eucaryotic Gene Expression*, **3**, 1.
8. Mueller, P. R. and Wold, B. (1989). *Science*, **246**, 780.
9. Jost, J. P., Jiricny, J., and Saluz, H. P. (1989). *Nucleic Acids Res.*, **17**, 2143.

# 15

# Ligation-mediated PCR

PAUL A. GARRITY, BARBARA WOLD, and
PAUL R. MUELLER

## 1. Introduction

Using PCR, a DNA fragment can be exponentially amplified with two flanking primers of known sequence. However, for many applications it is desirable to amplify a DNA fragment containing only one end of known sequence or, for genomic sequencing and *in vivo* footprinting, a family of fragments with one end in common. In ligation-mediated PCR (LMPCR), such substrates are made suitable for PCR by the ligation of a sequence of defined length and sequence content to the undefined end. PCR is then performed using one primer specific for the gene of interest and one primer specific for the ligated sequence (1). LMPCR differs from other single-sided exponential PCR methods in that the relative lengths of the starting DNA fragments are retained with base-pair resolution in the final product, making it suitable for use in genomic sequencing and *in vivo* footprinting (1, 2).

Although LMPCR will be discussed in the context of genomic sequencing and *in vivo* footprinting, it can potentially be used for any application where one wishes to amplify a fragment, but knows the sequence only at one end. The only restrictions are that the substrate DNA must be cleaved and must contain a 5′-phosphate at the cleavage site. Suitable cleavage agents include DNase I, restriction endonucleases, and Maxam–Gilbert sequencing reagents (such as DMS/piperidine). The preparation of DNA samples for DMS *in vivo* footprinting (ref. 3; Chapter 14), DNase I *in vivo* footprinting (4), genomic sequencing (ref. 5; Chapter 14), and cloning after restriction endonuclease digestion (6) have been described elsewhere.

The general strategy of LMPCR is outlined in *Figure 1*. In the first strand synthesis reaction, a gene-specific primer (primer 1) is hybridized to the cleaved DNA substrate. Annealed primer 1 is then extended with a DNA polymerase to create a blunt end at the site of the original cleavage event. To this blunt end is ligated a staggered linker (with one blunt end) of defined length and sequence. After the ligation reaction, the DNA is exponentially amplified using a primer that recognizes the ligated sequence (linker primer) and a gene-specific primer (primer 2). After exponential amplification, the

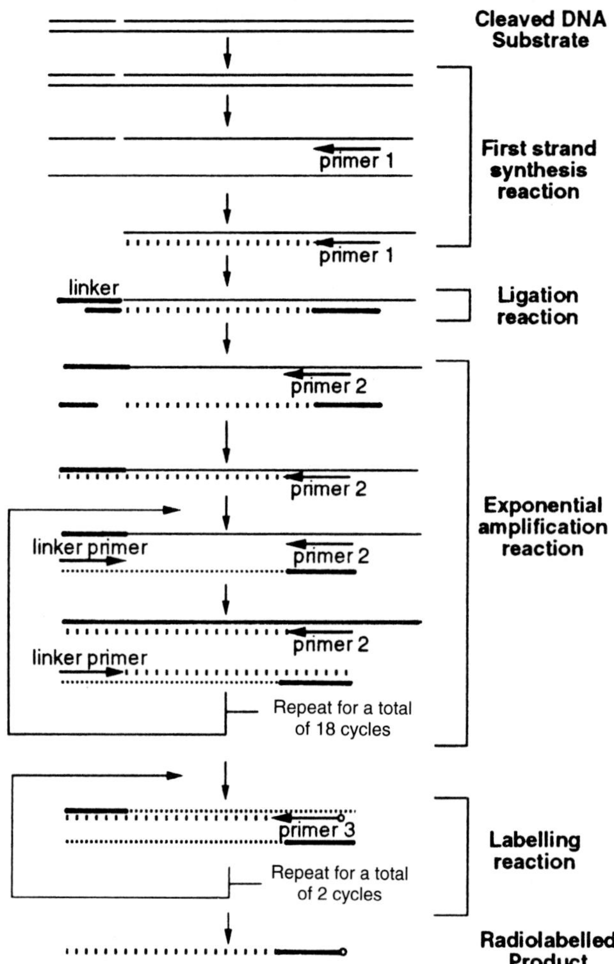


**Figure 1.** Scheme of ligation-mediated PCR. Thin solid lines indicate the original sample DNA, thick solid lines indicate staggered linker, and thick solid lines ending in arrows indicate oligonucleotide primers. Dotted lines indicate new DNA that has been synthesized via primer extension and PCR. The circle at end of primer 3 represents the radioactive label at its 5′-end. See text for detailed description.

LMPCR product is visualized by primer extension of a radioactively labelled third gene-specific primer (primer 3).

## 2. Procedures for LMPCR

### 2.1 Preparation of linker and radiolabelled primer 3

The two strands of the linker are synthesized by standard automated synthesis and then annealed under controlled temperature conditions. Their sequences and the annealing procedure are described in *Protocol 1*. End-labelling of primer 3, also required for LMPCR, is described in *Protocol 2*.

## **Protocol 1.** Preparation of unidirectional linker

### *Equipment and reagents*

- Heating block or waterbath (70°C)
- LMPCR 1: a DNA oligonucleotide of the following sequence (5′ → 3′): GCGGTGACC CGGGAGATCTGAATTC[a]
- LMPCR 2: a DNA oligonucleotide of the following sequence (5′ → 3′): GAATTCAGATC
- 1 M Tris–HCl pH 7.7
- Preparative DNA sequencing gel (see *Protocol 4 reagents*) and electrophoresis equipment and buffers

### *Method*

1. Purify LMPCR 1 and LMPCR 2 on a preparative DNA sequencing gel to ensure that all material is full length.
2. Determine the concentration of each gel-isolated oligonucleotide.[b]
3. Combine LMPCR 1 and LMPCR 2 with 1 M Tris–HCl pH 7.7 and water for a final concentration of 20 μM of each oligonucleotide in 250 mM Tris–HCl pH 7.7.
4. Denature for 5 min at 95°C.
5. Transfer to a 70°C heat-block and gradually cool to room temperature over a period of 1 h by carefully removing the metal block from the heater and putting it on the benchtop.[c]
6. Leave at room temperature for 1 h.
7. Gradually cool to 4°C by putting the metal block into a 4°C refrigerator. Leave at 4°C overnight.
8. Store the linker in aliquots at −20°C. Always thaw and handle the linker on ice to prevent denaturation.

[a] To facilitate potential cloning of LMPCR products, LMPCR 1 contains restriction sites for *Bst*EII, *Sma*I, *Bgl*II, and *Eco*RI. These sites are not used in genomic sequencing or *in vivo* footprinting applications.

[b] We commonly estimate oligonucleotide concentrations spectroscopically assuming that a concentration of 37 μg/ml yields $A_{260} = 1$.

[c] Alternatively, instead of the metal block, one can use a beaker containing 200 ml water drawn from a 70°C water bath.

**Protocol 2.** End-labelling of primer 3

*Equipment and reagents*

- Plexiglass shield, gloves, protective eyewear, and hand-held radioactivity monitor: the labelling mixture contains a substantial amount of $^{32}P$ and hence it is essential that all manipulations involving this reagent be performed using appropriate **safety precautions**.
- 10 × kinase buffer: 700 mM Tris–HCl pH 7.6, 100 mM $MgCl_2$, 50 mM DTT. Store at −20°C.
- TE buffer: 10 mM Tris–HCl pH 7.5, 1 mM EDTA
- Primer 3: 10 pmol/μl of gel-isolated oligonucleotide in TE
- [$\gamma$-$^{32}P$]ATP: 6000 Ci/mmol, 150 mCi/ml (e.g., DuPont NEN)
- T4 polynucleotide kinase: 10 units/μl (New England Biolabs)
- Nensorb-20 column (Dupont NEN; NL-022) and solutions detailed by the manufacturer
- 50% ethanol
- Rotary evaporator

*Method*

1. Combine the following in a microcentrifuge tube on ice (scale accordingly with amount of primer 3 desired):
   - 10 × kinase buffer 2 μl
   - primer 3 5 μl (50 pmol)
   - [$\gamma$-$^{32}P$]ATP 6.7 μl
   - water 4.3 μl
   - T4 polynucleotide kinase 2 μl
2. Incubate at 37°C for 30 min.
3. Separate labelled primer 3 from unincorporated ATP using the Nensorb-20 nucleic acid purification system according to the manufacturer's instructions, using 50% ethanol to elute the oligonucleotide.
4. Freeze on solid $CO_2$ and dry in a rotary evaporator.
5. Resuspend in TE at 1 pmol/μl (assume 100% recovery of the oligonucleotide).

## 2.2 First strand synthesis and ligation of the linker

These steps are carried out as described in *Protocol 3*. Note that, after the first strand has been synthesized and the linker has been ligated, the sample is stable at −20°C (*Protocol 3*, step 11) indefinitely. Thus one can either pause at this point or proceed directly to *Protocol 4*.

## 2.3 The LMPCR amplification reaction

PCR amplification of DNA and visualization of the ligated material by gel electrophoresis and autoradiography are described in *Protocol 4*.

**Protocol 3.** Synthesis of first strand and ligation of unidirectional linker

*Equipment and reagents*

- Automated thermal cycler (or three waterbaths at 60°C, 76°C, and 95°C)
- Siliconized microcentrifuge tubes
- Lid-locks (Intermountain Scientific C-32701) for microcentrifuge tubes
- 17°C water bath
- 5 × first strand buffer: 200 mM NaCl, 50 mM Tris–HCl pH 8.9, 25 mM $MgSO_4$, 0.05% gelatin (Sigma Chemical G1393). Store at −20°C
- TE: 10 mM Tris–HCl pH 7.5, 1 mM EDTA
- Primer 1: 1 pmol/μl in TE. Store at −20°C
- 25 mM dNTP mix. To prepare, combine equal volumes of 100 mM stocks of dATP, dCTP, dGTP, and dTTP (Pharmacia LKB 27–2050, 27–2060, 27–2070, and 27–2080). Store at −20°C
- Vent DNA polymerase: 2 units/μl (New England Biolabs; 254S)
- Unidirectional linker: 20 pmol/μl in 250 mM Tris–HCl pH 7.7 (see *Protocol 3*)
- T4 DNA ligase: 3 Weiss units/μl (Promega)
- 75% and 100% ethanol
- 0.4 μg/μl cleaved mouse genomic DNA in TE[a]
- First strand synthesis mix (25 μl per reaction). Prepare fresh by combining on ice (per reaction) 6.0 μl 5 × first strand buffer, 0.3 μl (0.3 pmol) primer 1, 0.25 μl 25 mM dNTP mix, and 18.2 μl water. Then add 0.25 μl (0.5 U) Vent DNA polymerase.
- Vent dilution solution (20 μl per reaction). Prepare fresh by combining on ice (per reaction) 2.2 μl 1 M Tris–HCl pH 7.5, 0.35 μl 1 M $MgCl_2$, 1.0 μl 1 M DTT, 0.25 μl 10 mg/ml DNase-free bovine serum albumin, and 16.2 μl water.
- Ligation solution (25 μl per reaction). Prepare fresh by combining on ice (per reaction) 0.25 μl 1 M $MgCl_2$, 0.5 μl 1 M DTT, 0.75 μl 100 mM rATP, 0.13 μl 10 mg/ml DNase-free bovine serum albumin and 17.37 μl water. Chill these components for a few minutes, then add 5.0 μl (100 pmol) unidirectional linker in 250 mM Tris–HCl pH 7.7.[b] Finally, add 1.0 μl (3 U) T4 DNA ligase.[c]
- Precipitation solution (9.4 μl per reaction). Prepare fresh by combining on ice (per reaction) 8.4 μl 3 M sodium acetate (pH 7.0) and 1.0 μl 10 mg/ml yeast tRNA.

*Method*

1. Add 5 μl (2 μg) cleaved genomic DNA sample to a microcentrifuge tube. Chill on ice.
2. Add 25 μl ice-cold first strand synthesis mix to the DNA sample.
3. Denature for 5 min at 95°C. Anneal for 30 min at 60°C. Extend for 10 min at 76°C.[d,e]
4. Transfer the sample to ice. Recover any condensation that has formed by spinning at 4°C in a microcentrifuge for a few seconds. Return to ice.[f]
5. Add 20 μl ice-cold Vent dilution solution.[g]
6. Add 25 μl ice-cold ligation solution.[h]
7. Incubate the sample at 17°C for 12–16 h.
8. After ligation, recover condensation by spinning at 4°C in a microcentrifuge for a few seconds. Put the sample on ice.
9. Add 9.4 μl ice-cold precipitation solution to the sample.
10. Add 220 μl ice-cold 100% ethanol and mix by inversion.
11. Precipitate at −20°C.

## Protocol 3. *Continued*

[a] It is the number of genomes, not the mass of DNA used that is the important parameter in LMPCR. The mass of DNA used may need to be adjusted depending on the genome size of the organism under study and on the particular application (see Section 4).

[b] The unidirectional linker should be thawed and handled on ice.

[c] The Tris–HCl contained in the linker solution provides the buffer for the ligation solution.

[d] This step is most conveniently performed in an automated thermal cycler. If three water baths are used instead, care should be taken to transfer samples between water baths rapidly.

[e] Microcentrifuge tubes commonly pop open at this step. This can be prevented by using Lid-Locks to clamp the tube lids shut.

[f] Vent DNA polymerase is not inactivated after the first strand synthesis. As Vent activity could interfere with the subsequent ligation, it is important to minimize its activity in steps 6–9 by keeping the solutions on ice as much as possible when handling.

[g] This converts the first strand reaction mixture into a solution that is optimal for T4 DNA ligase activity.

[h] As it is a component of the ligation solution (see *Equipment and reagents*), the unidirectional linker is added at this step.

## Protocol 4. PCR amplification and radioactive visualization of ligated material

### *Equipment and reagents*

- Plexiglass shield, gloves, protective eyewear, and hand-held radioactivity monitor. The labelling mix contains a substantial amount of $^{32}P$. It is therefore essential that all manipulations involving this reagent be performed using appropriate **safety precautions.**
- Automated thermal cycler
- 5 × amplification buffer: 200 mM NaCl, 100 mM Tris–HCl pH 8.9, 25 mM $MgSO_4$, 0.05% gelatin, 0.5% Triton X-100. Store at −20°C.
- 25 mM dNTP mix (see *Protocol 3*)
- 2 U/μl Vent DNA polymerase (New England Biolabs; 254S)
- 75% and 100% ethanol
- LMPCR 1: 10 pmol/μl LMPCR 1 (see *Protocol 1*) in TE
- Primer 2: 10 pmol/μl in TE
- End-labelled primer 3: 1 pmol/μl in TE (see *Protocol 2*)
- Amplification mix (30 μl per reaction). Prepare fresh by combining on ice (per reaction) 20 μl 5 × amplification buffer, 1.0 μl (10 pmol) LMPCR 1, 1.0 μl (10 pmol) primer 2, 0.8 μl 25 mM dNTP mix, and 7.2 μl water
- Diluted Vent DNA polymerase (3 μl per reaction). Prepare fresh by combining on ice (per reaction) 0.6 μl 5 × amplification buffer, 1.9 μl water. Then add 0.5 μl (1 U) Vent DNA polymerase
- Mineral oil
- Labelling mix (5 μl per reaction). Prepare fresh by combining on ice (per reaction) 1.0 μl 5 × amplification buffer, 2.3 μl (2.3 pmol) end-labelled primer 3, 0.4 μl 25 mM dNTP mix, and 0.8 μl water. Then add 0.5 μl (1 U) Vent DNA polymerase.
- Stop solution (300 μl per reaction): 10 mM Tris–HCl pH 7.5, 4 mM EDTA, 260 mM sodium acetate pH 7.0, 67 μg/ml yeast tRNA (added fresh from a 10 mg/ml stock stored at −20°C)
- Phenol/chloroform/isoamyl alcohol. Add 200 mg 8-hydroxyquinoline to 200 ml melted phenol. Equilibrate the phenol with 150 mM NaCl, 50 mM Tris–HCl pH 7.5, 1 mM EDTA. Prepare phenol/chloroform/isoamyl alcohol by mixing 50 vol. equilibrated phenol with 49 vol. chloroform and 1 vol. isoamyl alcohol. Store at −20°C. Discard after 6 months or if a brownish colour develops.
- Loading dye: 80% deionized formamide, 0.5 × TBE (44.5 mM Tris base, 44.5 mM boric acid, 1 mM EDTA)
- DNA sequencing gel. Prepare a 60 cm 6% (19:1) acrylamide/bis acrylamide, 7.7 M urea, 1 × TBE (89 mM Tris base, 89 mM boric acid, 2 mM EDTA) sequencing gel. Gels between 0.35 and 0.56 mm thick generally give better results than 0.2 mm gels. Combs that give lanes 8 mm in width with 2 mm between lanes are recommended.
- 12% methanol/10% glacial acetic acid (optional; can be used as a fixative for the DNA sequencing gel after electrophoresis (see step 24 and Section 5.5)
- Kodak XAR-5 X-ray film

*Method*

1. Allow ligated material from *Protocol 3* to precipitate at −20°C for at least 2 h.
2. Spin at 4°C in a microcentrifuge for 15 min.
3. Discard the supernatant. Add 500 μl of 75% ethanol to the tube, and invert to wash the walls of the tube.
4. Spin in a microcentrifuge at room temperature for 1 min to reposition the pellet.
5. Remove the supernatant and allow the pellet to air-dry.[a]
6. Resuspend the pellet in 70 μl water. With occasional mixing this is complete within 30 min.
7. Transfer the resuspended sample to ice. Add 30 μl ice-cold amplification mix and 3 μl ice-cold diluted Vent polymerase. Overlay with 90 μl mineral oil.
8. Perform 18 cycles of PCR. Denature for 4 min at 95°C in the first cycle, then for 1 min at 95°C in subsequent cycles. Anneal for 2 min at optimal hybridization temperature.[b] Extend for 3 min at 76°C in the first cycle and add 5 sec with each cycle. In the final cycle, extend for 10 min at 76°C.
9. Chill on ice and add 5 μl ice-cold labelling mix.
10. Perform 2 cycles of PCR. Denature for 4 min at 95°C in the first cycle and 1 min in the second. Anneal for 2 min at optimal hybridization temperature.[c] Extend to 10 min at 76°C.
11. Add 300 μl stop solution.
12. Phenol extract at room temperature using 400 μl phenol/chloroform/isoamyl alcohol.
13. Remove the aqueous phase to a fresh microcentrifuge tube.
14. Mix the aqueous phase thoroughly and split it into four aliquots of 94 μl each.[d]
15. Add 235 μl 100% ethanol to each aliquot and precipitate at −20°C for at least 2 h.[e]
16. Spin one aliquot in a microcentrifuge for 15 min at 4°C.
17. Discard the supernatant. Add 500 μl of 75% ethanol to the tube, and invert to wash the walls of the tube.
18. Spin in a microcentrifuge at room temperature for 1 min.
19. Remove the supernatant and allow the pellet to air-dry.[a]
20. Add 7 μl loading dye and vortex the sample. Check the extent of resuspension after 5 min. Draw the loading dye into a pipette tip and use a hand-held radiation monitor to compare the amount of radiation

**Protocol 4.** *Continued*

dissolved to the amount remaining in the tube. If > 10% remains in the tube, wash the tube walls with the loading dye until > 90% dissolves.

**21.** Denature the sample for 5 min at 85°C. Store on ice until ready for step 22.

**22.** Load the entire sample on one lane of the DNA sequencing gel. Electrophorese the gel at 50°C using standard conditions.

**23.** After electrophoresis, dry the gel (fixation with 12% methanol/10% glacial acetic acid is optional) and expose to autoradiographic film without an intensifying screen. A strong signal should be obtained within 12–24 h using XAR-5 film.

[a] A rotary evaporator can be used to speed the drying process.

[b] Usually it is 2–5°C above the calculated $T_m$ of primer 2 or the linker primer, whichever $T_m$ is lower (see text).

[c] Usually it is 2–5°C above the calculated $T_m$ of primer 3.

[d] This provides sufficient sample for multiple gel loadings to maximize the amount of readable sequence, and to serve as backups in case there is a problem with the sequencing gel.

[e] Though the specific activity of the product decreases with time, high quality results can be obtained even after one month of storage at this step.

# 3. Important criteria

## 3.1 Gene-specific primers

As noted in *Figure 1*, LMPCR uses three nested gene-specific primers. The use of multiple primers minimizes background. The following guidelines for primer selection have been empirically derived and have been generally successful.

- Primer 1, the primer for first strand synthesis, may or may not overlap primer 2, the amplification primer. If it does, the overlap should not exceed half of the length of primer 2.
- Primer 3, the labelling primer, *must* overlap at least the last half of primer 2; if it does not, the level of signal drastically decreases. This may reflect a need for primer 3 to exclude primer 2 from the template during the rounds of labelling.
- All primers should be gel-isolated prior to use to ensure that impurities are removed and that all oligonucleotides are full length.

Primers are usually designed such that the $T_m$ of primer 1<$T_m$ of primer 2<$T_m$ of primer 3. We use the formula

$$T_m = 59.5\,°C + 0.41\ (\%\ GC\ content) - 500/(length\ in\ nt)$$

to estimate $T_m$ in Vent DNA polymerase reaction buffer (7). Commonly, primer 1 has $T_m \geq 60\,°C$, primer 2 has $T_m >$ primer 1 and approximately the same as that of the linker primer. Primer 3 has $T_m >$ primer 2 to help it compete with primer 2 in the labelling step. A typical set of primer $T_m$s would be: primer 1 $T_m = 61\,°C$, and primer 2 $T_m = 64°C$, primer 3 $T_m = 67\,°C$. The hybridization temperatures commonly used are $T_m$ to $T_m+2\,°C$. However, even higher temperatures can be tried if background is a problem.

## 3.2 Unidirectional linker

The unidirectional linker has two key characteristics. First, it is not phosphorylated and thus it cannot be ligated to itself. Second, it has one blunt end as it is composed of two oligonucleotides of different lengths. Thus it can be ligated on in only one orientation. In addition, the duplex between the oligonucleotides in the linker has a low $T_m$. Thus the duplex is stable at the temperatures used in the ligation, but unstable at higher temperatures. Thus, the short strand will not compete for the linker primer (the long strand of the linker) during the higher temperatures of the PCR reaction. The sequence of the linker is probably not critical, but the linker that we use (sequence given in *Protocol 1*) works well, and contains several restriction enzyme sites that can be used to facilitate cloning.

## 3.3 Vent DNA polymerase

The choice of DNA polymerase is crucial to the success of LMPCR. Vent DNA polymerase yields LMPCR results superior to those observed when using the combination of Sequenase and *Taq* DNA polymerases in the original LMPCR protocol (2). It is important to note that LMPCR is extremely sensitive to the amount of Vent polymerase used. As little as a twofold increase in the amount of Vent polymerase above the optimal level can result in unacceptable levels of background. Usually, it is not necessary to test each batch of Vent polymerase, but if high background is encountered, it is a good idea to titrate the amount of this enzyme. The 'exonuclease-minus' version of Vent polymerase has not been tested in this procedure.

The buffers used in LMPCR are different from those suggested by the Vent polymerase supplier and have been optimized for used in LMPCR. The amount of $MgSO_4$ may require adjustment with some sets of primers, but the conditions given in *Protocol 4* have worked with all primers combinations tested. Finally, Vent polymerase is always added last to solutions and after they have been chilled on ice. The objective is to minimize the exonuclease activity of Vent polymerase, which is inhibited by cold and dNTPs.

# 4. Expected results

You should obtain a sequence ladder of relatively even intensity that extends ≥500 bases from the labelling primer. If size standards are used,

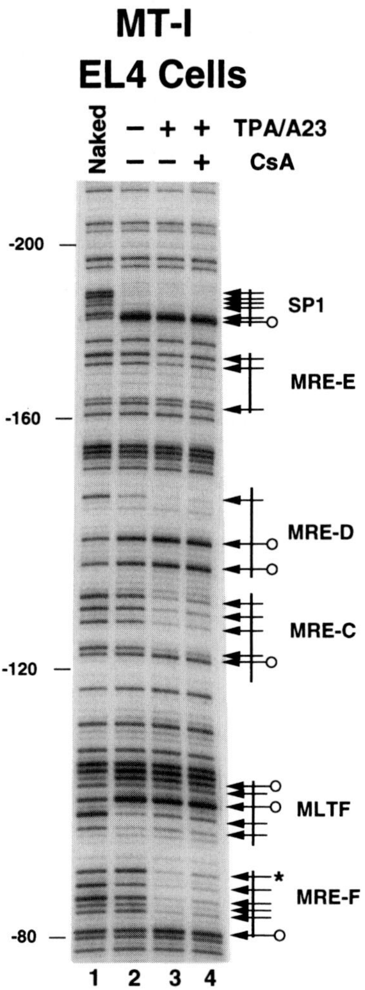


remember that all final products contain 25 nt of linker DNA at their 3′-ends. Since LMPCR is commonly used to visualize *in vivo* footprints, an example of an LMPCR *in vivo* footprint is shown in *Figure 2*. In an *in vivo* footprint, protein bound to a DNA sequence inside the cell can protect that site from being cut by a cleavage agent. In parallel, purified protein-free naked DNA is treated with the cleavage agent *in vitro*. The footprint is detected by comparing the two cleavage patterns. Sites of protein–DNA interaction appear either as bands of reduced intensity or, less frequently, bands of increased intensity in the genomic sequence ladder. A number of different cleavage agents can be used, each detecting a different spectrum of interactions. In *Figure 2*, dimethyl sulphate (DMS) was used as the cleavage agent

**Figure 2.** DMS *in vivo* footprint of the mouse metallothionein-I promoter visualized using LMPCR. The lanes contained 2 μg genomic DNA from *in vivo* DMS-treated EL4 cells that were either unstimulated (lane 2), stimulated with 10 ng/ml 12-*O*-tetradecanoyl phorbol 13-acetate (TPA) and 180 nM A23187 for 7 h (lane 3), or stimulated for 7 h in the presence of 0.5 μg/ml cyclosporin A (CsA) (lane 4). Lane 1 contained 2 μg of naked EL4 DNA, which had been treated wih DMS *in vitro*, for comparison. The location in the sequence ladder with respect to the major start site of MT-I transcription is indicated at the left margin and MT-I regulatory elements are indicated at the right margin. Bands marked by arrows were protected from reaction with DMS *in vivo*, and bands marked by arrows with circles at the end were made hypersensitive to DMS *in vivo*. An asterisk indicates a band that is DMS hypersensitive *in vivo* in unstimulated EL4 cells, but that is protected from DMS upon stimulation. The intensity of different interactions varies as a function of the percent of chromosomes at which an interaction is occurring, the stability of the interaction, and the degree to which an interaction inhibits or enhances the DMS reactivity of the binding site. The MT-I gene-specific primers used were: primer 1: CGGAGTAAGTGAGCAGAA-GGTACTC; primer 2: GGAGAAGGTACTCAGGACGTTGAAG; primer 3: GAAGGTACT-CAGGACGTTGAAGTCGTGG. LMPCR hybridization temperatures were: primer 1: 60 °C; primer 2: 66 °C; primer 3: 69 °C.

---

(see Chapter 14). DMS preferentially alkylates G residues, with subsequent piperidine treatment cleaving the DNA at the alkylated bases. DMS detects protein–DNA interactions in the major groove.

*Figure 2* shows a DMS *in vivo* footprint across a region of the mouse metallothionein-I (MT-I) promoter in EL4 thymoma cells. EL4 cells express MT-I RNA at a basal level that is increased upon treatment with the phorbol ester TPA and the calcium ionophore A23187. This induction is not prevented by treatment with the immunosuppressive drug cyclosporin A (CsA), which blocks other gene induction responses in these cells. Comparison of the cleavage ladders from naked and *in vivo* samples reveals multiple sites of *in vivo* protein–DNA interaction (*Figure 2*). The SP1 and MLTF recognition sites show high levels of occupancy in unstimulated cells, and this is relatively unaffected by TPA/A23 treatment. The metal-responsive elements (MREs), some of which are detectably occupied in unstimulated cells (e.g. MRE-D), show increased binding of factors upon TPA/A23 treatment. This increased binding is largely unaffected by CsA treatment. Thus TPA/A23 induction of MT-I RNA levels correlates with the binding of factors to elements in the MT-I promoter. These are the same elements that are occupied by factors in response to metal exposure and convey metal-inducibility upon the MT-I gene.

*In vivo* footprinting involves quantitative comparison of band intensities between different DNA samples. Thus an individual DNA sample must give identical results each time it is visualized by LMPCR. Such reproducibility requires, of course, that all reactions are performed in a uniform manner (for example, through the use of reagent cocktails). It also requires that the correct amount of DNA starting material be used. If too much DNA is used, the PCR reaction may depart from linearity as reagents become exhausted, potentially exaggerating small sample to sample differences in reaction components or conditions. (Similar effects are observed if too many rounds of PCR are used.) However, large sample to sample fluctuations in band intensities also arise if too little substrate DNA is used. Since LMPCR is an exponential amplification-based procedure, this does not reflect an inability to obtain a signal from small quantities of DNA, but rather that statistical sampling effects can begin to have a large impact when the size of a sampled population is small. For example, if, from a homogeneous mixture, aliquots were taken containing an average of two substrate molecules, some aliquots might contain no substrate molecules, some only one, and others three or more. Thus large fluctuations would be observed even though the parental mixture was uniform. If larger aliquots were taken instead, the percentage fluctuation would decrease. Similarly, a reliable footprint can only be detected if a sufficient quantity of starting DNA is used. We have empirically determined that for a DMS *in vivo* footprint, $6 \times 10^5$ haploid genomes is a sufficient quantity of starting DNA to minimize statistical variations. This is $\approx 2$ μg of mouse DNA, but for organisms of different genome size different masses of DNA will be required. Less quantitative uses of LMPCR will be more tolerant of statistical fluctuation and may be effectively performed using less DNA. Genomic sequencing can use moderately lower amounts of substrate, whereas cloning can often be done from much lower amounts of substrate.

Good *in vivo* footprints require not only that ladders be reproducibly obtained from a single sample, but that there be a general matching of band intensities between samples. In *Figure 2*, for example, note that band intensities outside regions of protein–DNA interaction are very similar in all lanes. It is especially important that bands above and below regions of footprints match identically (in *Figure 2* below −80 and above −190). This indicates both that the amount of DNA used in each lane is the same and that the degree of cleavage of each sample is similar. If either of these conditions are not met, footprint information cannot be interpreted with confidence.

# 5. Troubleshooting

## 5.1 High background

The most frequently encountered problem is high background. As noted in Section 3.3, this is often due to the use of too much Vent DNA polymerase. Try adjusting the amount of Vent DNA polymerase used, especially in the PCR amplification step. High background can also be caused by the gene-specific primers. Try increasing the annealing temperature at the amplification and labelling steps. In some cases it may be necessary to choose new primers. High background can also arise from problems at the electrophoresis step (see Sections 5.4 and 5.5).

### 5.2 Short ladders

If ladders substantially shorter than 500 nt are obtained, there are several common causes. An abrupt ending of the ladder may indicate that the denaturation temperatures used are too low. Thermal cyclers do not always accurately measure the temperature inside the reaction. Try increasing the denaturation temperatures by a couple of degrees. A premature, graded tailing off of the ladder may reflect overdigestion of the DNA with the cleavage agent, especially if products near the labelling primer are very intense.

### 5.3 No ladder/weak ladder

No ladder or a very weak ladder can be caused by a number of things:

- a primer combination that will not work (which is occasionally encountered)
- a bad reagent (such as old or low quality nucleotides)
- a hybridization temperature that is too high for one or more of the primers
- contaminants in the DNA preparation (such as residual piperidine)
- a DNA polymorphism across a primer binding site

It is often useful to try different primer sets with the suspect DNA samples and different DNA samples with the suspect primer set. One can also check the primers by doing traditional PCR. Finally, ensure that all reagents are properly stored and of the highest quality.

### 5.4 Spurious bands

If spurious bands are a problem, common causes are old cleavage agents (e.g. old DMS or piperidine), old loading dye (containing formamide that has not been recently deionized), or genomic DNA of poor quality.

### 5.5 Fuzzy ladders/difficult to see footprints

The use of sharkstooth combs or traditional combs with narrow teeth for the electrophoresis step can make it difficult to clearly see footprints. Use a comb that gives lanes 8 mm in width with 2 mm between lanes. Old or impure acrylamide gel reagents or loading dye can make the result appear fuzzy or to have high levels of background. Inefficient drying of the gel can also cause fuzziness. Fixing the gel in 12% methanol/10% glacial acetic acid can eliminate this problem. However, if the non-extended labelled primer still remains in the gel (i.e. has not been run off into the lower buffer well), it must be cut off prior to fixation. Otherwise, this very radioactive low-molecular weight material will diffuse to some extent into the fix solution and can potentially contaminate the rest of the gel.

## Acknowledgements

This work was supported by grants from the NIH and Muscular Dystrophy Association to B.W. P.A.G. and P.R.M. were supported by a National Research Service Award from the USPHS.

## References

1. Mueller, P. R. and Wold, B. (1989). *Science*, **246**, 780.
2. Garrity, P. A. and Wold, B. J. (1992). *Proc. Natl Acad. Sci. USA*, **89**, 1021.
3. Mueller, P. R., Salser, S. J., and Wold, B. (1988). *Genes Dev.*, **2**, 412.
4. Rigaud, G., Roux, J., Pictet, R., and Grange, T. (1991). *Cell*, **67**, 977.
5. Saluz, H. P. and Jost, J. P. (1987). In *A laboratory guide to genomic sequencing: the direct sequencing of native uncloned DNA*. Birkhauser, Boston.
6. Fors, L., Saavedra, R. A., and Hood, L. (1990). *Nucleic Acids Res.*, **18**, 2793.
7. Wahl, G. M., Berger, S. L., and Kimmel, A. R. (1987). In *Methods in enzymology* (ed. S. L. Berger and A. R. Kimmel), Vol. 152, p. 399. Academic Press, New York.

# A1

# Suppliers of specialist items

**Aldrich Chemical Company Ltd**, The Old Brickyard, New Road, Gillingham, Dorset, SP8 4JL, UK.

**Amersham International plc**, Lincoln Place, Green End, Aylesbury, Buckinghamshire HP20 2TP, UK.

**Amersham North America**, 2636 South Clearbrook Drive, Arlington Heights, IL 60005, USA.

**Anachem Ltd**, 20 Charles Street, Luton, Bedfordshire, LU2 0EB, UK.

**Applied Biosystems Division of Perkin-Elmer**, 850 Lincoln Center Drive, Foster City, CA 94404, USA.

**Applied Biosystems Division of Perkin-Elmer Ltd**, Kelvin Close, Birchwood Science Park North, Warrington, Cheshire WA3 7PB, UK.

**Appligene SA**, Parc d'Innovation, BP 72, F-767402 Illkirch, France.

**Ateliers Cloup**, BP 60, 94502 Champigny Cedex, France.

**Becton Dickinson and Co.**, 2 Bridgewater Lane, Lincoln Park, NJ 07035, USA.

**Bio-Rad Laboratories Ltd**, Bio-Rad House, Maylands Avenue, Hemel Hempstead HP2 7TD, UK.

**Bio-Rad Laboratories**, Alfred Nobel Drive, Hercules, CA 94547, USA.

**BIOS Laboratories**, 5 Science Park, New Haven, CT 06511, USA

**Biotecx Labs Inc.**, 6023 South Loop East, Houston, Texas 77033, USA.

**Boehringer Mannheim Corporation**, Biochemical Products, 9115 Hague Road, PO Box 50414, Indianapolis, IN 46250–0414, USA.

**Boehringer Mannheim (Diagnostics and Biochemicals) Ltd**, Bell Lane, Lewes, East Sussex BN17 1LG, UK.

**Boehringer Mannheim GmbH**, Biochemica, Sandhofer Str. 116, Postfach 310120 6800 Mannheim 31, Germany.

**British Biotechnology Products Ltd**, 4–10 The Quadrant, Barton Lane, Abingdon, Oxfordshire, OX14 3YS, UK.

**BRL**; see Gibco/BRL.

**Burdick & Jackson**, Division of Baxter Diagnostics Inc., 1953 South Harvey Street, Muskegon, MI 49442, USA.

**Cambridge Biosciences Ltd,** Stourbridge Common Business Centre, Swann's Road, Cambridge CB5 8LA, UK.

**Cambridge Research Biochemicals Inc.**, Fairfax Research Room 205, Wilmington, DE 19897, USA.

**Cambridge Research Biochemicals Ltd**, Gadbrook Park, Northwich, Cheshire CW9 7RA, UK.

**Clontech Laboratories Inc.**, 4030 Fabian Way, Palo Alto, California 94303–4607, USA.

**Clontech Laboratories Inc.**, in Europe, see Cambridge Biosciences Ltd.

**Cruachem Ltd**, Todd Campus, West of Scotland Science Park, Acre Road, Glasgow G20 0UA, UK

**Dupont (UK) Ltd**, NEN, Diagnostics and Biotechnology Systems, Medical Products, Wedgewood Way, Stevenage, Hertfordshire, SG1 4QN, UK.

**DuPont NEN**, 549 Albany Street, Boston, MA 02118, USA.

**Dynal Inc.**, 475 North Station Plaza, Great Neck, NY 11021, USA.

**Dynal International**, PO Box 158 Skøyen, N0212, Oslo, Norway.

**Dynal UK Ltd**, Station House, 26 Grove Street, New Ferry, Wirral, Merseyside, L62 5A2, UK.

**Eastman Kodak**, Acorn Field, Knowsley Industrial Park North, Liverpool L33 72X, UK.

**Eastman Kodak**, Rochester, New York 14650, USA.

**Falcon**; Falcon is a registered trademark of Becton Dickinson and Co.

**Fisher Scientific**, 711 Forbes Ave., Pittsburgh, PA, USA.

**Flowgen Instruments Ltd**, Broad Oak Enterprise Village, Broad Road, Sittingbourne, Kent ME9 8BR, UK.

**FMC BioProducts (Europe)**, Risingevej 1, DK-2665 Vallensbaek Strand, Denmark.

**FMC BioProducts Inc.**, 5 Maple Street, Rockland, ME 04841, USA.

**FMC BioProducts Inc.**; distributed in UK by Flowgen Instruments Ltd.

**Genset**, 11 passage Delaunay, 75011 Paris, France.

**Glen Research/Cambio**, 34 Millington Road, Cambridge CB3 9HP, UK

**Gibco/BRL**, PO Box 68, Grand Island, NY 14072–0068, USA.

**Gibco/BRL**; for UK see Life Technologies Ltd.

**Hoefer Scientific Instruments**, 654 Minnesota Street, PO Box 77387, San Francisco CA 94107-0387, USA.

**Hoefer Scientific Instruments Ltd.,** Croft Road Workshops, Croft Road, Off Hempstalls Lane, Newcastle-under-Lyme, Staffs ST5 0TW.

**Hybaid**, 111–113 Waldegrave Road, Teddington, Middlesex, TW11 8LL, UK.

**Hybaid**, National Labnet Corporation, PO Box 841, Woodbridge, N.J. 07095, USA.

**ICN Biochemicals Inc.,** 3300 Hyland Ave., Costa Mesa, CA 92626, USA.

**ICN Biochemicals Ltd.,** Unit 18, Thame Park Business Centre, Wenman Road, Thame, Oxon OX9 3XA, UK.

**Intermountain Scientific**, 1610 South Main, Bountiful, UT 84010, USA.

**Invitrogen Corporation**; distributed in UK by British Biotechnology Products Ltd.

**Invitrogen Corporation**, 3985 B Sorrento Valley Building, San Diego, CA 92121, USA.

**J.T Baker Inc.**, 222 Red School Lane, Phillipsburg, NJ 08865, USA.

**Kodak**; see Eastman Kodak.

**Life Sciences International**

**Life Technologies Ltd**, Trident House, Renfrew Road, Paisley PA3 4EF, UK.

**Mallinckrodt Inc.**, Paris. KT. 40361, USA.

**Merck,** 6100 Darmstadt 1, Postfach 4119, Germany.

**Millipore (UK) Ltd**, The Boulevard, Blackmoor Lane, Watford, Hertfordshire WD1 8YW, UK.

**New England Biolabs**, 32 Tozer Road, Beverley, MA 01915, USA.

**New England Biolabs/C.P. Laboratories**, PO Box 22, Bishop's Stortford, Hertfordshire, CM23 3DH, UK.

**NUNC**; distributed in UK by Life Technologies Ltd.

**Perkin-Elmer**, 761 Main Avenue, Norwalk, CT 06856, USA.

**Perkin-Elmer**, Maxwell Road, Beaconsfield, Buckinghamshire HP9 1QA, UK.

**Perkin-Elmer Holding GmbH**, Bahnhofstrasse 30, D-8011 Vaterstetten, Munich, Germany.

**Pharmacia LKB Biotechnology AB**, Pharmacia Biotech Europe, Rue de la Fusée 66, B-1130 Brussels, Belgium.

**Pharmacia LKB Biotechnology Inc.**, 800 Centennial Avenue, PO Box 1327, Piscataway, NJ 08855–1327, USA.

**Pharmacia LKB Biotechnology Ltd**, 23 Grosvenor Road, St Albans, Hertfordshire AL1 3AW, UK.

**Pierce**, Pierce Europe BV, Box 1512, 3260 BA Ous-Beijerland, The Netherlands

**Pierce**, PO Box 117, Rockford, Illinois 61105, USA.

**Promega Corp.**, 2800 Woods Hollow Road, Madison, WI 53711–5399, USA.

**Promega Ltd**, Delta House, Enterprise Road, Chilworth Research Centre, Southampton, SO1 7NS, UK.

**QIAGEN GmbH,** Max-Volmer-Str. 4, 40724 Hilden, Germany.

**QIAGEN Inc.**, 9259 Eton Avenue, Chatsworth, CA 91311, USA.

**Research Genetics Inc.,** 2130 Memorial Parkway, Hunteville, Alabama 35801, USA.

**Sarstedt Ltd**. 68 Boston Road, Beaumont Leys, Leicester LE4 1AW, UK.

**Savant Instruments Inc.,** Farmingdale, New York, USA.

**Savant;** distributed in UK by Stratech Scientific Ltd.

**Schleicher and Schüll Inc.**, 10 Optical Avenue, Keene, NH 03431, USA.

**Sepracor**, 35 Av. Jean Jaures, 92395 Villeneuve la Garenne Cedex, France.

**Sigma Chemical Company**, 3050 Spruce Street, PO Box 14508, St Louis, MO 63178–9916 USA.

**Sigma Chemical Company (UK)**, Fancy Road, Poole, Dorset, BH17 7NH, UK.

**Stratagene**, 11099 North Torrey Pines Road, La Jolla, CA 92037, USA.
**Stratagene Ltd**, Cambridge Innovation Centre, Cambridge Science Park, Milton Road, Cambridge, CB4 4GF, UK.
**Stratech Scientific Ltd.,** 61–63 Dudley Street, Luton, Beds LU2 0NP.
**Techne Ltd,** Duxford, Cambridge CB2 4PZ, UK.
**Uniscience Ltd,** Beamont Close, Banbury, Oxon OX16 7RG, UK.
**Vector Laboratories**, 16 Wulfrie Square, Bretton, Peterborough, PE3 8RF, UK.
**Wheaton Inc.**, 1501 N. Tenth Street, Millville, NJ 08332, USA.

# Index